# OCR gateway

# GCSE chemistry

**Authors**

Angela Saunders

Nigel Saunders

# Contents

## C4 The periodic table 126



## C5 Quantitative analysis 164



## C6 Chemistry out there 202

# How to use this book

**Welcome to your Gateway Chemistry course.** This book has been specially written by experienced teachers and examiners to match the 2011 specification.

On these two pages you can see the types of pages you will find in this book, and the features on them. Everything in the book is designed to provide you with the support you need to help you prepare for your examinations and achieve your best.

## Module openers

**Specification matching grid:** This shows you how the pages in the module match to the exam specification for GCSE Chemistry, so you can track your progress through the module as you learn.

**Why study this module:** Here you can read about the reasons why the science you're about to learn is relevant to your everyday life.

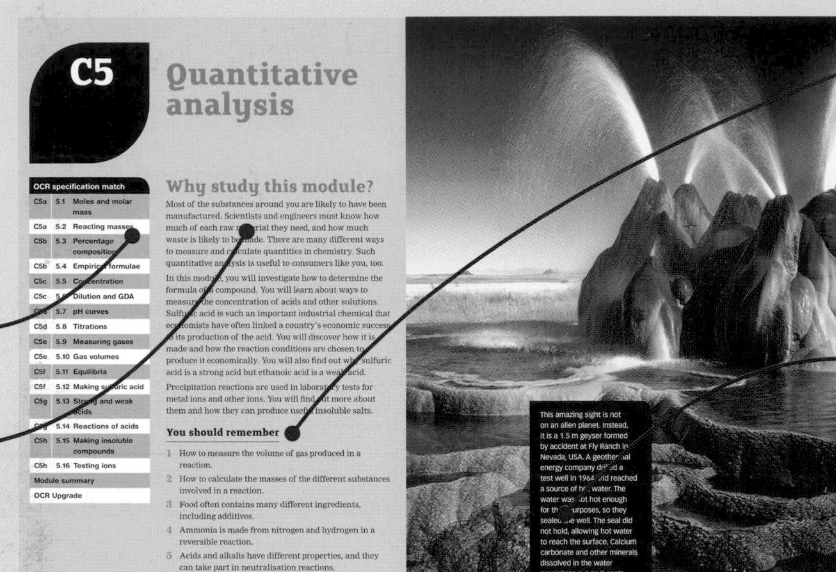

**You should remember:** This list is a summary of the things you've already learnt that will come up again in this module. Check through them in advance and see if there is anything that you need to recap on before you get started.

**Opener image:** Every module starts with a picture and information on a new or interesting piece of science that relates to what you're about to learn.

## Main pages

**Learning objectives:** You can use these objectives to understand what you need to learn to prepare for your exams. Higher Tier only objectives appear in pink text.

**Key words:** These are the terms you need to understand for your exams. You can look for these words in the text in bold or check the glossary to see what they mean.

**Questions:** Use the questions on each spread to test yourself on what you've just read.

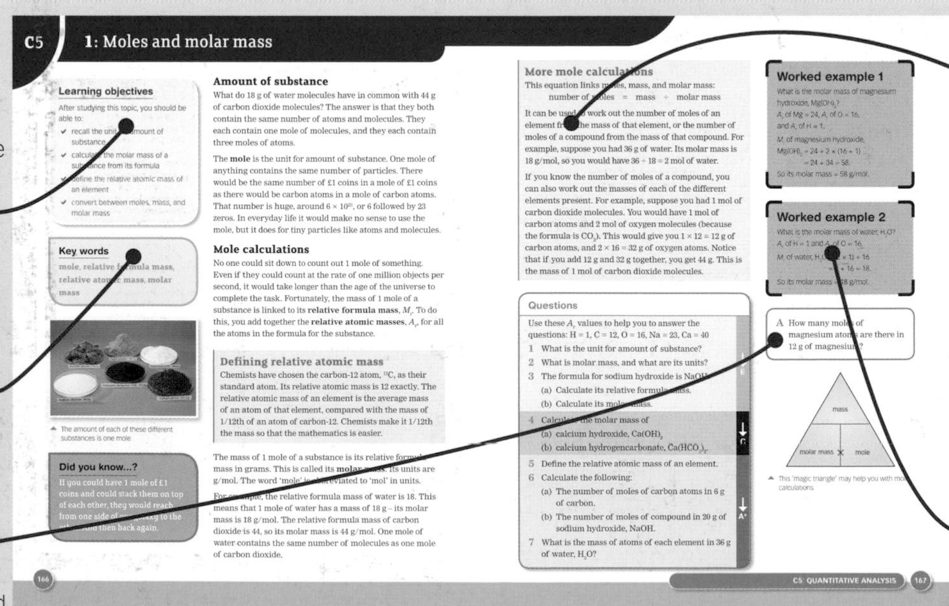

**Higher Tier content:** Anything marked in pink is for students taking the Higher Tier paper only. As you go through you can look at this material and attempt it to help you understand what is expected for the Higher Tier.

**Worked examples:** These help you understand how to use an equation or to work through a calculation. You can check back whenever you use the calculation in your work.

## Summary and exam-style questions

Every summary question at the end of a spread includes an indication of how hard it is. You can track your own progress by seeing which of the questions you can answer easily, and which you have difficulty with.

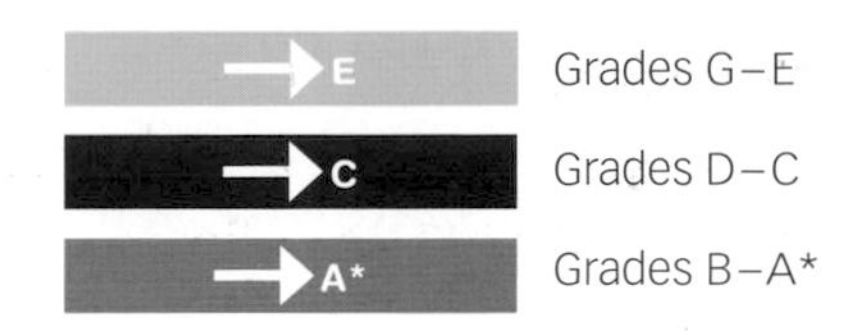


When you reach the end of a module you can use the exam-style questions to test how well you know what you've just learnt. Each question has a grade band next to it, so you can see what you need to do for the grade you are aiming for.

**Revision checklist:** This is a summary of the main ideas in the module. You can use it as a starting point for revision, to check that you know about the big ideas covered.

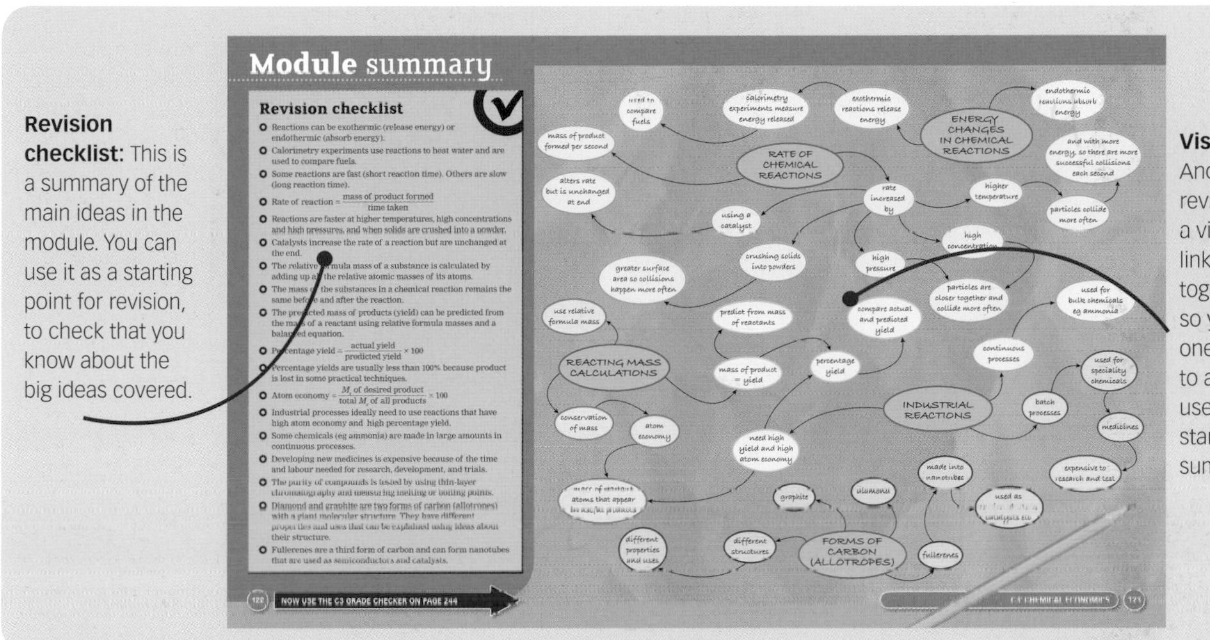


**Visual summary:** Another way to start revision is to use a visual summary, linking ideas together in groups so you can see how one topic relates to another. You can use this page as a start for your own summary.

**Module summaries**

**Upgrade:** Upgrade takes you through an exam question in a step-by-step way, showing you why different answers get different grades. Using the tips on the page you can make sure you achieve your best by understanding what each question needs.

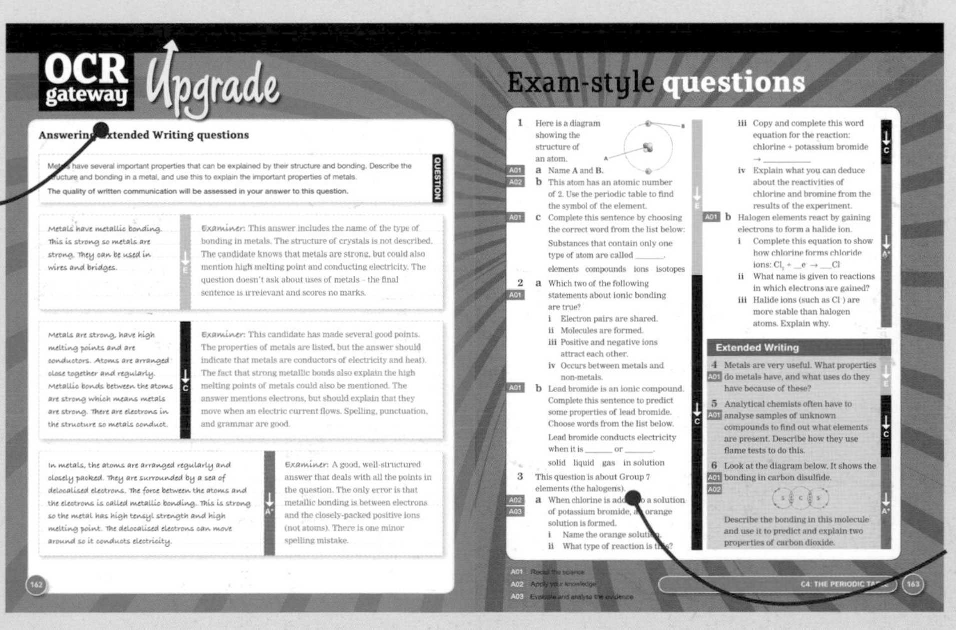


**Exam-style questions:** Using these questions you can practice your exam skills, and make sure you're ready for the real thing. Each question has a grade band next to it, so you can understand what level you are working at and focus on where you need to improve to get your target grade.

**OCR Upgrade**

# Routes and assessment

## Matching your course

The modules in this book have been written to match the specification for **OCR GCSE Gateway Science Chemistry B**.

In the diagram below you can see that the modules can be used to study either for **GCSE Chemistry** or as part of **GCSE Science** and **GCSE Additional Science** courses.

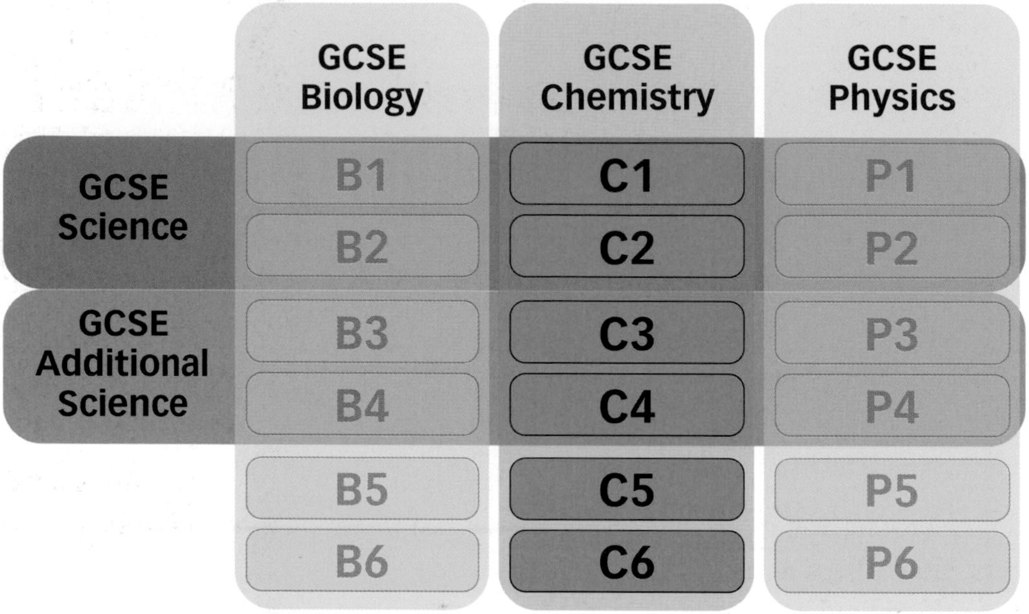

| | GCSE Biology | GCSE Chemistry | GCSE Physics |
|---|---|---|---|
| GCSE Science | B1 | C1 | P1 |
| | B2 | C2 | P2 |
| GCSE Additional Science | B3 | C3 | P3 |
| | B4 | C4 | P4 |
| | B5 | C5 | P5 |
| | B6 | C6 | P6 |

## GCSE Chemistry assessment

The content in the modules of this book match the different exam papers you will sit as part of your course. The diagram below shows you which modules are included in each exam paper. It also shows you how much of your final mark you will be working towards in each paper.

| Unit | Modules tested | % | Type | Time | Marks available |
|---|---|---|---|---|---|
| B741 | C1 C2 C3 | 35% | Written exam | 1hr 15 | 75 |
| B742 | C4 C5 C6 | 40% | Written exam | 1hr 30 | 85 |
| B743 | Controlled Assessment | 25% | | 4hrs | 48 |

# Understanding exam questions

The list below explains some of the common words you will see used in exam questions.

**Calculate**
Work out a number. You can use your calculator to help you. You may need to use an equation. The question will say if your working must be shown. (Hint: don't confuse with 'Estimate' or 'Predict')

**Compare**
Write about the similarities and differences between two things.

**Describe**
Write a detailed answer that covers what happens, when it happens, and where it happens. Talk about facts and characteristics. (Hint: don't confuse with 'Explain')

**Discuss**
Write about the issues related to a topic. You may need to talk about the opposing sides of a debate, and you may need to show the difference between ideas, opinions, and facts.

**Estimate**
Suggest an approximate (rough) value, without performing a full calculation or an accurate measurement. Don't just guess – use your knowledge of science to suggest a realistic value. (Hint: don't confuse with 'Calculate' and 'Predict')

**Explain**
Write a detailed answer that covers how and why a thing happens. Talk about mechanisms and reasons. (Hint: don't confuse with 'Describe')

**Evaluate**
You will be given some facts, data or other information. Write about the data or facts and provide your own conclusion or opinion on them.

**Justify**
Give some evidence or write down an explanation to tell the examiner why you gave an answer.

**Outline**
Give only the key facts of the topic. You may need to set out the steps of a procedure or process – make sure you write down the steps in the correct order.

**Predict**
Look at some data and suggest a realistic value or outcome. You may use a calculation to help. Don't guess – look at trends in the data and use your knowledge of science. (Hint: don't confuse with 'Calculate' or 'Estimate')

**Show**
Write down the details, steps or calculations needed to prove an answer that you have been given.

**Suggest**
Think about what you've learnt and apply it to a new situation or a context. You may not know the answer. Use what you have learnt to suggest sensible answers to the question.

**Write down**
Give a short answer, without a supporting argument.

**Top tips**

Always read exam questions carefully, even if you recognise the word used. Look at the information in the question and the number of answer lines to see how much detail the examiner is looking for.

You can use bullet points or a diagram if it helps your answer.

If a number needs units you should include them, unless the units are already given on the answer line.

As part of the assessment for your GCSE Chemistry course you will undertake a Controlled Assessment task. This section of the book includes information designed to help you understand what Controlled Assessment is, how to prepare for it, and how it will contribute towards your final mark.

## Understanding Controlled Assessment

### What is Controlled Assessment?

Controlled Assessment has taken the place of coursework for the new 2011 GCSE Sciences specifications. The main difference between coursework and Controlled Assessment is that you will be supervised by your teacher when you carry out some parts of your Controlled Assessment task.

### What will I have to do during my Controlled Assessment?

The Controlled Assessment task is designed to see how well you can:

- develop hypotheses
- plan practical ways to test hypotheses
- assess and manage risks during practical work
- collect, process, analyse, and interpret your own data using appropriate technology
- research, process, analyse, and interpret data collected by other people using appropriate technology
- draw conclusions based on evidence
- review your method to see how well it worked
- review the quality of the data.

### How do I prepare for my Controlled Assessment?

Throughout your course you will learn how to carry out investigations in a scientific way, and how to analyse and compare data properly. These skills will be covered in all the activities you work on during the course.

In addition, the scientific knowledge and understanding that you develop throughout the course will help you as you analyse information and draw your own conclusions.

### How will my Controlled Assessment be structured?

Your Controlled Assessment is a task divided into three parts. You will be introduced to each part of the task by your teacher before you start.

### What are the three parts of the Controlled Assessment?

Your Controlled Assessment task will be made up of three parts. These three parts make up an investigation, with each part looking at a different part of the scientific process.

| | What skills will be covered in each part? |
|---|---|
| **Part 1** | Research and collecting secondary data |
| **Part 2** | Planning and collecting primary data |
| **Part 3** | Analysis and evaluation |

### Do I get marks for the way I write?

Yes. In two of the three parts of the Controlled Assessment you will see a pencil symbol (✎). This symbol is also found on your exam papers in questions where marks are given for the way you write.

These marks are awarded for quality of written communication. When your work is marked you will be assessed on:

- how easy your work is to read
- how accurate your spelling, punctuation, and grammar are
- how clear your meaning is
- whether you have presented information in a way that suits the task
- whether you have used a suitable structure and style of writing.

## Part 1 – Research and collecting secondary data

At the beginning of your task your teacher will introduce Part 1. They will tell you:

- how much time you have – for Part 1 this should be about 2 hours, either in class or during your homework time
- what the task is about
- about the material you will use in Part 1 of the task
- the conditions you will work under
- your deadline.

The first part of your Controlled Assessment is all about research. You should use the stimulus material for Part 1 to learn about the topic of the task and then start your own research. Whatever you find during your research can be used during later parts of the Controlled Assessment.

### Sources, references, and plagiarism

For your research you can use a variety of sources including fieldwork, the Internet, resources from the library, audio, video, and others. Your teacher will be able to give you advice on whether a particular type of source is suitable or not.

For every piece of material you find during your research you must make sure you keep a record of where you found it, and who produced it originally. This is called referencing, and without it you might be accused of trying to pass other people's work off as your own. This is known as plagiarism.

### Writing up your research

At the end of Part 1 of the Controlled Assessment you will need to write up your own individual explanation of the method you have used. This should include information on how you carried out your own research and collected your research data.

This write up will be collected in by your teacher and kept. You will get it back when it is time for you to take Part 3.

## Part 2 – Planning and collecting primary data

Following Part 1 of your Controlled Assessment task your teacher will introduce Part 2. They will tell you:

- how much time you have – for Part 2 this should be about 2 hours for planning and 1 hour for an experiment
- what the task is about
- about the material you will use in Part 2 of the task
- the conditions you will work under
- your deadline.

Part 2 of the Controlled Assessment is all about planning and carrying out an experiment. You will need to develop your own hypothesis and plan and carry out your experiment in order to test it.

### Risk assessment

Part of your planning will need to include a risk assessment for your experiment. To get the maximum number of marks, you will need to make sure you have:

- evaluated all significant risks
- made reasoned judgements to come up with appropriate responses to reduce the risks you identified

- manage all of the risks during the task, making sure that you don't have any accidents and that there is no need for your teacher to come and help you.

### Working in groups and writing up alone

You will be allowed to work in groups of no more than three people to develop your plan and carry out the experiment. Even though this work will be done in groups, you need to make sure you have your own individual records of your hypothesis, plan, and results.

This write up will be collected in by your teacher and kept. You will get it back when it is time for you to take Part 3.

## Part 3 – Analysis and evaluation

Following Part 2 of your Controlled Assessment task your teacher will introduce Part 3. They will tell you:

- how much time you have – for Part 3 this should be about 2 hours
- what the task is about
- about the answer booklet you will use in Part 3
- the conditions you will work under.

Part 3 of the Controlled Assessment is all about analysing and evaluating the work you carried out in Parts 1 and 2. Your teacher will give you access to the work you produced and handed in for Parts 1 and 2.

For Part 3 you will work under controlled conditions, in a similar way to an exam. It is important that for this part of the task you work alone, without any help from anyone else and without using anyone else's work from Parts 1 and 2.

### The Part 3 answer booklet

For Part 3 you will do your work in an answer booklet provided for you. The questions provided for you to respond to in the answer booklet are designed to guide you through this final part of the Controlled Assessment. Using the questions you will need to:

- evaluate your data
- evaluate the methods you used to collect your data
- take any opportunities you have for using mathematical skills and producing useful graphs
- draw a conclusion
- justify your conclusion.

# C1 Carbon chemistry

| OCR specificaton match | | |
|---|---|---|
| C1a | 1.1 | Making crude oil useful |
| C1a | 1.2 | Making crude oil more useful |
| C1b | 1.3 | Burning hydrocarbons |
| C1b | 1.4 | Using carbon fuels |
| C1c | 1.5 | Clean air |
| C1c | 1.6 | Air pollution |
| C1d | 1.7 | Alkanes and alkenes |
| C1d | 1.8 | Making polymers |
| C1e | 1.9 | Designer polymers |
| C1e | 1.10 | Polymer problems |
| C1f | 1.11 | Cooking |
| C1f | 1.12 | Food additives |
| C1g | 1.13 | Smells |
| C1g | 1.14 | Nail varnish and testing |
| C1h | 1.15 | Paints and pigments |
| C1h | 1.16 | Smart paints |
| Module summary | | |
| OCR Upgrade | | |

## Why study this module?

You use many different materials in your everyday life. Many of these are made from crude oil or depend on it for their manufacture. Crude oil is vital to modern living, yet stories of the problems caused when things go wrong are often in the news.

In this module you will investigate how crude oil is changed into useful products like fuels, plastics, cosmetics, and paints. You will also explore the causes of air pollution and how they may be prevented, and issues surrounding testing products on animals.

Everyone needs to eat, so you will examine the chemical reactions that happen in cooking. You will also find out about food additives and some of the issues concerned with their use.

### You should remember

1. There are differences between elements, mixtures, and compounds.
2. Compounds show characteristic properties and patterns in their properties.
3. Mixtures may be separated in different ways, depending on the mixture.
4. New substances are made in chemical reactions.
5. There are benefits and drawbacks in the use of fossil fuels.
6. Advances in science and technology can lead to the production of new materials with desirable properties.
7. Human activity and natural processes can lead to changes in the environment.

This 1.2 metre diameter pipeline can carry up to 3.4 billion litres of crude oil per day across Alaska. It is 1288 kilometres long, and crosses 34 major rivers and nearly 500 streams. The pipeline is so long that it takes nearly 12 days for oil to travel south from the oilfields to the port of Valdez, ready to be shipped to refineries across the USA.

# 1: Making crude oil useful

## Learning objectives

After studying this topic, you should be able to:

- explain what fossil fuels are
- describe how crude oil is separated into fractions
- describe the main fractions obtained from crude oil
- explain why crude oil can be separated by fractional distillation

Crude oil is a thick liquid

## The fossil fuels

**Crude oil** is a **fossil fuel**. It was formed from the remains of living things that lived in the sea millions of years ago. These became buried deep in the seabed after they died. Chemical reactions happened that eventually turned them into crude oil.

Crude oil is a **non-renewable resource**. We are using it up much faster than more can form. Coal and natural gas are also fossil fuels and non-renewable resources. All the fossil fuels are **finite resources**. They will run out one day if we continue to use them.

**A** Name three fossil fuels.

**B** Explain why the fossil fuels are non-renewable resources.

## Separating crude oil

Crude oil is a mixture of many different compounds called hydrocarbons. The hydrocarbons in crude oil are separated from each other by **fractional distillation**. This happens at an oil refinery in a tower called a fractionating column. Fractional distillation works because different hydrocarbons have different boiling points.

Oil refineries run continuously

The crude oil is heated. Its vapours are piped into the bottom of the fractionating column, which has a temperature gradient.

The fractionating column is hot at the bottom and gradually gets colder towards the top. The vapours cool as they rise through the column. They condense to form a liquid when they reach a part that is cold enough. The liquid falls into a tray and is piped out of the column.

- Hydrocarbons with the highest boiling points leave at the bottom of the column.
- Hydrocarbons with the lowest boiling points reach the top without cooling enough to condense, leaving as gases.

The substances leaving the column are called **fractions** because they are just a part of the crude oil. Each fraction contains hydrocarbons with similar boiling points. Different fractions have different uses, depending on their properties.

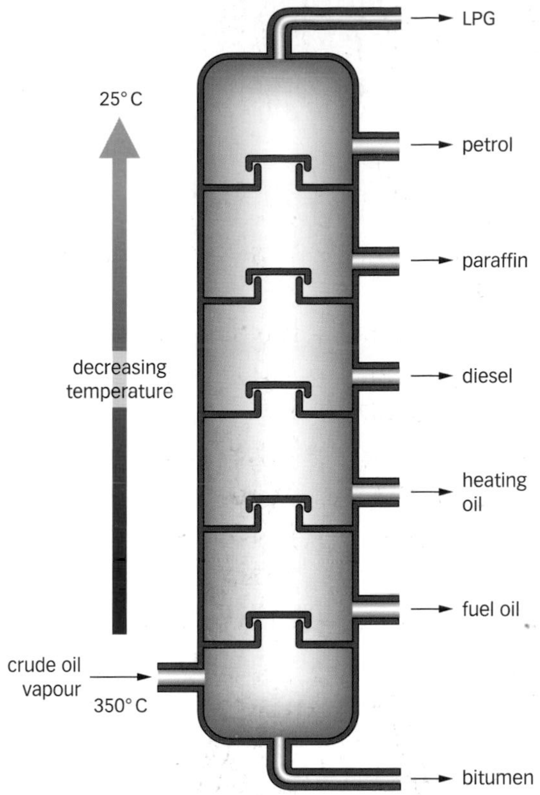

▲ The different fractions from a fractionating column. LPG contains propane and butane gases.

## Intermolecular forces

The atoms in hydrocarbon molecules are joined to each other by covalent bonds. These are much stronger than the intermolecular forces that attract hydrocarbon molecules towards each other. Intermolecular forces are broken when crude oil is boiled, rather than covalent bonds. This causes the hydrocarbon molecules to separate from each other, forming a vapour.

The larger the hydrocarbon molecules, the stronger the intermolecular forces, and the higher the hydrocarbon's boiling point. The hydrocarbons with the smallest molecules are gases and those with the largest molecules are solids.

### Key words

**crude oil, fossil fuel, non-renewable resource, finite resource, fractional distillation, fraction**

### Questions

1. What property of hydrocarbons lets fractional distillation of crude oil work?
2. State the names of three fractions from crude oil. Include a solid, a liquid, and a gas.

↓ E

3. Describe how fractional distillation works.
4. Where does the fraction with the lowest boiling point leave the fractionating column? Name two hydrocarbons that this fraction contains.

↓ C

5. Explain, in terms of intermolecular forces, why crude oil can be separated by fractional distillation.

↓ A*

### Exam tip OCR

✔ Remember that crude oil vapours must be cooled and condensed so they can leave the fractionating column as a liquid.

# 2: Making crude oil more useful

## Learning objectives

After studying this topic, you should be able to:

- explain how oil is obtained and some of the environmental problems involved
- describe how large hydrocarbon molecules may be converted into smaller ones
- explain why cracking is carried out
- discuss the problems associated with oil as a finite resource

## Key words

**impermeable**, **oil well**, **oil slick**, **petrochemical**, **cracking**, **catalyst**

## Did you know...?

There are oil wells on every continent except for Antarctica. The Madrid Protocol became law in 1998. It bans drilling for oil in Antarctica for 50 years.

A seabird covered in crude oil from an oil slick

## Drilling for oil

As crude oil forms, the seabed above it gradually turns to rock. The oil is forced up through the rock. In some parts of the world it eventually reaches the surface. The hydrocarbons with low boiling points evaporate, leaving thick tar behind. Tar was used in the past to waterproof wooden ships. In other parts of the world, layers of **impermeable** rock trap the oil so it cannot escape to the surface. **Oil wells** are drilled through the rock to reach the oil below. The oil is pumped to the surface so that it can be transported to oil refineries around the world.

An oil rig in the North Sea

## Problems with oil

Exploiting crude oil can cause environmental problems. For example, oil may be spilled at the oil well or during unloading at an oil refinery. Ocean-going oil tankers are used to transport oil. Sometimes these sink or run aground, and oil leaks out. This causes **oil slicks** on the surface of the sea, and these can come ashore and damage beaches. The oil also damages birds' feathers, stopping them flying, and the birds become very ill or die. It takes a long time for an oil slick to break down fully. Detergents may be sprayed on an oil slick to make it break up, but these chemicals may also damage wildlife.

**A** Explain how oil is obtained.

**B** Describe how exploiting crude oil may cause environmental problems.

## Energy security

The exploitation of crude oil is also associated with political problems. For example, countries in the Middle East are the largest producers, but the USA and European countries are the largest consumers. Like many other countries, they have to import a lot of the oil they use. If the oil producers decide to reduce supplies or to raise prices, there may be little the oil consumers can do to prevent it.

At some point in the future, normal oil wells will run dry. There are other sources of oil, such as tar sands, but these are more difficult to exploit. Most oil is used for fuel, but around 10% of it is used for making plastics and other **petrochemicals**. Alternative fuels are being developed, but oil cannot easily be replaced as a raw material for petrochemicals.

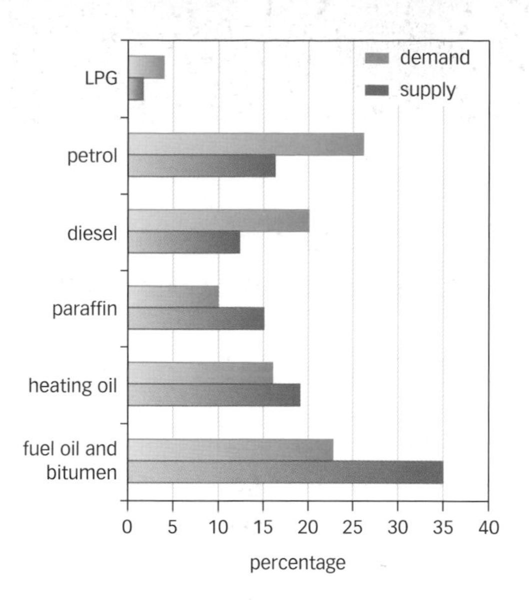


▲ An example of the supply of crude oil fractions and customer demand for them

## Cracking fractions

The supply of some fractions may be greater than the demand for them. **Cracking** is a chemical process that converts large hydrocarbon molecules into smaller, more useful, hydrocarbon molecules. It involves heating oil fractions to a high temperature and passing them over a **catalyst**. For example:

$$\text{octane} \rightarrow \text{hexane} + \text{ethene}$$

$$C_8H_{18} \rightarrow C_6H_{14} + C_2H_4$$

Cracking helps to match the supply of fractions to the demand for them creating more of the widely used fractions such as petrol. Octane and hexane are both alkanes, while ethene is an alkene. Alkenes are useful because they can be used to make polymers.

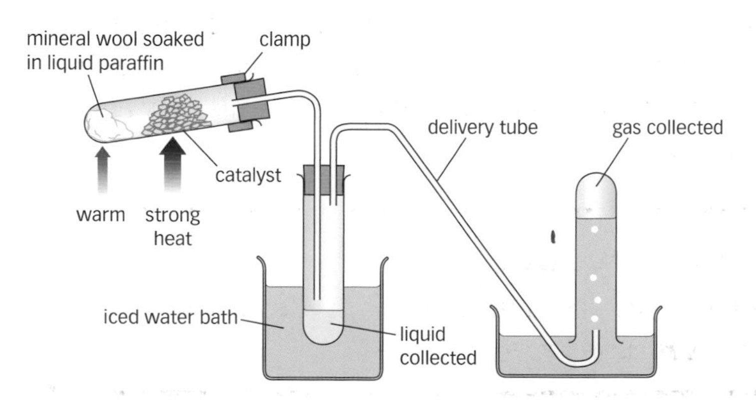


▲ Liquid paraffin can be cracked in the laboratory using this apparatus

### Questions

1. What is cracking?
2. Describe the conditions needed for cracking.

E

3. Use the bar chart to help you answer these questions.
   (a) Identify three fractions where demand is greater than supply.
   (b) Explain how cracking could help to satisfy the demand for petrol.
4. Why are the products of cracking large hydrocarbon molecules more useful than the original large molecules?

C

5. Crude oil is paid for internationally in US dollars. Explain how variations in currency exchange rates between the pound and the dollar might affect the cost of petrol and diesel in the UK.

A*

# 3: Burning hydrocarbons

## Learning objectives

After studying this topic, you should be able to:

- describe what happens when hydrocarbon fuels burn
- describe how the products of combustion can be identified
- write balanced symbol equations for complete combustion

Flames from a propane burner keep the air in this balloon hot

**A** What are the products of complete combustion of a hydrocarbon?

## Key words

**fuel, combustion, complete combustion, limewater**

## Burning fuels

A **fuel** is a substance that reacts with oxygen to release useful energy. This is mostly heat energy. The oxygen needed for burning or **combustion** usually comes from the air. If there is a plentiful supply of air, **complete combustion** happens.

Hydrocarbon molecules contain only hydrogen and carbon atoms. When a hydrocarbon burns completely:

- the carbon atoms react with oxygen to produce carbon dioxide
- the hydrogen atoms react with oxygen to produce water.

Here is the general word equation for the complete combustion of a hydrocarbon:

hydrocarbon + oxygen → carbon dioxide + water

It does not matter which hydrocarbon fuel is being burned. If it reacts completely with oxygen, water and carbon dioxide are produced. For example:

ethane + oxygen → carbon dioxide + water

## A burning experiment

The complete combustion of a hydrocarbon can be investigated in the laboratory. The diagram shows how this could be done for natural gas, which is mostly methane. As the gas burns, the products of combustion escape into the air. They are drawn through the apparatus using a water pump. They pass first into a boiling tube kept cold with iced water, then through **limewater** in a second boiling tube. A clear, colourless liquid collects in the first boiling tube. It looks like water and a simple test shows that it really is water. A piece of blue cobalt chloride paper turns pink when it is dipped in the liquid, showing that water is there.

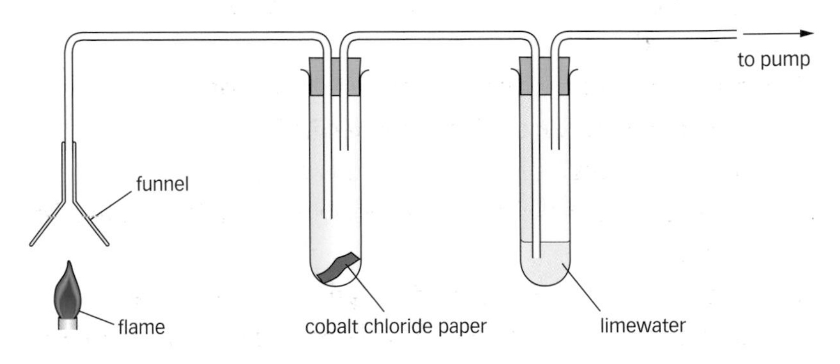


Showing the products of combustion

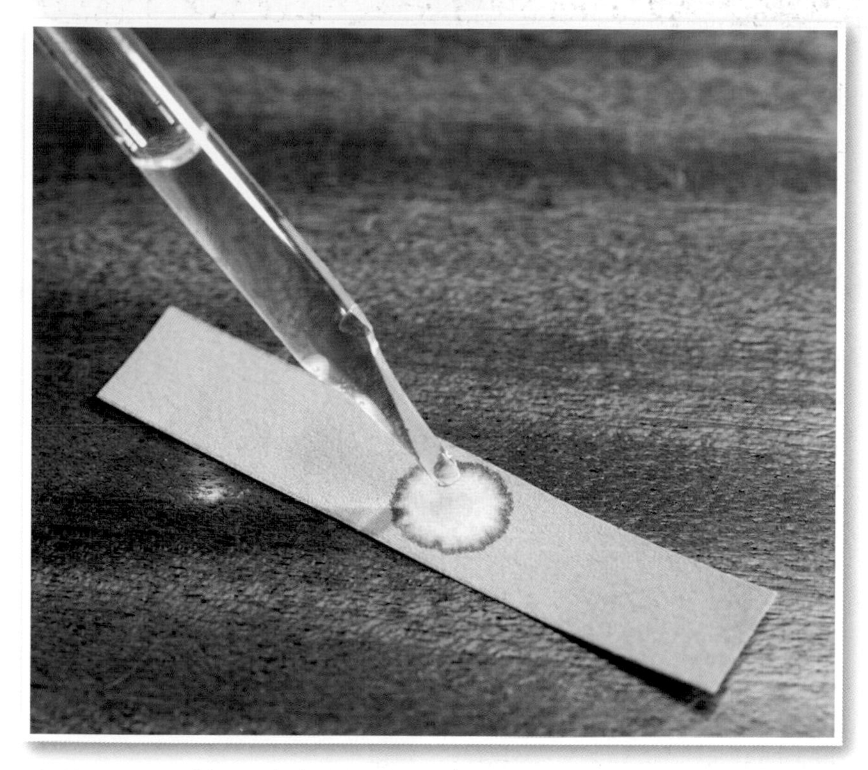

Water turns blue cobalt chloride paper pink

Carbon dioxide turns limewater milky white

The limewater in the second test tube gradually turns milky white. This shows that carbon dioxide is there. The apparatus will also become hot. This shows that heat energy is released during combustion.

## Balanced equations

Complete combustion of a hydrocarbon can be described using a balanced symbol equation. For example, for ethane:

$$2C_2H_6 + 7O_2 \rightarrow 4CO_2 + 6H_2O$$

There are four easy steps involved in writing an equation like this.

1. Write down the formulae for the hydrocarbon and oxygen $O_2$. Do the same for carbon dioxide $CO_2$ and water $H_2O$.
2. Count the number of carbon atoms in the hydrocarbon. Write this number in front of $CO_2$.
3. Count the number of hydrogen atoms in the hydrocarbon. Divide this by two. Write this number in front of $H_2O$.
4. Count the total number of oxygen atoms on the right-hand side. Divide this by two. Write this number in front of $O_2$.

### Exam tip OCR

✔ Make sure you can describe an experiment to show that complete combustion of a hydrocarbon produces water and carbon dioxide.

### Questions

1 Which gas is needed for hydrocarbons to burn?

2 What is a fuel?

E

3 Write a word equation for the complete combustion of octane.

4 Describe an experiment to show that complete combustion of a hydrocarbon produces carbon dioxide and water.

C

5 Write balanced symbol equations for the complete combustion of:

(a) pentane, $C_5H_{12}$

(b) butene, $C_4H_8$

A*

C1

# 4: Using carbon fuels

## Learning objectives

After studying this topic, you should be able to:

- explain the advantages of complete combustion
- explain the factors involved in choosing a fuel
- write balanced symbol equations for incomplete combustion

## Key words

**incomplete combustion, carbon monoxide, soot, pollution, toxicity**

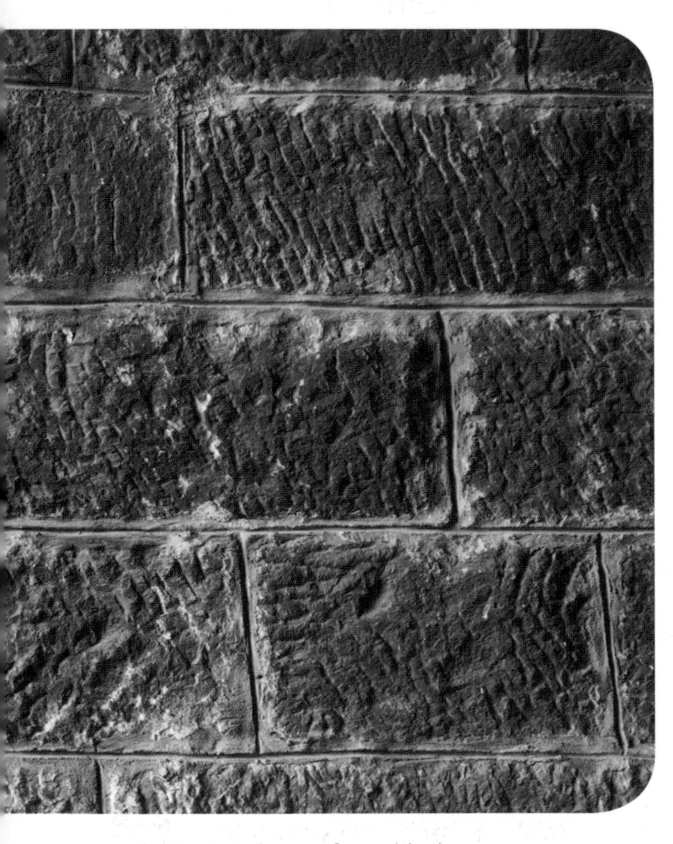

Soot makes surfaces black

## Incomplete combustion

Complete combustion happens when there is a plentiful supply of air. On the other hand, **incomplete combustion** happens when there is a shortage of air. You can easily see the difference using a Bunsen burner.

When the air hole is open, air mixes with the gas in the chimney. The Bunsen burner has a blue flame. When the air hole is closed, the Bunsen burner has a yellow flame. The blue flame from complete combustion releases more energy than the yellow flame from incomplete combustion.

Complete combustion

Incomplete combustion

During incomplete combustion, **carbon monoxide** and carbon are produced rather than carbon dioxide. Here is the general word equation for the incomplete combustion of a hydrocarbon:

hydrocarbon + oxygen → carbon monoxide + carbon + water

The carbon forms tiny black particles called **soot**.

## A poisonous product

Carbon monoxide is a poisonous gas. It is also colourless and tasteless, and it has no smell. So you cannot tell if you are breathing it in. If you breathe in a lot of carbon monoxide, you can fall ill and even die. This is why gas appliances like gas fires and boilers should be serviced regularly. They are checked to make sure they are getting enough air for complete combustion. This helps the appliance release more heat and less soot, without producing carbon monoxide.

### Balanced equation

Incomplete combustion of a hydrocarbon can be described using a balanced symbol equation. For example, for ethane:

$$C_2H_6 + 2O_2 \rightarrow CO + C + 3H_2O$$

## Which is the best fuel?

There are seven factors to consider when choosing a fossil fuel. These are: availability, cost, energy value, **pollution**, ease of use, **toxicity**, and storage. The table shows these factors for three fuels.

| Factor | | Coal | Fuel oil | Natural gas |
|---|---|---|---|---|
| Availability | How easy is it to get? | Easy | Easy | Needs pipes |
| Cost | How expensive is it? | Low | High | High |
| Energy value | How much energy does it release? | Medium | High | High |
| Pollution | Does it contribute to acid rain or the greenhouse effect? | High | Medium | Medium |
| Ease of use | How easy is it to use? | Difficult | Easy | Easy |
| Toxicity | How poisonous is it? | Low | Medium | Very low |
| Storage | How easy is it to store? | Easy | Easy | Difficult |

The amount of fossil fuels being burnt is increasing because of an increasing world population, and increased use in newly industrialised countries such as India and China.

A household carbon monoxide alarm

**A** List the factors involved in choosing a fossil fuel.

### Exam tip OCR

✔ You can remember the factors in the table from their first letters, ACE PETS.

### Questions

1. What are the products of incomplete combustion of hydrocarbons?
2. Describe the differences, apart from colour, between blue and yellow Bunsen burner flames.

E

3. Write a word equation for the incomplete combustion of octane.
4. Explain the importance of servicing gas appliances.

C

5. Compare the fuels in the table. Which fuel would be best for power stations, and why?

A*

## Learning objectives

After studying this topic, you should be able to:

- recall the percentage of the main gases in the air
- describe how the present-day atmosphere evolved

The atmosphere is the Earth's outer layer

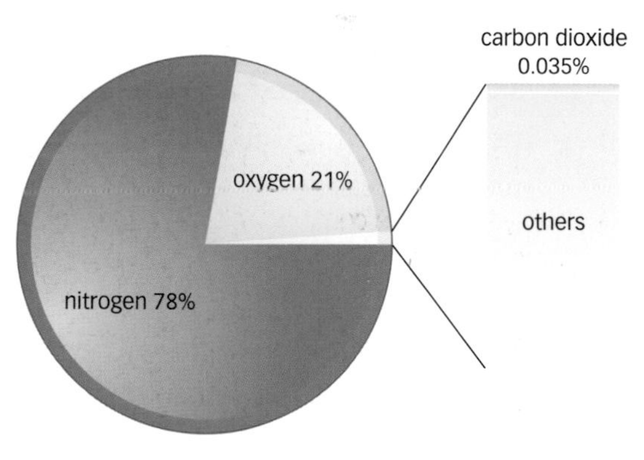


This pie chart shows the proportions of the main gases in air

**A** Outline the main differences between the Earth's early atmosphere and its atmosphere today.

## The atmosphere today

The Earth's **atmosphere** is a mixture of gases surrounding the planet. The Earth's gravity stops these gases escaping into space. The lower part of the atmosphere is the air that we breathe. It has stayed very much the same for the last 200 million years. The original atmosphere was formed by gases escaping from the Earth's interior. Just two gases, nitrogen and oxygen, make up about 99% of the air. Today, there are smaller proportions of other gases in the air, such as carbon dioxide, water vapour, and noble gases such as argon.

### Forming the atmosphere

Scientists have several theories about how the atmosphere formed. It is difficult to be certain which one, if any, is correct, because no-one was around to record events as they happened. One possible theory involves volcanoes and plants.

The Earth is about 4.5 billion years old. There was a lot of **volcanic activity** during its first billion years. When they erupt, volcanoes release huge volumes of gases. These are mainly water vapour and carbon dioxide, with smaller proportions of other gases such as ammonia and methane. It is likely that the Earth's early atmosphere came from gases released from inside the Earth's crust by volcanoes. The water vapour condensed to form the oceans, leaving an atmosphere of mostly carbon dioxide. There would have been little or no oxygen in the early atmosphere.

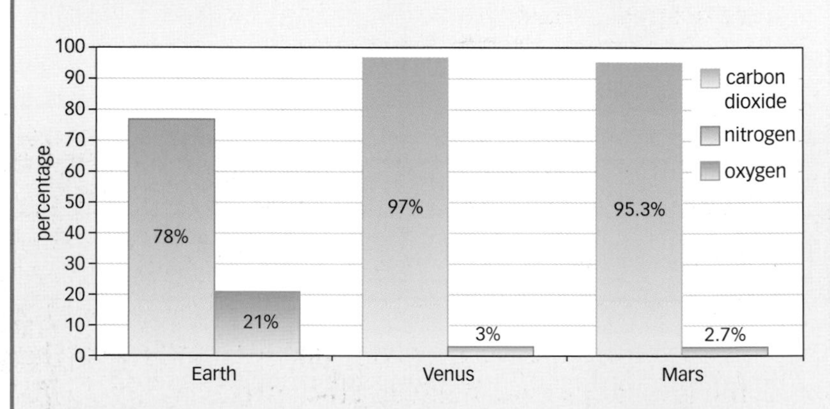


The Earth's early atmosphere was probably like the present-day atmospheres of Mars and Venus, shown above with Earth for comparison

## An evolving atmosphere

Plants make their own food using **photosynthesis**. As part of this process, they use carbon dioxide from the atmosphere and release oxygen. When plants evolved, their photosynthesis reduced the amount of carbon dioxide in the atmosphere. Photosynthesis also increased the amount of oxygen until it reached today's level.

Photosynthesis was not the only reason why carbon dioxide levels decreased. Carbon dioxide is a soluble gas and large amounts of it dissolved in the oceans. It was used by marine organisms to form the chemicals needed for their shells and skeletons. Limestone is a sedimentary rock formed from these shells and skeletons. Carbon dioxide is locked away inside limestone, mainly as calcium carbonate. More carbon dioxide was locked away in fossil fuels, which formed from the remains of ancient plants and animals.

Ammonia was broken down by sunlight and by its reaction with oxygen, forming nitrogen. Nitrogen is an unreactive gas. Once released into the atmosphere, nitrogen reacts with other substances only with difficulty. So the levels of nitrogen in the atmosphere increased until they reached today's level.

### Questions

1 What are the approximate proportions of nitrogen, oxygen, and carbon dioxide in the atmosphere today?

2 How is the early atmosphere thought to have formed?

E

3 Explain why the level of oxygen in the atmosphere increased.

C

4 Explain why the level of nitrogen in the atmosphere increased.

5 Describe three processes responsible for reducing the level of carbon dioxide in the early atmosphere.

A*

### Key words

**atmosphere**, **volcanic activity**, **photosynthesis**

### Did you know...?

At current rates of photosynthesis, it would take living things just 2000 years to make all the oxygen found in the Earth's atmosphere today.

Shelly limestone contains the fossilised remains of ancient marine organisms

### Exam tip OCR

✔ Remember that oxygen and nitrogen in the atmosphere increased over time. Carbon dioxide, ammonia, and methane decreased.

# 6: Air pollution

## Learning objectives

After studying this topic, you should be able to:

- explain some of the processes affecting the composition of the air
- describe the origin and effects of some atmospheric pollutants

Burning coal releases carbon dioxide

**A** Describe a simple carbon cycle involving respiration, combustion, and photosynthesis.

## Key words

**respiration**, **global warming**, **deforestation**, **pollutant**, **acid rain**, **photochemical smog**, **catalytic converter**

## Carbon dioxide

About 0.035% of the Earth's atmosphere is carbon dioxide. Plants absorb this gas to make their own food by photosynthesis. However, the levels of carbon dioxide are gradually increasing year by year.

Carbon dioxide is released by **respiration**, and oxygen is used up. Respiration is a process that can increase the level of carbon dioxide in the air and decrease the level of oxygen. Combustion can do this, too. Large amounts of carbon dioxide are released through our use of fossil fuels such as coal, oil, and natural gas.

The processes that release carbon dioxide ought to be balanced by the processes that remove it, but they are not.

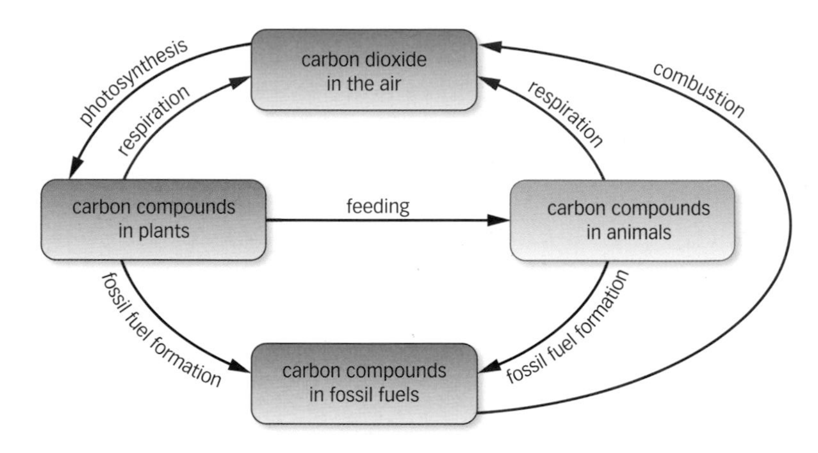


Processes releasing and absorbing carbon dioxide form part of the carbon cycle

Increasing levels of carbon dioxide are thought to be one of the reasons for **global warming**. Human activities are releasing carbon dioxide faster than it can be removed. Most electricity production and transport uses fossil fuels. The human population is increasing rapidly. This adds to the demand for energy, so more fossil fuels are used. It also adds to the demand for land for buildings, roads, and farms.

Forests may be cut down to make way for human use. Such **deforestation** has a big impact on carbon dioxide levels. The trees may be burnt to clear the land quickly, and this releases carbon dioxide. With fewer trees left, less carbon dioxide can be removed by photosynthesis.

## Atmospheric pollutants

**Pollutants** are substances released into the environment that may damage living things, including us. Pollutants may be released into the water or soil, but atmospheric pollutants are released into the air. Some are produced as a result of burning fossil fuels.

Carbon monoxide is produced if fuels burn in a poor supply of oxygen. This can happen in car engines. Carbon monoxide is a poisonous gas. It can cause sickness and even death.

> Nitrogen and oxygen are the main gases in air. Normally they do not react together. But the high temperature and pressure inside an engine allows them to react to produce $NO_x$.

Photochemical smog contains many unpleasant atmospheric pollutants that can damage health

Fossil fuels naturally contain small amounts of sulfur compounds. These impurities form sulfur dioxide when the fuel is burnt. This gas dissolves in the water in clouds, forming acids. The acids make rain more acidic than normal. Such **acid rain** erodes stonework and corrodes metals. It can kill trees, and living things in rivers and lakes. $NO_x$ causes acid rain, too. It can also react with other atmospheric pollutants, particularly in sunlight, to produce **photochemical smog**. This can cause breathing problems.

## Controlling pollutants

Vehicles release large amounts of atmospheric pollutants. The levels of these must be controlled to reduce damage to human health and the environment. **Catalytic converters** are fitted to vehicle exhaust systems. They convert carbon monoxide into carbon dioxide:

$$\text{carbon monoxide} + \text{nitrogen oxide} \rightarrow \text{nitrogen} + \text{carbon dioxide}$$

$$2CO + 2NO \rightarrow N_2 + 2CO_2$$

Notice that catalytic converters can also control oxides of nitrogen.

This forest has been badly damaged by acid rain

**B** What are atmospheric pollutants?

### Questions

1. Why is it important to control the levels of atmospheric pollutants?

2. What is photochemical smog and how does it form?
3. What is acid rain, how does it form, and why does it cause problems?

4. Explain how carbon monoxide forms, why it is dangerous, and how emissions from vehicle exhausts may be controlled.

5. Outline why deforestation may have a big impact on carbon dioxide levels.

C1

# 7: Alkanes and alkenes

## Learning objectives

After studying this topic, you should be able to:

- ✔ describe the structures of alkanes and alkenes
- ✔ recognise alkanes and alkenes from their molecular or displayed formula
- ✔ describe the nature of covalent bonds
- ✔ describe a laboratory test for unsaturation

## Key words

**hydrocarbon, alkane, displayed formula, covalent bond, alkene, saturated, unsaturated, addition reaction**

**A** What do — and ═ mean in a displayed formula?

**B** Draw the displayed formula for butane.

## The alkanes

**Hydrocarbons** are compounds that contain only hydrogen atoms and carbon atoms. The **alkanes** are a 'family' of hydrocarbons. The table shows information about three alkanes.

| Name of alkane | Chemical formula |
|---|---|
| methane | $CH_4$ |
| ethane | $C_2H_6$ |
| propane | $C_3H_8$ |
| butane | $C_4H_{10}$ |

Notice that the names of alkanes end in 'ane'. The first part tells you how many carbon atoms each molecule contains. 'Meth' means one, 'eth' means two, and 'prop' means three. The alkanes have the same general formula: $C_nH_{2n+2}$ where $n$ stands for the number of carbon atoms.

A **displayed formula** shows how the atoms in a molecule are joined together. Each atom is shown by its chemical symbol, and each chemical bond by a straight line. Hydrogen atoms make one bond but carbon atoms make four bonds. In alkanes, the carbon atoms are joined together by single **covalent bonds**.

```
     H            H   H            H   H   H
     |            |   |            |   |   |
 H — C — H    H — C — C — H    H — C — C — C — H
     |            |   |            |   |   |
     H            H   H            H   H   H
  methane        ethane           propane
```

▲ Displayed formulae for three alkanes

## The alkenes

The **alkenes** are another 'family' of hydrocarbons. Unlike the alkanes, alkenes contain a double covalent bond between carbon atoms. The table on the next page shows information about three alkenes.

| Name of alkene | Chemical formula |
|---|---|
| ethene | $C_2H_4$ |
| propene | $C_3H_6$ |
| butene | $C_4H_8$ |

Notice that the names of alkenes end in 'ene'. The first part again tells you how many carbon atoms each molecule contains. A double bond is shown by two lines in a displayed formula.

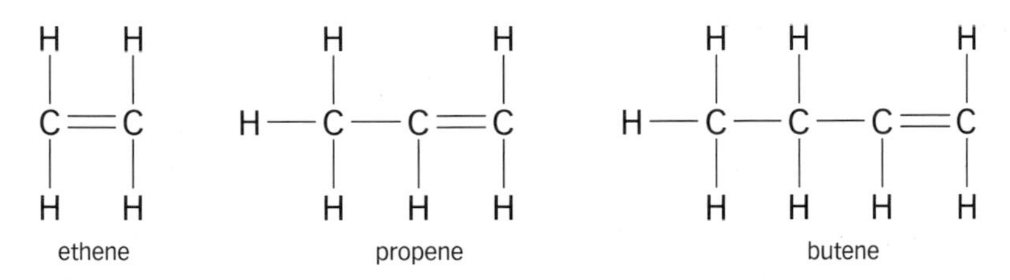


Displayed formulae for three alkenes

Bromine reacts with alkenes but not with alkanes. Bromine water, bromine dissolved in water, is orange. It turns colourless when it is mixed with alkenes, but not when it is mixed with alkanes.

## Unsaturation

A covalent bond is a shared pair of electrons. In hydrocarbons, a hydrogen atom shares its electron with an electron from a carbon atom, forming a covalent bond. Alkanes are **saturated** compounds. They contain only single covalent bonds between their carbon atoms. On the other hand, alkenes are **unsaturated** compounds. They contain a double covalent bond between two of their carbon atoms. The reaction between bromine and an alkene is an **addition reaction** used to test for unsaturation.

A colourless dibromo compound forms when bromine reacts with an alkene

Alkenes, like other unsaturated hydrocarbons, decolorise bromine water. Alkanes cannot decolorise bromine water.

### Exam tip OCR

✔ Write a and e clearly so you do not muddle up alkanes and alkenes.

### Questions

1. What are hydrocarbons?
2. Explain, with examples, what alkanes and alkenes are.

E

3. What is the general formula for alkenes?
4. Describe the differences between alkanes and alkenes. Include displayed formulae in your answer.

C

5. Explain what saturated and unsaturated hydrocarbons are, and how they may be distinguished.

A*

# 8: Making polymers

## Learning objectives

After studying this topic, you should be able to:

- describe the relationship between a monomer and its polymer
- describe how polymerisation works
- construct displayed formulae of monomers and polymers

These items are all made from polymers

Separate plastic beads

## Polymers

The molecules in plastics are called polymer molecules. **Polymers** are very large molecules. They are made when very many smaller molecules, called **monomers**, join together. For example:

- poly(ethene) is a polymer made from ethene monomers
- poly(propene) is a polymer made from propene monomers.

Notice that the name of the polymer comes from the name of its monomer. You probably know poly(ethene) as polythene. It is the plastic used for carrier bags and bottles. Polystyrene is much tougher than polythene, so it is used for packaging.

**A** Name the polymer made from chloroethene monomers.

**B** Name the monomer needed to make polystyrene.

## Polymerisation

The chemical reaction needed to make a polymer from monomers is called **polymerisation**. A high pressure and a catalyst are usually needed for polymerisation to happen. During polymerisation, the monomer molecules join together end to end. Think of a necklace made of lots of beads joined together. The beads are like the monomer molecules, and the necklace is like the polymer molecule they make.

A necklace made from plastic beads joined together

The displayed formula for a polymer is based on the displayed formula of its monomer.

## Addition polymerisation

The type of polymerisation involving alkene monomers is called **addition polymerisation**. During addition polymerisation, unsaturated monomer molecules react together to form a saturated polymer molecule.

You can easily draw the displayed formula of an addition polymer from its monomer:

- draw the monomer but change the double bond to a single bond
- draw two bonds either side of where the double bond was
- put brackets round the molecule, passing through the outer bonds.

Reverse the process to draw the displayed formula of a monomer from its polymer.

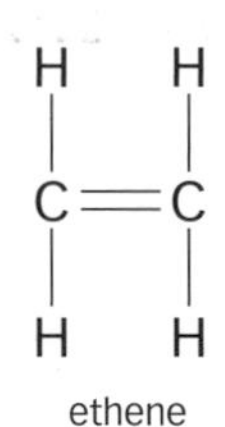


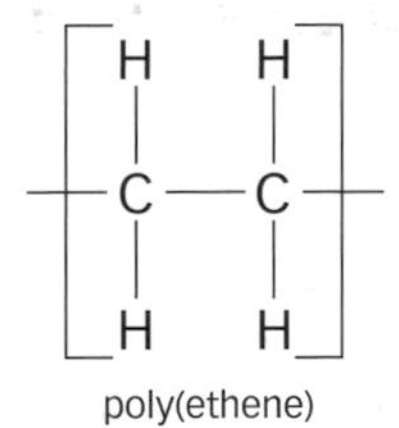


▲ The displayed formulae of a monomer and the polymer it forms

### Did you know...?

The polymers poly(ethene), PVC, and Teflon were all discovered by accident.

### Key words

**polymer**, **monomer**, **polymerisation**, **addition polymerisation**

### Questions

1 Explain what monomers and polymers are. E

2 What feature do monomer molecules have in common?

3 Describe what polymerisation is and the conditions needed for it to happen.

4 Explain which of the two displayed formulae belongs to a polymer. C

H H
C=C
H $CH_3$

A

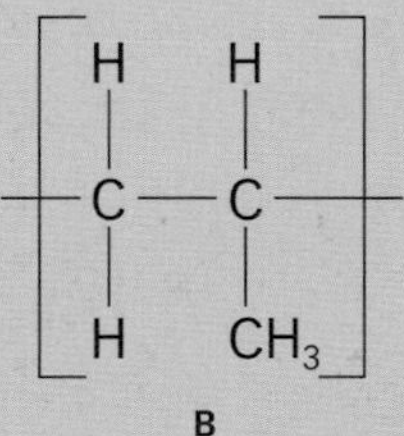


5 Draw the displayed formula of chloroethene, $CH_2$=CHCl, and its polymer. A*

### Exam tip OCR

✔ Make sure you can recognise the displayed formula for a polymer.

# 9: Designer polymers

## Learning objectives

After studying this topic, you should be able to:

- use information to explain the uses of polymers
- describe the properties of waterproof and breathable materials
- explain how the properties of plastics are determined by their structure

Plastic outdoor furniture

**A** What property, other than stiffness, makes HDPE more suitable than LDPE for making outdoor furniture?

### Key words

**cross-link**, **waterproof clothing**, **breathable clothing**, **laminated**

## Using polymers

Different polymers have different properties. The uses of a particular polymer depend upon its properties. For example, LDPE and HDPE are two types of poly(ethene). The table compares some of their properties.

| | LDPE | HDPE |
|---|---|---|
| Density in g/cm³ | 0.92 | 0.95 |
| Maximum usable temperature in °C | 85 | 120 |
| Strength in MPa | 11.8 | 31.4 |
| Relative flexibility | flexible | stiff |

LDPE is more suitable for making plastic bags because it is flexible. On the other hand, HDPE is more suited to making outdoor furniture because it is stiff.

Polystyrene is used in protective packaging and as insulation. Nylon and polyester fibres are used to make clothing.

### Forces and cross-links

The polymer molecules in many plastics are attracted to each other only by weak intermolecular forces. These are easily overcome, so such plastics have low melting points. They are also easily stretched because the molecules can slide past one another.

The polymer molecules in other plastics have strong covalent bonds between them. These **cross-links** stop the molecules moving apart or past one another. Such plastics have high melting points. They are rigid and cannot be easily stretched.

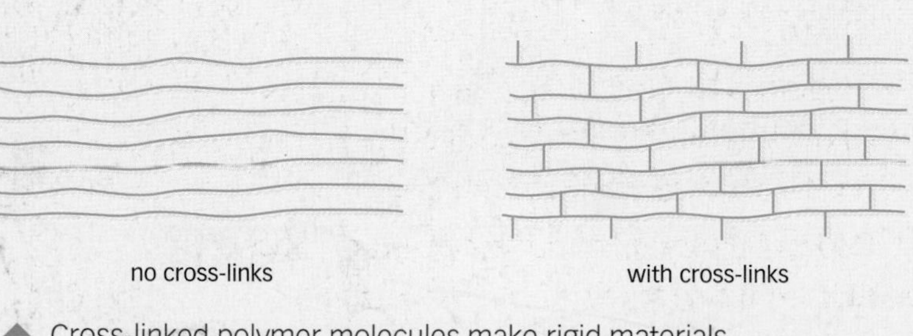


Cross-linked polymer molecules make rigid materials

## Polymers outdoors

**Waterproof clothing** stops water getting in when it rains. Unfortunately, it also stops your sweat getting out. **Breathable clothing** stops the rain water getting in *and* lets sweat out.

Nylon is waterproof, tough, lightweight, and blocks UV light. But it does not let water vapour through it. Sweat condenses inside nylon clothing, making you damp and uncomfortable. Gore-Tex® has the same properties as nylon but it is also breathable. So you stay dry in the rain, even if you are exercising hard.

Breathable clothing helps active outdoor people cope with sweat

### A layered material

Gore-Tex® and other breathable materials contain different layers. One layer is made from polyurethane, or a polymer called PTFE. This contains many tiny holes. Each one is too small for water droplets to pass through. However, they are large enough for water vapour to pass through. This is why rain cannot get in but sweat can get out. The PTFE layer is fragile, so it is **laminated** with nylon.

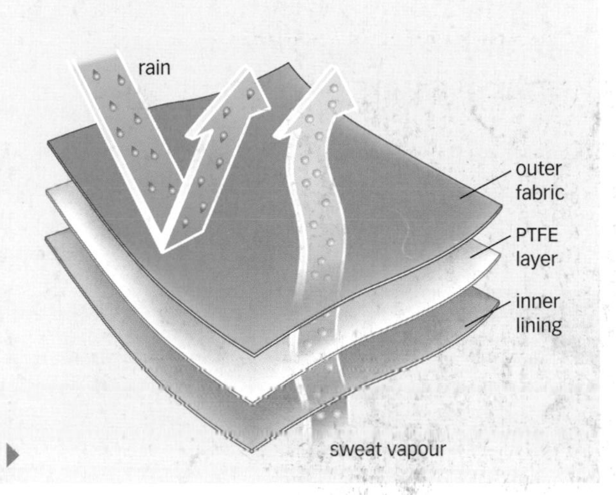


Breathable materials contain several layers

### Questions

1. Use the information in the table to explain why HDPE is better for making disposable cups than LDPE.
2. Describe the advantages of waterproof and breathable clothing.

E

3. Explain why breathable materials are suitable for outdoor clothing.

C

4. Suggest why melamine has a high melting point and is rigid, whereas poly(ethene) has a low melting point and is stretchy.
5. Describe how the construction of Gore-Tex® makes it waterproof and breathable.

A*

### Did you know...?

The PTFE layer in Gore-Tex® contains around 1.4 billion holes per square centimetre.

# 10: Polymer problems

## Learning objectives

After studying this topic, you should be able to:

- explain that many polymers are non-biodegradable
- describe and explain some problems with disposing of polymers
- explain why chemists are developing new types of polymers

## Key words

**non-biodegradable**, **landfill site**, **biodegradable**

## Did you know...?

It has been estimated that there are around 17 000 pieces of plastic per square kilometre of ocean, and that plastic bags take 1000 years to degrade. The chemical energy contained in ten carrier bags could take a family car 1 km down the road.

**A** What does non-biodegradable mean?

**B** Explain why many polymers are difficult to dispose of.

## Litter, litter, everywhere

Many polymers are unreactive. They do not react easily with acids, alkalis, or many other chemicals. They are often tough and strong. These properties are some of the reasons why polymers are so very useful. For example, we can store shampoo and water in plastic bottles without worrying that the bottle will dissolve. We can bury plastic gas pipes deep underground, knowing that they will serve us for many years. Unfortunately, these very benefits make polymers difficult to dispose of.

Decay bacteria help substances rot away. But many polymers are **non-biodegradable**. They cannot be decomposed by bacterial action. Plastic waste litters the Earth almost everywhere.

## Disposing of polymers

When you throw out a used plastic item, the chances are it will end up in a **landfill site**. These are places where local authorities take waste to be buried. In the UK, around 85% of waste is dumped in landfill sites. These eventually become full of waste. It is difficult to find land for a new landfill site. Local people may object to a new landfill site, for example, because it will be unsightly. Although the plastic itself may not smell, other materials rotting away are smelly and may attract rats and gulls.

▲ Landfill sites waste valuable land

Crude oil is the raw material for most polymers. It is also the source of fuels such as petrol. Waste plastic can be burnt to release heat energy. This can be used to generate electricity or to heat buildings. The plastic must be burnt at a high temperature to stop toxic gases being made. Unless the heat energy released is used, disposal by burning is a waste of a valuable resource.

It is possible to recycle polymers. For example, PET is the tough plastic used to make drinks bottles. It can be recycled to make fibres for clothing. Unfortunately, plastic waste is usually a mixture of different polymers. It must be sorted into separate types of polymer, often by hand. This is difficult and expensive to do.

## Chemists to the rescue

Chemists are developing new types of polymers. **Biodegradable** polymers can be broken down by microorganisms. In one type, starch is added to poly(ethene) or other polymers during manufacture. Once the polymer gets wet, bacteria break down the starch, causing the plastic item to crumble into very small pieces. Other new polymers dissolve in water. For example, dishwasher detergent tablets are wrapped in a plastic that dissolves once it gets wet in the dishwasher.

### Questions

**E**

1. Describe three ways that waste polymers can be disposed of.
2. Explain why recycling symbols are put on plastic items.

**C**

3. Describe the benefits of biodegradable polymers.
4. Explain why burying waste plastics and burning them is a waste of valuable resources.

**A***

5. In 2002, Ireland introduced a tax on plastic carrier bags. Sales of carrier bags fell by 90% but sales of other bags increased by 400%.
   (a) Suggest why the tax was introduced.
   (b) Suggest why plastic bag sales changed in the way they did.

▲ A waste incinerator

**Did you know...?**

It takes 25 PET bottles to make enough fibre for a fleece jacket.

PET

low-density poly(ethene)

high-density poly(ethene)

poly(propene)

PVC

polystyrene

▲ Recycling symbols found on plastic items

**Exam tip** **OCR**

✔ Make sure you can describe the drawbacks of different ways to dispose of waste polymers.

# 11: Cooking

## Learning objectives

After studying this topic, you should be able to:

- explain why cooking is a chemical change
- explain the changes observed during cooking
- describe what happens when baking powder is heated

Uncooked meat

Cooked meat

**A** State three signs that a chemical reaction has taken place.

## Cooking up a chemical change

Cooking involves **chemical changes**. You can tell that a chemical change happens if

- a new substance is made
- the change is irreversible
- an energy change happens.

Eggs and meat are good sources of **protein**. When they are cooked, they absorb heat. Their appearance and **texture** or feel changes too. These changes cannot be reversed. While eggs and meat cook, their protein molecules change shape. This is called **denaturing**.

### Cooking changes

The texture and appearance of egg white, egg yolk, and meat change when they are cooked. This is because their protein molecules change shape when heated. Once a protein has been denatured, it cannot be changed back again, so the process is irreversible.

Potatoes contain a lot of **starch**, a complex **carbohydrate**. Raw potato is tough and bitter-tasting. It is difficult for us to digest. Cooked potato, however, is soft and sweet-tasting. Starch grains are trapped inside potato cells. When a potato is cooked, the cell walls break. This makes the potato lose its rigid structure and become softer. It also releases the starch grains. These absorb water and swell up. Cooked potato is easier to digest than raw potato.

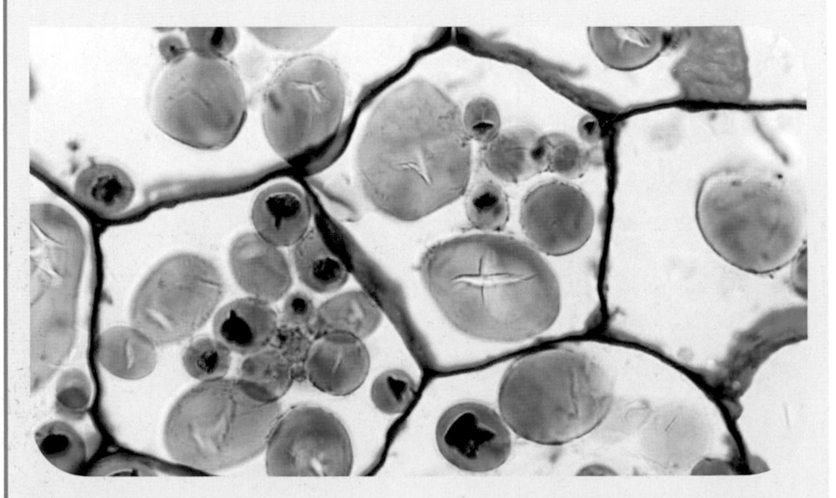

Starch grains in potato cells, as seen through a microscope (× 250)

## Baking powder

Baking powder is added to cake mixes to help the cake rise during cooking. It contains sodium hydrogencarbonate. When this is heated, it breaks down and releases carbon dioxide. Here are the equations for the reaction:

sodium hydrogencarbonate → sodium carbonate + carbon dioxide + water

$$2NaHCO_3 \rightarrow Na_2CO_3 + H_2O + CO_2$$

Baking powder helps cakes rise in the oven

The reaction is called a **decomposition** reaction because the sodium hydrogencarbonate decomposes or breaks down. The carbon dioxide released expands in the heat, helping the cake rise. Limewater turns milky white when carbon dioxide is bubbled through it.

### Questions

1 Describe the changes that happen when an egg is cooked.

2 Describe how baking powder helps cakes rise in the oven.

E

3 Describe what happens to the protein molecules in meat as it cooks.

4 Write an equation for the decomposition of sodium hydrogencarbonate, and describe a chemical test for carbon dioxide.

C

5 Explain the changes that happen when eggs and potatoes are cooked.

A*

### Exam tip OCR

✔ You need to be able to write the word equation for the decomposition of sodium hydrogencarbonate. If you are taking the Higher Tier paper, you need to be able to write the balanced symbol equation too.

**B** Why is the reaction that happens when sodium hydrogencarbonate is heated called a decomposition reaction?

**C** Limewater contains calcium hydroxide. Calcium carbonate and water are formed when this reacts with carbon dioxide. Write a word equation for the reaction.

### Key words

**chemical change**, **protein**, **texture**, **denaturing**, **starch**, **carbohydrate**, **decomposition**

## Learning objectives

After studying this topic, you should be able to:

- recall the main types of food additive and describe their function
- describe what emulsifiers are
- describe how emulsifiers work

## Key words

**food additive, antioxidant, food colour, flavour enhancer, emulsifier, hydrophilic, hydrophobic**

**A** Give an example of an additive used as an antioxidant and explain what it does.

**B** Suggest why potato crisps and other fatty foods gradually 'go off' and develop an unpleasant taste after their bag is opened.

## Did you know...?

Food additives are given E-numbers if the European Food Safety Authority considers them safe. Even vinegar has an E-number. It is E260.

## Food additives

**Food additives** are substances added to food to improve its properties. Many food additives are natural substances such as vinegar. This preserves pickles and other food, and it gives the food a pleasant flavour. There are several different types of food additive. The table describes the main ones: **antioxidants**, **food colours**, **flavour enhancers**, and **emulsifiers**.

| Type of additive | Reason for adding it | Example |
|---|---|---|
| antioxidant | stops food reacting with oxygen and going off | vitamin C |
| food colour | gives the food an improved colour to make it more attractive | caramel |
| flavour enhancer | improves the flavour of the food | monosodium glutamate |
| emulsifier | stops oil and water mixed together in food from separating | egg yolk lecithin |

Food exposed to the air will gradually react with oxygen. The fats in food are particularly likely to do this. When fats are oxidised, they form substances with an unpleasant rancid taste. Antioxidants stop these reactions happening.

Food additives can improve the shelf-life of food, helping to prevent waste

## Emulsifiers

You can make a simple salad dressing by whisking a mixture of vegetable oil and vinegar together. If it is left to stand, the oil and vinegar separate into two layers. This is because oil and water do not mix.

▲ Oil and water do not mix

Emulsifiers are substances that help oil and water to mix, and then not separate. Mayonnaise contains an emulsifier, often egg yolk lecithin. This improves the texture of the mayonnaise, making it thicker than the oil or vinegar on their own, and makes the mixture less likely to separate.

Emulsifier molecules have a water-loving part (the **hydrophilic** end) and an oil-loving part (the **hydrophobic** end).

▲ Emulsifier molecules have two different parts

### Emulsifier molecules

The hydrophobic end of an emulsifier molecule bonds with oil molecules. The hydrophilic end bonds to water molecules. This stops droplets of oil or water forming and joining together, so helping to keep the mixture from separating.

▲ Emulsifiers help to stop the oil and water in mayonnaise separating

**Exam tip**

✔ You do not need to recall any examples of food additives or their E-numbers for the exam.

### Questions

1 List the main types of food additive.

2 Describe why emulsifiers are needed in foods like mayonnaise.

3 Describe the functions of the main types of food additive.

↓ E

4 Describe the structure of an emulsifier molecule.

↓ C

5 Describe how an emulsifier works.

↓ A*

# 13: Smells

## Learning objectives

After studying this topic, you should be able to:

- recall some examples of natural and synthetic cosmetics
- describe how to make an ester
- describe the properties needed by a perfume
- explain evaporation in terms of the kinetic particle theory

Moisturisers are cosmetics that are intended to prevent dry skin and hide imperfections

Fragrant rose oil can be extracted from rose petals

## Cosmetics

**Cosmetics** are substances used to change a person's appearance or smell, hopefully for the better. They include skin creams, deodorants, lipstick, nail varnish, and perfume. Some cosmetics are made from natural sources. For example, skin creams may contain glycerine to keep the skin moist and plant oils to stop it drying out. Cosmetics may also contain artificial or **synthetic** substances. For example, skin creams may contain polyethylene glycol rather than plant oils or glycerine.

## Perfumes

**Perfumes** are substances with pleasant smells. They may be used on their own, or they may be used to improve the smell of other products such as cosmetics and air fresheners. Perfumes need certain properties so that they work properly. The table describes some of these properties and why they are needed.

| Desirable physical property | Reason why the perfume needs the property |
|---|---|
| Evaporates easily | The perfume particles must reach the nose easily. |
| Non-toxic (not poisonous) | It must not poison the user or make them ill. |
| Does not react with water | It must not react with perspiration. |
| Does not irritate the skin | It could not be applied to the skin if it would cause harm. |
| Insoluble in water (does not dissolve in water) | It must not be washed off easily. |

Perfumes can be obtained from natural sources. These are often plant materials such as fruits, flowers, or certain woods, for example sandalwood.

**A** Explain why a perfume should not react with water or dissolve in it.

## Volatility

**Volatility** is a measure of how easily a liquid evaporates. Very volatile liquids evaporate easily. The molecules in a liquid are attracted to one another. They must overcome these forces so that they can escape from the liquid, allowing the liquid to evaporate. Molecules with lots of energy can do this, which is why liquids evaporate faster when they are warmed up. Perfumes evaporate easily because the attractions between their molecules are very weak and easily overcome.

## Esters

**Esters** are compounds with pleasant smells. They are used in perfumes. Esters are found naturally in fruit but they can be made in the laboratory, too. This is done by reacting an alcohol with an **organic acid** such as methanoic acid or ethanoic acid. In general:

alcohol + acid → ester + water

Ethyl ethanoate is an ester that smells of pear drops. It is made by reacting ethanol with ethanoic acid:

ethanol + ethanoic acid → ethyl ethanoate + water

The reaction happens if some concentrated sulfuric acid is added (to act as a catalyst) and the mixture is warmed to around 60 °C.

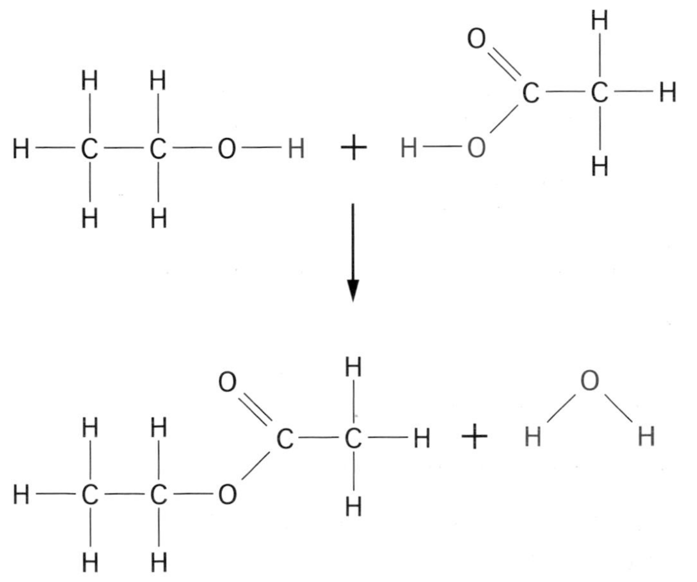

▲ Esters are produced by reacting an alcohol with an organic acid. Water is produced as well.

### Did you know...?

Women in ancient Rome used crocodile dung to lighten the colour of their faces.

### Key words

cosmetics, synthetic, perfume, volatility, ester, organic acid

**B** What two types of substance react together to make an ester?

**C** What does ethyl ethanoate smell like?

### Questions

1. Name two natural sources of perfume.
2. Describe five properties needed by a perfume.

E

3. Ethyl methanoate smells of raspberries. Describe how you could produce this ester using ethanol and methanoic acid. Include an equation in your answer.
4. Explain why perfumes should evaporate easily and not irritate the skin.

C

5. Explain, in terms of particles, why a perfume may smell stronger once it has been applied to the skin.

A*

# 14: Nail varnish and testing

## Learning objectives

After studying this topic, you should be able to:

- describe a solution in terms of the substances it contains
- interpret information on the effectiveness of solvents
- explain why water will not dissolve nail varnish
- explain why testing cosmetics on animals has been banned in the EU
- describe some advantages and disadvantages of animal testing

Sugar is soluble in water. It dissolves to form sugar solution.

## Key words

**solution, solvent, solute, soluble, insoluble**

Nail varnish remover dissolves nail varnish colours

## Solutions

A **solution** is a mixture of a solvent and a solute. The **solvent** is the liquid that does the dissolving, and the **solute** is the substance that becomes dissolved. For example, water and sugar mix completely to form a solution that does not separate out. The water is the solvent and the sugar is the solute.

Nail varnish is a coloured coating painted onto fingernails or toenails

**A** Sea water contains dissolved salt. Name the solvent and solute involved.

If a substance can dissolve in a particular liquid, we say that it is **soluble** in that liquid. If the substance cannot dissolve, we say that it is **insoluble** in that liquid. For example, sand is insoluble in water. Nail varnish is insoluble in water. Different solvents are needed to remove it from nails.

## Nail varnish remover

Esters such as ethyl ethanoate are useful as solvents. They dissolve some substances that are insoluble in water. For example, they are an ingredient of nail varnish remover. This liquid dissolves nail varnish so that it can be removed from a nail.

**B** Suggest why nail varnish should be insoluble in water.

## Attractive particles

Whether or not a substance will dissolve in a particular solvent depends upon the relative strength of the attractions between

- the particles of the substance and the solvent molecules
- the particles of the substance itself
- the solvent molecules.

The attractions between the particles in nail varnish and water molecules are weaker than the attractions between the particles in nail varnish. They are also weaker than the attractions between water molecules. So nail varnish will not dissolve in water.

## Testing cosmetics

Cosmetics are meant to go on the body, so they should not contain substances that cause illness or allergic reactions. It is important that cosmetics are tested to make sure they are safe to use. In some countries, cosmetics and their ingredients are tested on animals. This used to happen in the European Union (EU) but it was banned in 2009. Instead, manufacturers use ingredients that have already been tested on animals, use human volunteers, or avoid using new or untested ingredients.

In the UK, cosmetics can only be tested on animals if they also have medicinal properties. There are advantages to testing medicines on animals. For example, it helps researchers find new medicines to improve the health of humans and other animals. It also helps to ensure that new medicines are safe to use. On the other hand, it is expensive to look after the animals. Around 60 million animals are used in research or testing around the world each year. Some people believe that this is morally wrong.

Artificial human skin can be used to test mascara

### Did you know...?

The UK laws controlling experiments on animals are very strict. For example, cosmetics testing was banned in 1998. Three types of licence are needed before scientists can start their research. These are only granted if the likely results are important enough to justify the use of animals, if the research cannot be done without them, and if a minimum number will be used.

### Exam tip OCR

✔ For the exam, you need to be able to describe one advantage and one disadvantage of animal testing.

### Questions

1 Copper sulfate crystals are dissolved in water. Name the solution, the solvent, and the solute.

2 Why must cosmetics be tested?

E

3 Butyl ethanoate is an ester. Suggest why it might be an ingredient of nail varnish remover.

4 Describe one advantage and one disadvantage of testing on animals.

C

5 Explain why nail varnish is soluble in ethyl ethanoate but it is insoluble in water.

A*

# 15: Paints and pigments

## Learning objectives

After studying this topic, you should be able to:

- ✔ recall and explain the functions of the main ingredients in paint
- ✔ describe and explain what happens when paints dry
- ✔ explain why the components of a colloid do not separate

**A** What is a pigment?

**B** Suggest why some powdered rocks can be used as pigments in paints.

▲ Modern pigments are bright and do not fade easily

## Pigments

A **pigment** is a coloured substance used in **paints** or dyes. In the past, pigments came from natural sources such as plants, animals, and powdered rocks. Most of these pigments were not brightly coloured and they faded easily over time. Most modern pigments are synthetic – they have been made by chemists.

## Paints

Paints are used to decorate or protect surfaces such as wood or metal. Paints have three main ingredients. These are described in the table.

| Ingredient | Function |
|---|---|
| pigment | the substance that gives the paint its colour |
| binding medium | sticks the pigments to the surface being painted |
| solvent | thins the paint so it spreads more easily |

Many paints are applied as a thin layer using a brush, roller, or spray. The paint dries as its solvent evaporates. **Emulsion paints** are water-based paints. Their ingredients are dissolved in water. As the water evaporates, the paint dries and the other ingredients join together to form a tough, coloured layer on the painted surface. In an **oil paint** the pigment is dispersed in an oil, which may be dissolved in a solvent.

▲ Emulsion paints are often applied with a roller

## Oil paints

Oil paints are used to decorate doors and window frames with a tough, shiny layer of paint. The solvent in oil paints is a hydrocarbon oil. This evaporates as the paint dries, but something else happens, too. Oxygen in the air oxidises the hydrocarbon molecules. Eventually they join together to form a very tough layer on the painted surface.

## Colloids

A **colloid** is a type of mixture in which one substance is spread or **dispersed** evenly in another substance. Colloids are different from solutions because none of the substances is dissolved. Paints are colloids. Their solid particles are mixed and dispersed with the particles of liquid. The 'solvents' they contain are there to spread the paint thinly rather than to dissolve it.

## Too small

The components of a colloid do not separate out or settle to the bottom. This is because the particles are dispersed throughout the mixture and are too small to settle.

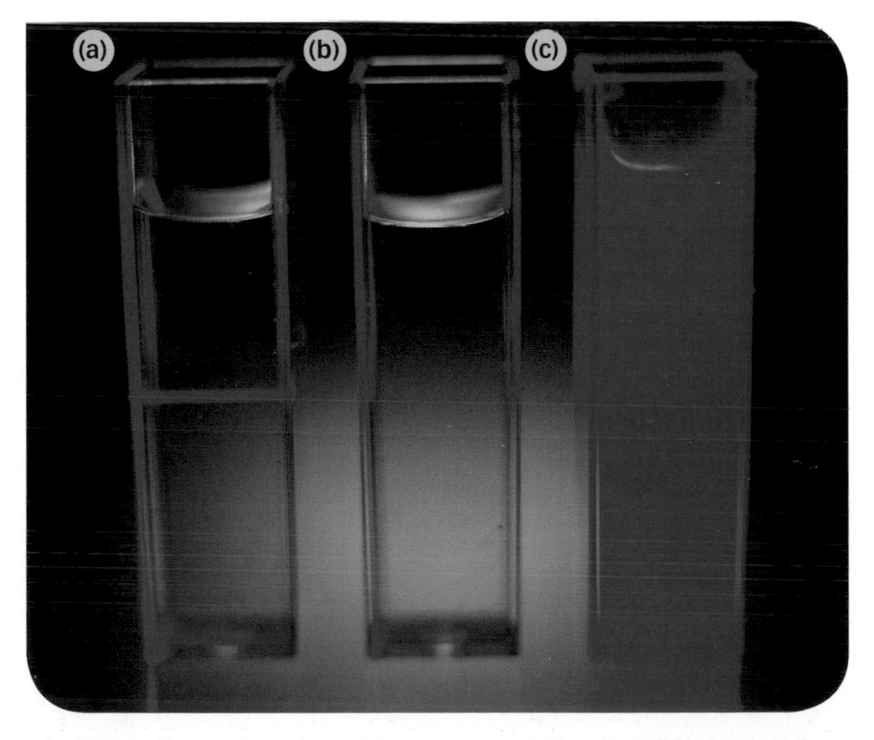


▲ Impurities in tap water scatter laser light a little (a), but pure water does not scatter it (b). A colloid scatters laser light a lot (c).

### Key words

**pigment, paint, emulsion paint, oil paint, colloid, dispersed**

### Did you know...?

Tyrian purple was a dye used by the ancient Romans. It was extracted from sea snails. At the time, Tyrian purple was more expensive than gold, so it was only used to dye the robes of the most wealthy people.

### Questions

1 State the main ingredients of paint.

E

2 Why do emulsion paints contain a solvent?

3 Describe the functions of each ingredient in a paint.

4 Describe the differences between a colloid and a solution.

C

5 Explain why the components of a colloid do not settle out, and what happens when an oil paint dries.

A*

C1

# 16: Smart paints

## Learning objectives

After studying this topic, you should be able to:

- ✔ describe the properties and uses of thermochromic pigments and phosphorescent pigments
- ✔ explain why phosphorescent pigments are safer than radioactive substances

## Did you know...?

Radium-based paint was used to paint military dials in the 1920s in America and Canada. The women who painted the dials used to lick their paintbrushes to get a narrow point. Many of them later suffered from the effects of exposure to radiation. They developed anaemia and 'radium jaw', a condition involving bleeding gums and bone cancer. Radium is no longer used for glow-in-the-dark paints.

**A** What is a thermochromic pigment?

**B** Suggest why a thermochromic strip may be attached to an aquarium containing tropical fish.

## Key words

**thermochromic**, **dye**, **phosphorescent**, **radioactive**, **radiation**

## Colour by temperature

**Thermochromic** pigments are sensitive to temperature. They change colour when they are heated up or cooled down. Liquid crystals are used in calculator displays. Some liquid crystals are thermochromic. They have a different colour at different temperatures. This makes them useful for making strip thermometers and mood rings. These change colour according to the temperature of the body.

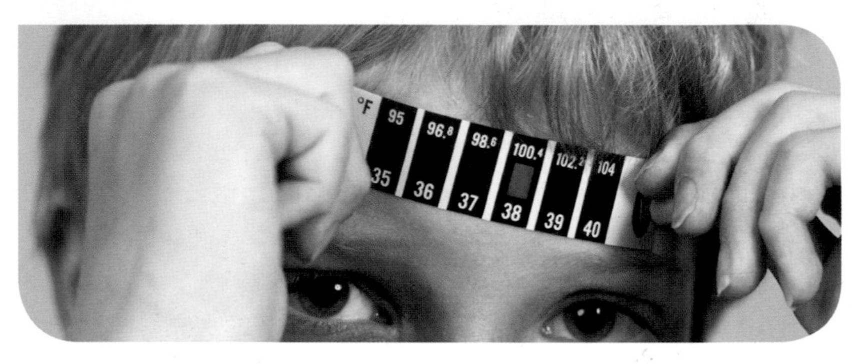


▲ This strip thermometer shows a raised body temperature of 38 °C

Other thermochromic pigments are heat-sensitive **dyes**. They are usually colourless at one temperature and become coloured as the temperature is changed. Thermochromic pigments are useful as general indicators of temperature, often as a warning that something is hot. For example, kettles and mugs coated with thermochromic pigments can give a warning that they contain a hot liquid. In a similar way, babies' spoons and bath toys warn parents if the food or bath water is too hot.

▲ Thermochromic pigments on this mug warn if the drink inside is hot

Thermochromic pigments can be mixed with acrylic paints to give even more colour changes.

## A glow in the dark

**Phosphorescent** pigments can glow in the dark. When they are in the light, they absorb energy. This energy is released as light over a period of time. It is difficult to see this happening in daylight, but you can see the pigments glow in the dark. They stop glowing after a while when they have released their stored energy. Phosphorescent pigments are useful for 'glow in the dark' warning signs and clock faces.

### A safer alternative

**Radioactive** substances can be used to make glow-in-the-dark paints. For example, radium is a radioactive metal. If radium is mixed with substances called phosphors, the **radiation** it produces makes the phosphor glow brightly. Tritium is another radioactive substance that can be used in this way. However, radiation can damage our cells and may cause cancer. Modern phosphorescent pigments are not radioactive and are much safer.

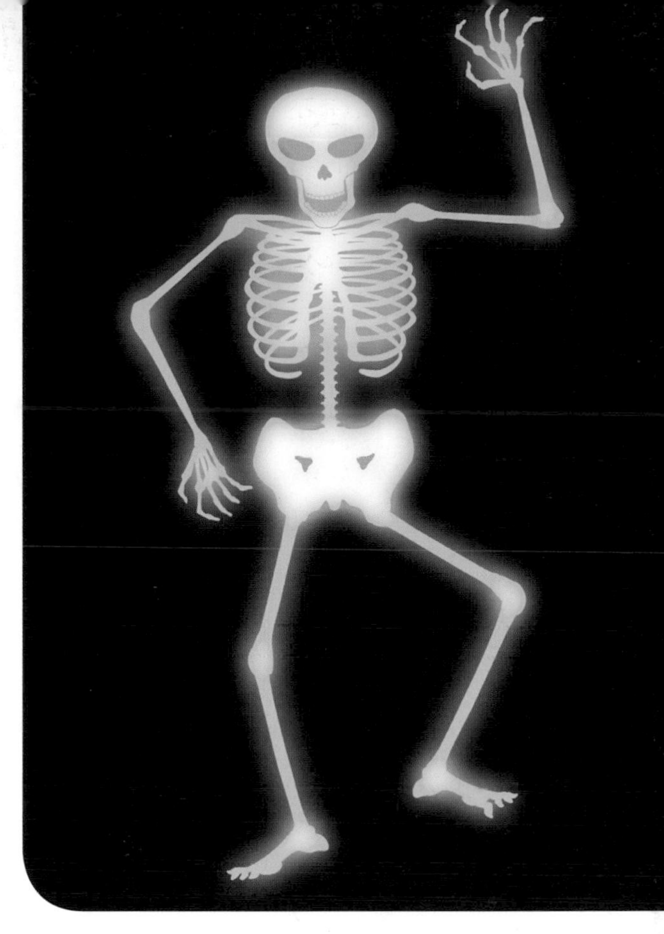

▲ Phosphorescent pigments are used in novelty items such as this spooky skeleton

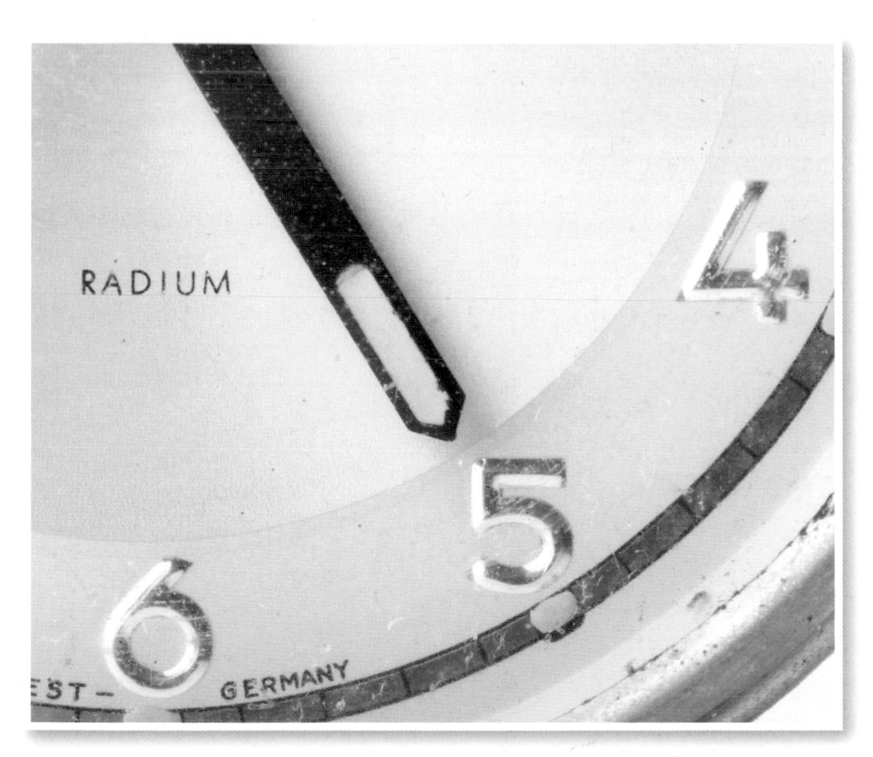


▲ This old alarm clock has radium-based paint on its hands and dial

### Questions

1. What is the difference between a thermochromic pigment and a phosphorescent pigment? (E)
2. Describe two uses of thermochromic pigments.
3. Briefly describe how phosphorescent pigments work. (C)
4. Explain how thermochromic pigments can be used to give a range of colour changes.
5. Explain why phosphorescent pigments are safer than alternative radioactive substances. (A*)

# Module summary

## Revision checklist

- Chemical reactions during cooking produce new substances.
- Balanced equations describe what happens in reactions.
- Food additives are added to food to improve its properties. Emulsifiers stop oil and water from separating.
- Esters are used in perfumes. They evaporate easily and have pleasant smells.
- Different solvents are used in paints and cosmetics to dissolve different solutes (varnish, pigments, or oils).
- Paint is a colloid (one substance is dispersed in another).
- The pigments in paint can be made sensitive to light or heat (phosphorescent or thermochromic).
- Crude oil is a non-renewable resource. It is a mixture of hydrocarbons (separated by fractional distillation).
- Distillation uses differences in boiling points. Larger hydrocarbons have stronger intermolecular forces and higher boiling points.
- Exploiting crude oil can cause environmental problems.
- Cracking using heat and a catalyst breaks down long-chain alkanes into smaller molecules, including alkenes.
- Alkanes are saturated hydrocarbons which contain only single bonds. Alkenes are unsaturated hydrocarbons which contain C=C double bonds.
- Alkenes are used in polymerisation reactions. Many small monomer molecules join to form one large polymer molecule.
- The uses of a polymer depend on its properties. Disposal of non-biodegradable polymers causes environmental problems.
- Alkanes in crude oil are used as fuels. The equation for complete combustion is: hydrocarbon + oxygen $\rightarrow$ carbon dioxide + water. Heat energy is released.
- Burning hydrocarbon fuels releases carbon dioxide (greenhouse effect), carbon monoxide (toxic), sulfur oxides and nitrogen oxides (acid rain), and particulates (global dimming).
- The Earth's atmosphere has a stable composition.
- The composition of the Earth's atmosphere has developed and changed due to: volcanic eruption, condensing of water, dissolving of gases in seas, and photosynthesis.

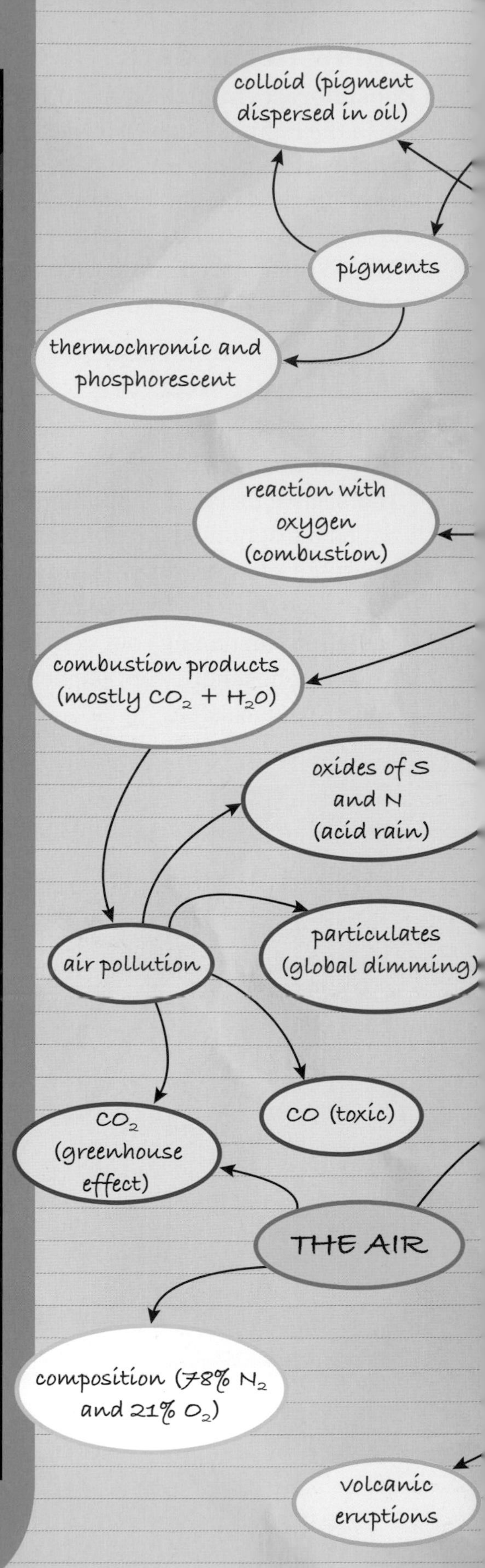




NOW USE THE C1 GRADE CHECKER ON PAGE 240

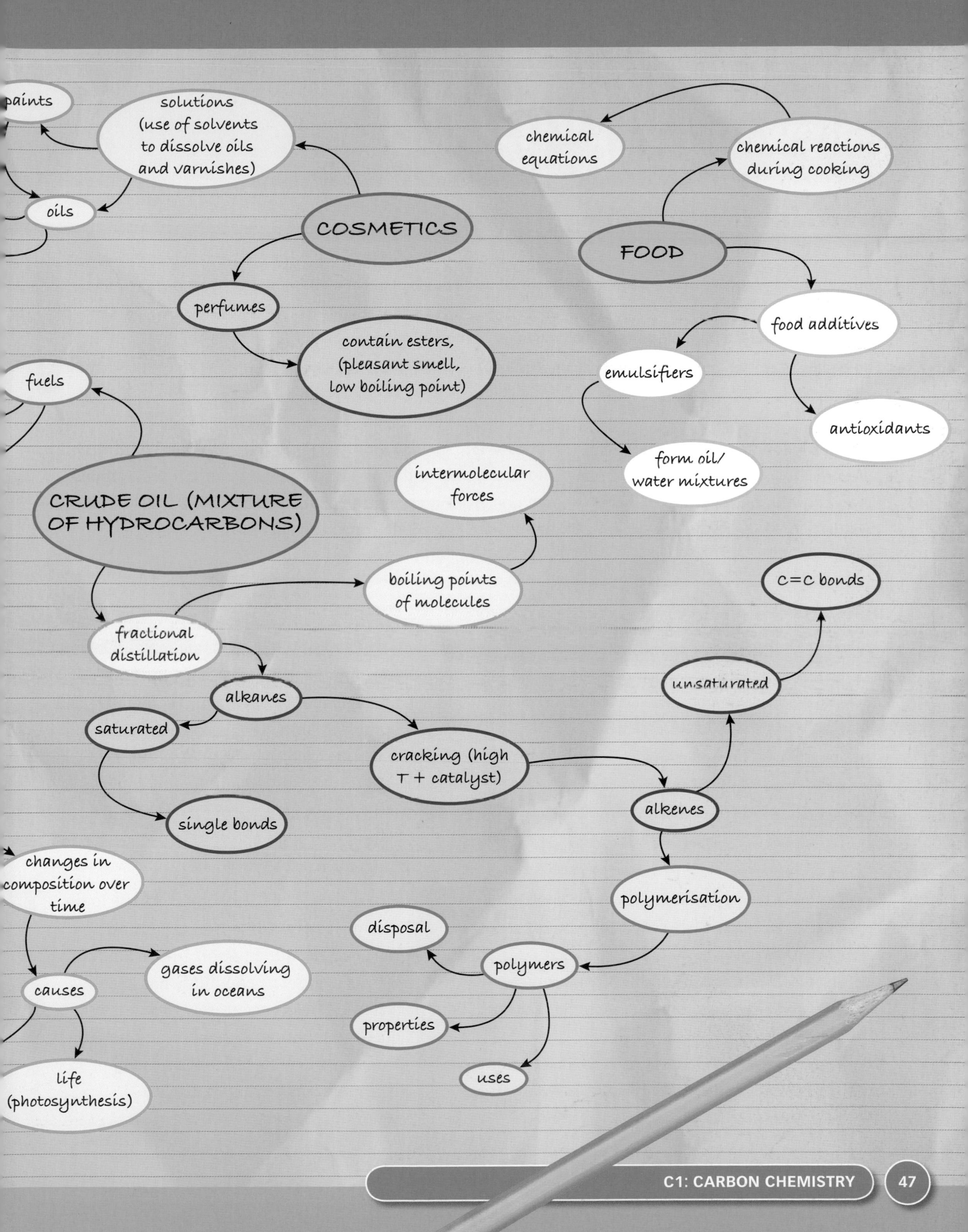
paints
solutions (use of solvents to dissolve oils and varnishes)
oils
COSMETICS
perfumes
contain esters, (pleasant smell, low boiling point)
chemical equations
chemical reactions during cooking
FOOD
food additives
emulsifiers
antioxidants
form oil/ water mixtures
fuels
CRUDE OIL (MIXTURE OF HYDROCARBONS)
intermolecular forces
boiling points of molecules
fractional distillation
alkanes
saturated
single bonds
cracking (high T + catalyst)
alkenes
unsaturated
C=C bonds
polymerisation
polymers
disposal
properties
uses
changes in composition over time
causes
gases dissolving in oceans
life (photosynthesis)

## Answering Extended Writing questions

**QUESTION**

Some shops now supply carrier bags made of cornstarch instead of polythene. Cornstarch is made from maize plants.

Outline the benefits and problems of cornstarch bags, compared to polythene bags.

**The quality of written communication will be assessed in your answer to this question.**

Cornstarch and polythene both protect food. Cornstarch break down natrally, but polythene doesnt. Polythene bags kill wales.

*Examiner:* The candidate knows that cornstarch is biodegradable and that polythene bags can kill sea animals, but has given no scientific detail and has not used scientific terms, such as 'biodegradable'. There are several spelling, punctuation, and grammar errors.

Bacteria can break down cornstarch bags, but it cannot brake down polythene bags. Cornstarch is a renewable resource. Bacterias can only brake down cornstarch bags in good conditions. Cornstarch is biodegradable. Polythene is not a renewable resource. It is made from oil.

*Examiner:* The candidate has made two relevant comparisons, and has also pointed out that biodegradable bags are only broken down by bacteria under certain conditions. Scientific terms have been used correctly. However, the answer is poorly organised and includes errors of spelling and grammar.

Cornstarch is biodegradable – it can be broken down by bacteria. Polythene is not biodegradable. Some waste polythene bags end up as litter, and may suffocate animals. Other waste polythene bags are burned in incinerators, so toxic gases go into the atmosphere.

Cornstarch is made from maize, which is a renewable resource. However, some people think it is wrong to use land to grow maize for cornstarch, not food.

*Examiner:* This answer covers the main scientific points. It is well organised and the spelling, punctuation, and grammar are accurate. The candidate has used scientific terms accurately. The answer does not mention that polythene is made from a finite resource, but this is implied by the statement that maize is a renewable resource.

# Exam-style questions

**1** Burning fuels can cause pollution of the atmosphere. Complete this table which shows how different pollutants form.
A01

| Pollutant | How formed |
|---|---|
| | Incomplete combustion of petrol |
| Oxides of nitrogen | |
| | Burning of fuels containing sulfur impurities |

**2** One of the pollutants produced by cars is carbon monoxide. Modern cars have been designed so that carbon monoxide is converted into carbon dioxide.

A01 **a** Give the formulae of carbon monoxide and carbon dioxide.

A01 **b** Name the device that converts carbon monoxide to carbon dioxide.

A01 **c** Explain why it is important to control the emission of carbon monoxide.

**3** The displayed formula of chloroethene is:

H, Cl, C=C, H, H

A02 **a** **i** Why is this molecule described as unsaturated?

**ii** A student adds bromine water to a small sample of this substance. What would she observe?

**iii** Complete this balanced equation to show the reaction when the bromine is added:

$C_2H_3Cl + Br_2 \rightarrow$ ________

**b** Chloroethene is a monomer used in industry to manufacture the polymer, poly(chloroethene).

A02 **i** Draw out the displayed formula of poly(chloroethene).

A01 **ii** What conditions are needed for a polymerisation reaction?

E ↓ C ↓ A*

## Extended Writing

**4** Crude oil is often transported by sea.
A01 Describe some of the environmental problems that can be caused by transporting crude oil in this way.

**5** Polymers are a very important type of
A01 substance in today's society. However, many people are worried about how to dispose of them.
Explain why disposing of polymers is a particular problem, and describe how chemists are trying to solve the problem of disposal.

**6** The atmosphere of the Earth today
A01 contains 21% oxygen, 78% nitrogen, and small amounts of other gases.
Scientists think that the atmosphere of the early Earth, billions of years ago, was very different to this.
Describe how the early atmosphere may have been different to today's, and explain the processes that caused the atmosphere to evolve into today's atmosphere.

E ↓ C ↓ A*

A01 Recall the science
A02 Apply your knowledge
A03 Evaluate and analyse the evidence

# Chemical resources

| OCR specification match | |
|---|---|
| C2a | 2.1 The structure of the Earth |
| C2a | 2.2 Plate tectonics and igneous rocks |
| C2b | 2.3 Construction materials |
| C2b | 2.4 Concrete |
| C2c | 2.5 Copper |
| C2c | 2.6 Alloys |
| C2d | 2.7 Corrosion |
| C2d | 2.8 Making cars |
| C2e | 2.9 Making ammonia |
| C2e | 2.10 Reversible reactions |
| C2f | 2.11 Making salts – 1 |
| C2f | 2.12 Making salts – 2 |
| C2g | 2.13 Fertilisers and crop yields |
| C2g | 2.14 Making fertilisers |
| C2h | 2.15 Chemicals from salt |
| C2h | 2.16 Electrolysis of brine |
| Module summary | |
| OCR Upgrade | |

## Why study this module?

Earthquakes and volcanoes are often in the news. Although scientists understand the Earth's structure, it is still not possible to predict exactly when and where these natural events will happen. In this module you will study the theory of plate tectonics and how the Earth's surface has changed over time.

The Earth's crust is a source of valuable raw materials. These include rocks for making concrete, and metal ores for producing steel and other useful metals. You will investigate how these materials are extracted and used, and the environmental problems this causes.

Salt and the air around us are also important sources of useful materials. This includes hydrogen from salt solution and nitrogen from air. These substances are used to make ammonia, the raw material for fertilisers, dyes, and explosives. You will discover how chemists choose the conditions needed for industrial chemical processes like making ammonia. You will also examine the benefits and risks of using fertilisers.

### You should remember

1. Geological activity is caused by chemical and physical processes.
2. Different materials have different physical and chemical properties.
3. Rocks and metals are useful building materials.
4. The air is a mixture of gases, including nitrogen and oxygen.
5. Iron and steel rust when exposed to air and water.
6. Acids react with metals and bases.
7. Plants need minerals to grow well.
8. Some substances conduct electricity while others do not.

The Millau Viaduct in France has the highest road deck in Europe. It soars 343 metres above the ground, 19 metres higher than the Eiffel Tower, and is almost 2.5 kilometres long. The bridge also has the highest towers and pylons in the world. It took just three years to build and is designed to last for 125 years. It contains 206 000 tonnes of concrete and 36 000 tonnes of steel, all produced from the Earth's crust.

# 1: The structure of the Earth

## Learning objectives

After studying this topic, you should be able to:

- describe the structure of the Earth
- explain how molten rock reaches the Earth's surface
- describe the theory of plate tectonics

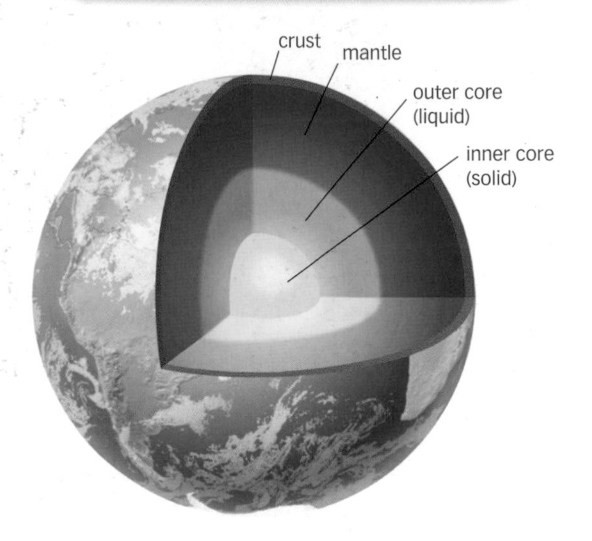


▲ The structure of the Earth

**A** Describe the structure of the Earth.

### Did you know...?

The deepest hole ever drilled is the Kola Superdeep Borehole in Russia. It reached 12.261 km into the crust in 1989. Conditions deep down became too hot for drilling to continue.

**B** What is magma?

## The structure of the Earth

The Earth is a sphere containing several different layers. The main ones are:

- the **crust**, which is rocky and thin compared with the other layers
- the **mantle**
- the **core**, which contains iron and is very hot.

The radius of the Earth is about 6370 km. Imagine that you could ride in a jumbo jet right down to the centre of the Earth. The journey would take seven hours at top speed, and there would be just enough fuel to get back to the surface again. But it is not really possible to study the Earth's structure directly. The crust is too thick to drill through and the deepest mine is only 3.5 km deep. So instead, information is used from seismic waves produced by earthquakes or explosions.

## Tectonic plates

Taken together, the crust and the outer part of the mantle are called the **lithosphere**. This is rigid and relatively cold. It is broken up into huge parts called **tectonic plates**. These are less dense than the mantle, so they float on top of it. Tectonic plates move very slowly, only about 2.5 cm per year.

Earthquakes happen because of the movements of tectonic plates. Volcanic activity happens where the plates meet. Molten rock or **magma** gets to the surface through weak spots in the crust. The magma can rise through the crust because it is less dense than the crust.

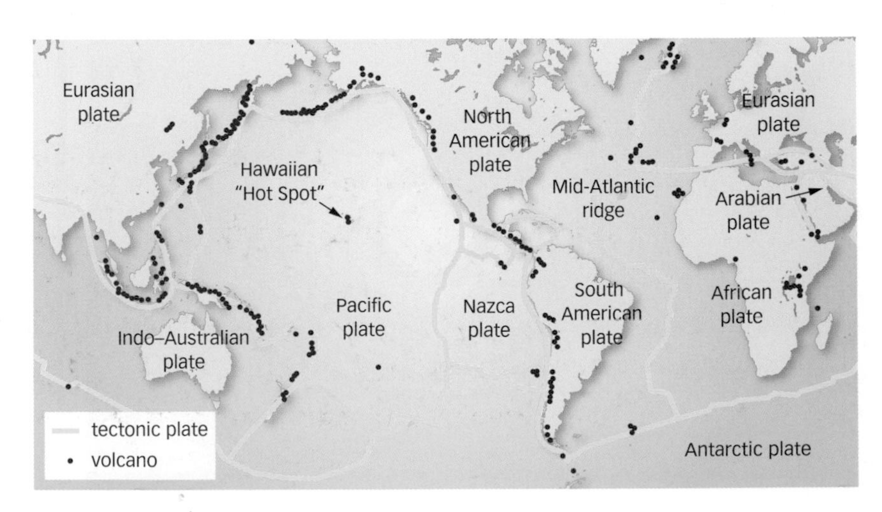


▲ The red spots show where there is volcanic activity

## The theory of plate tectonics

The mantle is the zone between the crust and the core. It is relatively cold and rigid near to the crust. Deeper down, the mantle becomes hotter and less rigid, and it flows slowly. Huge convection currents are set up. Hot material flows upwards, cools near the crust, then sinks again. These currents push the tectonic plates around the Earth.

The oceanic plates are denser than the continental plates. At ocean margins the oceanic plates cool. If they collide with a continental plate, the cooler oceanic plate is pulled under the continental plate. This is called **subduction**. The subducted oceanic plate partially melts and volcanoes can occur.

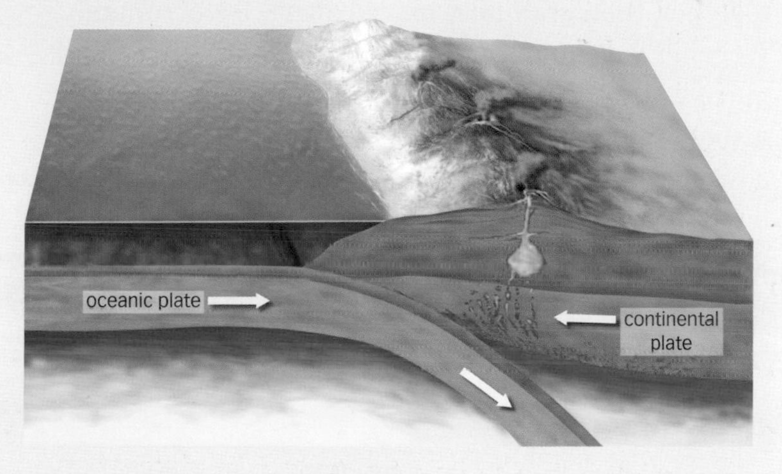


Subduction also builds mountains

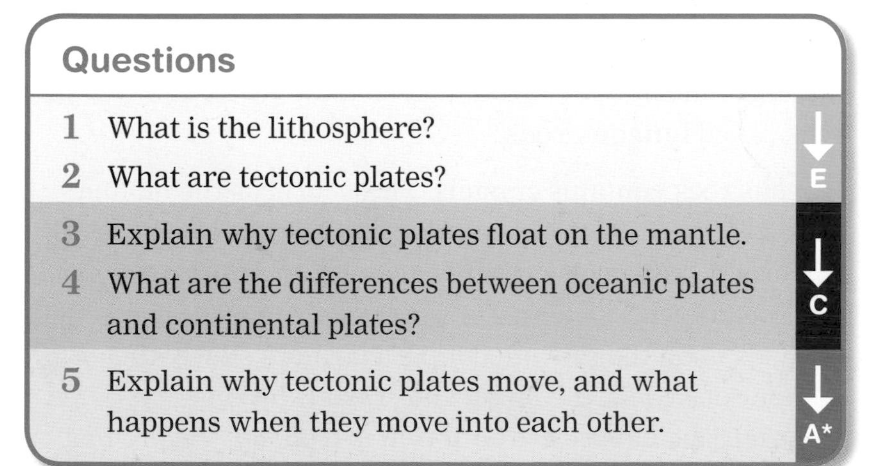

### Questions

1 What is the lithosphere?

2 What are tectonic plates?

E

3 Explain why tectonic plates float on the mantle.

4 What are the differences between oceanic plates and continental plates?

C

5 Explain why tectonic plates move, and what happens when they move into each other.

A*

An active volcano

### Key words

**crust, mantle, core, lithosphere, tectonic plate, magma, subduction**

### Did you know...?

Tectonic plates move about as fast as your thumbnail grows.

### Exam tip OCR

✔ It might help to think of the Earth as an egg. The shell is the crust, the white is the mantle, and the yolk is the core.

# 2: Plate tectonics and igneous rocks

## Learning objectives

After studying this topic, you should be able to:

- describe how the theory of plate tectonics developed
- explain how igneous rocks form
- describe the work of geologists in studying volcanoes

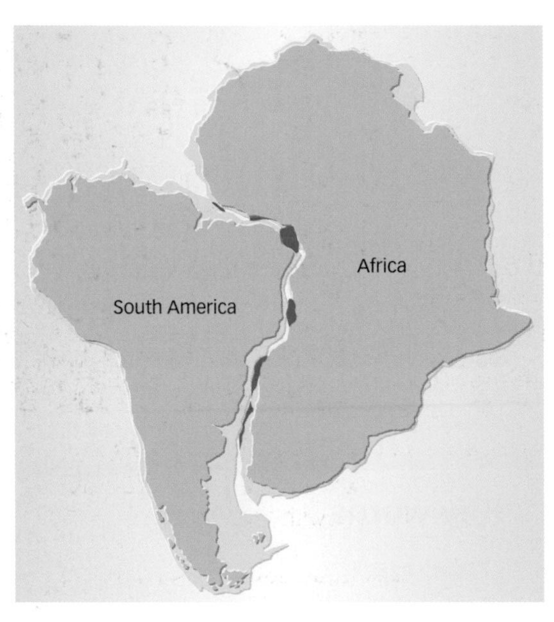


▲ The eastern coastline of South America matches the western coastline of Africa

**A** Why does rhyolite have small crystals, but granite large crystals?

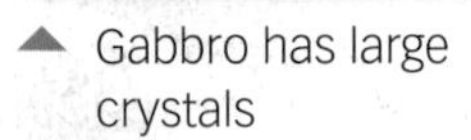

▲ Gabbro has large crystals

▲ Rhyolite is silica-rich with small crystals

## Why plate tectonics?

Many theories have been put forward to explain the nature of the Earth's surface, but Earth scientists or **geologists** now accept the theory of plate tectonics. This is because it explains a wide range of evidence in the world around us, and has been discussed and tested by scientists worldwide over many years.

### Continental drift

Alfred Wegener was a German scientist who published his theory about 'continental drift' in 1914. He noticed that the shapes of the continents fitted together. He suggested that the continents were all once joined together, then gradually moved apart. Unfortunately, Wegener was unable to explain how continents could move, and he was not a geologist. So his theory was disbelieved by most scientists until the 1960s, when advances in science began to explain how continents move. Scientists discovered convection currents in the mantle. They discovered evidence that tectonic plates were moving apart under the Atlantic Ocean, leading to sea-floor spreading, and they discovered subduction.

## Igneous rocks

Rock melts when it is heated to a high enough temperature. Liquid or molten rock under the Earth's surface is called magma. On the other hand, molten rock that erupts from a volcano is called **lava**. When molten rock cools down, either underground or above ground, it solidifies to form a type of rock called **igneous rock**.

Igneous rock contains crystals. These form as the molten rock solidifies. They lock together randomly to form the solid rock. Lava cools quickly and magma cools slowly inside the Earth's crust. The size of the crystals in an igneous rock depends upon the rate of cooling. The more slowly the molten rock cools, the larger the crystals become.

- Basalt and gabbro are both rich in iron, but gabbro has larger crystals because the molten rock cooled slowly.
- Rhyolite and granite are both rich in a mineral called silica, but granite has larger crystals.

Granite has large crystals. It is often used for kitchen worktops

## Did you know...?

There are around 60 volcanic eruptions on the Earth's surface each year, but this varies a lot. There are many more deep under the sea.

**B** Suggest why people might still live near an active volcano, even though there is the danger of an eruption.

## Volcanoes and eruptions

Volcanic soil is very fertile. It is rich in minerals that plants need to grow well. Some people choose to live near volcanoes because of this. This may be dangerous. Not all volcanoes are the same. Some erupt runny lava, others erupt thick lava. Those that erupt thick lava erupt more violently and are a greater risk to those around them.

Some geologists study volcanoes. They want to find information about the Earth's structure, and to see whether they can predict future eruptions.

Although they might get a good idea of when a volcano might erupt, geologists are still unable to give accurate predictions.

### Different lava, different rock

Basalt and rhyolite are both formed as a result of volcanic eruptions. Basalt is iron-rich and forms from runny lava. This flows steadily and is regarded as relatively safe. Rhyolite is different. It is silica-rich and forms from thick lava. This explodes out of the volcano, producing lava bombs and volcanic ash.

The Giant's Causeway in Northern Ireland. It consists of hexagonal columns of basalt, an iron-rich rock with small crystals.

## Key words

**geologists, lava, igneous rock, eruption**

## Questions

1 What is the difference between magma and lava?

2 How do igneous rocks form?

E

3 What determines the size of the crystals in igneous rocks?

4 What do the igneous rocks basalt and gabbro have in common? Explain why basalt has smaller crystals than gabbro.

C

5 Describe how the theory of plate tectonics developed, and how it supported Alfred Wegener's earlier theory of continental drift.

A*

# 3: Construction materials

## Learning objectives

After studying this topic, you should be able to:

- ✔ explain the use of limestone in construction
- ✔ describe the environmental, social, and economic effects of mining limestone

▲ Modern buildings use many different construction materials

## Did you know...?

Dark granite is harder than light-coloured granite, which contains more of a mineral called quartz, reducing its strength.

**B** Describe two uses of limestone as a construction material.

## Mining limestone

**Construction materials** are the substances used to make buildings, bridges, and roads. Many of these materials are manufactured from rocks found in the Earth's crust. For example, aluminium and iron are metals extracted from metal **ores**. Bricks are made by baking clay, and glass is made by heating sand to very high temperatures. Sometimes rocks are used without any further chemical processing. **Limestone**, marble, and granite can be used this way.

## Rocks in construction

Limestone and other rocks must be removed from the ground. A **quarry** is a place where this is done. It is a large but shallow hole in the ground, where large amounts of the desired rock can be found. Explosives are usually used to break up the rock so that it can be removed. There are around 1300 quarries in the UK.

**A** What is a quarry?

Limestone is a particularly important raw material. Over 80 million tonnes of it are quarried in the UK each year. Although limestone from some quarries is crumbly, a lot of it is tough enough to be used as a construction material. Limestone can be cut into blocks, rather like large bricks, to make buildings. It can also be crushed to make **aggregate**. This consists of lumps of limestone. Aggregate is used as a firm base underneath railway lines, and as the foundation for roads before the tarmac is laid over the top.

▲ Limestone aggregate is used for railways

## Quarry queries

There are benefits and problems caused by quarrying limestone and other rocks.

Quarries are often sited in the countryside. They provide jobs in places where they may be scarce. This helps local families, and also shops and schools. The products from the quarries are valuable, and have many uses. For example, the UK's quarries produce materials that are worth several billion pounds per year. Although the UK imports some limestone, it exports even more. This helps the country's economic success.

On the other hand, quarries are often sited in attractive areas such as national parks. This may damage the tourist industry. Quarries are noisy and dusty, and so may affect the quality of life for local people. Using lorries to carry the rocks from the quarry to the customers produces extra traffic, which may have to pass through small towns or villages.

Quarries take up land space, making it unavailable for other uses such as farming and recreation. They destroy the original landscape, so this must be restored when the quarry closes. However, this process usually takes place in stages during the lifetime of a modern quarry. The land may be restored as farmland or a nature reserve, for example.

### Hard rocks

Limestone is a **sedimentary rock** formed from the shells and skeletons of sea creatures. Marble is a **metamorphic rock** formed from limestone. Large-scale movements of the Earth's crust can bury rocks, putting them under high pressures and temperatures. The rocks do not melt when this happens, but their crystal structure changes. Marble is harder than limestone. It has small grains that are joined together more strongly than the grains in limestone.

Granite is an igneous rock. It is harder than marble. Its crystals are joined together more strongly than the grains in marble.

▲ Quarrying for limestone

### Key words

construction material, ore, limestone, quarry, aggregate, sedimentary rock, metamorphic rock

### Questions

1. How is limestone quarried?
2. Name three construction materials and describe their use in buildings.
3. Describe three benefits and three drawbacks of a limestone quarry.

↓ E

4. Construction materials are manufactured from substances found in the Earth's crust. What are the substances used to manufacture aluminium, brick and glass?

↓ C

5. Correctly order granite, limestone, and marble from softest to hardest. Explain the reasons for the differences in hardness.
6. To what extent do you think that a limestone quarry is a good thing? Consider its environmental, social, and economic effects in your answer.

↓ A*

## Learning objectives

After studying this topic, you should be able to:

- ✔ explain how metal carbonates may be decomposed by heating
- ✔ describe how cement and concrete are made
- ✔ describe the advantages of reinforced concrete

▲ Calcium oxide reacting with water

**A** Write a word equation for the thermal decomposition of copper carbonate.

### Key words

**thermal decomposition, cement, mortar, concrete, reinforced concrete, composite material**

## Thermal decomposition

Limestone and marble are both forms of calcium carbonate, $CaCO_3$. When calcium carbonate is heated strongly, it breaks down to form calcium oxide and carbon dioxide:

calcium carbonate → calcium oxide + carbon dioxide

$$CaCO_3 \rightarrow CaO + CO_2$$

This sort of reaction is called **thermal decomposition**. It is a reaction in which one substance is chemically changed, when heated, to produce two or more new substances.

Calcium carbonate is not the only metal carbonate that decomposes when heated. Other metal carbonates do so, too. In general:

metal carbonate → metal oxide + carbon dioxide

For example, green copper carbonate decomposes to form black copper oxide and carbon dioxide when heated.

The carbonates of very reactive metals need higher temperatures to decompose than can be achieved with a Bunsen burner. For example, potassium and sodium are reactive metals in Group 1 of the periodic table. Potassium carbonate and sodium carbonate do not decompose when heated with a Bunsen burner.

## Alkalis from limestone

Limestone contains calcium carbonate. It forms calcium oxide when it is heated strongly. This is carried out industrially in large ovens called lime kilns, where the temperature reaches 900 °C or more. The carbon dioxide produced in the reaction escapes into the atmosphere. The calcium oxide is an ingredient of **cement**.

It also reacts with water to produce calcium hydroxide:

calcium oxide + water → calcium hydroxide

$$CaO + H_2O \rightarrow Ca(OH)_2$$

Calcium hydroxide is an alkali. It can be used to neutralise acids. For example, farmers may spread it on their fields to reduce excess acidity in the soil, helping the crops to grow. It may also be used to treat lakes damaged by acid rain.

You may have used a solution of calcium hydroxide to test for carbon dioxide. This solution is called limewater. It turns cloudy white when carbon dioxide is bubbled through it.

## Cement and concrete

Cement is made by heating powdered limestone with clay. Cement is mixed with sand and water to make **mortar**. This is the mixture used by bricklayers to stick bricks together. As it sets, it absorbs carbon dioxide from the air, forming tiny crystals of calcium carbonate, which lock together with the other ingredients.

Cement is mixed with sand, aggregate, and water to make **concrete**. This is the tough construction material that is widely used in buildings, bridges, and other man-made structures.

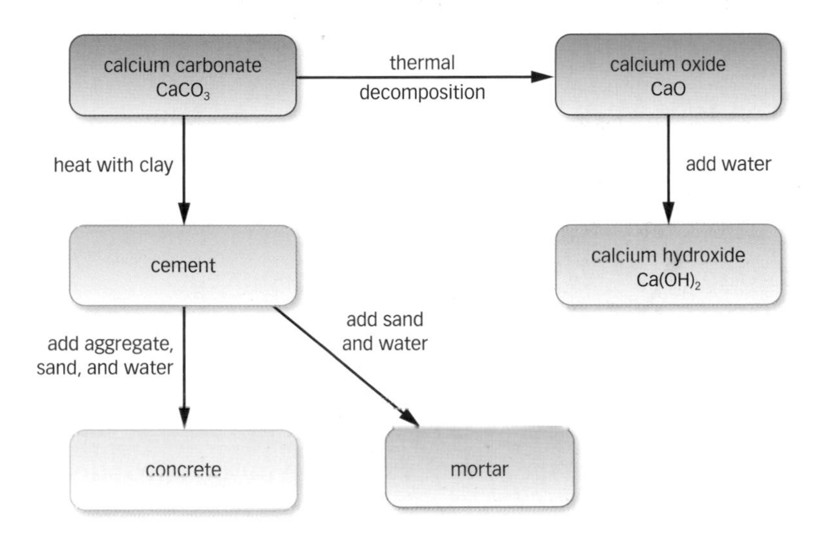


This flow chart summarises the uses of calcium carbonate and its products

## Reinforced concrete

Concrete can be strengthened by letting it set around a supporting mesh of solid steel. Such **reinforced concrete** is a **composite material**.

It has useful properties from both the materials it contains:

- concrete is hard and strong in compression (when squashed)
- steel is flexible and strong in tension (when stretched).

Reinforced concrete is widely used in the construction industry. For example, it is used to make bridges, dams, and the floors of tall buildings.

### Did you know...?

The world manufactures 2.8 billion tonnes of cement every year.

### Exam tip OCR

✔ Make sure you know how concrete is made, starting with limestone.

### Questions

1 What is produced when limestone decomposes?

2 Describe how concrete is made.

E

3 Write word equations for the thermal decomposition of magnesium carbonate and zinc carbonate. Which product do the two reactions have in common?

C

4 Explain why reinforced concrete is a better construction material than non-reinforced concrete.

5 Explain why the manufacture of cement is a major source of carbon dioxide, a greenhouse gas. In your answer, consider how the raw materials might be heated and any reactions involved.

A*

Wet concrete being poured into a steel mesh to make reinforced concrete

## Learning objectives

After studying this topic, you should be able to:

- describe how copper can be extracted and purified
- describe the benefits and drawbacks of recycling copper

Malachite

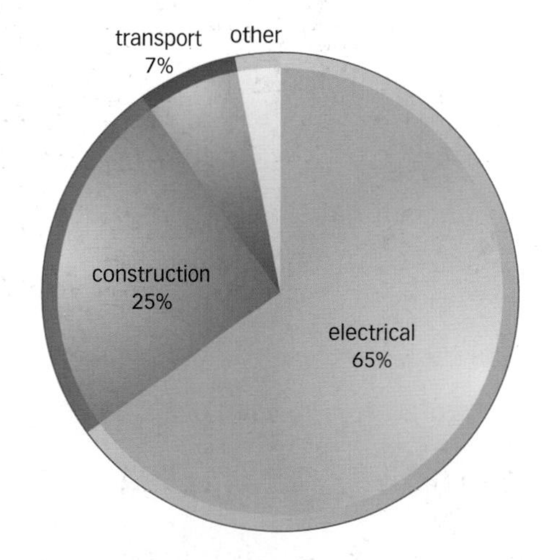


The main uses of copper

## Extracting copper

Copper is a valuable metal with properties that make it very useful. It is a good conductor of heat and electricity. Copper is ductile, which means it is easily made into wires. It is unreactive compared with most metals, which makes it useful for water pipes.

Copper is usually found combined with other elements to form compounds. Copper ores contain these compounds. Copper metal is extracted by heating crushed ore with carbon. This works because carbon is more reactive than copper. For example, malachite is a green mineral containing copper carbonate. It easily breaks down to form copper oxide when heated:

copper carbonate → copper oxide + carbon dioxide

$$CuCO_3 \rightarrow CuO + CO_2$$

The copper oxide is then heated with carbon. This is a **reduction** reaction because oxygen is removed:

copper oxide + carbon → copper + carbon dioxide

$$2CuO + C \rightarrow 2Cu + CO_2$$

**A** What percentage of copper is used for construction?

**B** How is carbon used to extract copper?

## Purifying copper

The extracted copper contains enough impurities to make it less useful. It must be purified. This is done using **electrolysis**, which involves passing an electric current through copper(II) sulfate solution.

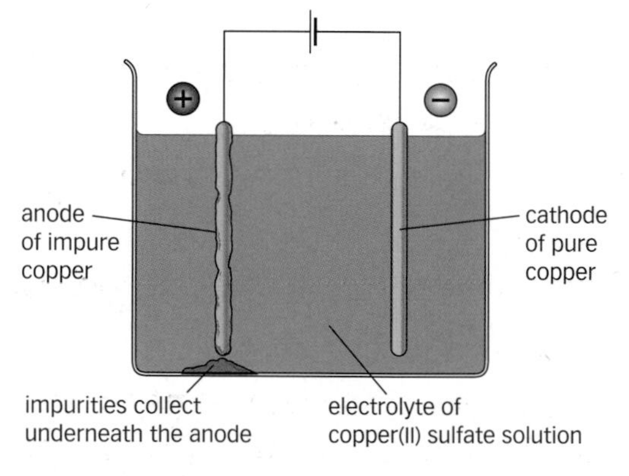


Purifying copper by electrolysis

## Electrolysis

Copper(II) sulfate solution is an **electrolyte**. It is a liquid that conducts electricity. The negatively charged **electrode**, the **cathode**, is pure copper. During electrolysis, copper is deposited and the cathode gains mass:

$$Cu^{2+} + 2e^{-} \rightarrow Cu \text{ (reduction)}$$

The positively charged electrode, the **anode**, is impure copper. During electrolysis, copper dissolves and it loses mass:

$$Cu - 2e^{-} \rightarrow Cu^{2+} \text{ (oxidation)}$$

This is an example of oxidation because electrons are lost.

## Recycling copper

The best copper ores have been used up. The remaining ores contain less copper and are more difficult to mine. It is cheaper to recycle copper than to extract it from its ores. Recycling makes better use of limited resources. It also reduces the amount of waste and pollution produced.

Scrap metal thieves realise how valuable copper is, but most people do not recycle it. They throw old electrical items away, for example. It is then difficult to recover the copper for recycling because the waste metals have to be sorted.

### Questions

1 Study the pie chart on the previous page.
   (a) What percentage of copper is used for 'other' uses?
   (b) One of these other uses is for coins. Why is copper suitable for making coins?

↓ E

2 Write equations to show the extraction of copper from copper carbonate.

3 Draw a labelled diagram to show the apparatus needed to purify copper.

4 Describe some of the problems of recycling copper.

↓ C

5 The sludge that collects underneath the anode during copper purification contains other metals, including silver and gold. How is it useful?

↓ A*

### Did you know...?

The ancient Romans had copper mines on the island of Cyprus. They called copper 'aes Cyprium', which became shortened to 'cuprum'. This is where the modern symbol for copper, Cu, comes from.

### Key words

**reduction, electrolysis, electrolyte, electrode, cathode, anode**

▲ Scrap copper roofing ready for recycling

### Exam tip OCR

✔ You should be able to label the apparatus needed to purify copper by electrolysis.

# 6: Alloys

## Learning objectives

After studying this topic, you should be able to:

- ✔ recognise some common alloys and their uses
- ✔ relate the properties of an alloy to a particular use
- ✔ explain the use of 'smart alloys'

## Key words

**alloy, shape-memory alloy**

▲ Brass is used to make door knockers

## Exam tip OCR

- ✔ Make sure you can state the uses of amalgam, brass, and solder, and a large-scale use of each one.

## Metal mixtures

Pure metals are often not very useful. For example, iron is too soft for most uses. So it is mixed with carbon and other elements to make steel. Copper is also too soft for some uses. It is mixed with tin to make bronze, which is much tougher. Steel and bronze are **alloys**.

An alloy is a mixture of two or more elements, at least one of which is a metal. Other alloys include amalgam, brass, and solder. The table shows information about these three alloys.

| Alloy | Main metals in the alloy | Typical large-scale use |
|---|---|---|
| amalgam | mercury | tooth fillings |
| brass | copper and zinc | musical instruments and coins |
| solder | lead and tin | joining electrical wires |

**A** Name two alloys that contain copper.

## The right alloy for the job

The properties of an alloy are different from the properties of the individual metals it contains. For example, solder is made from lead, which melts at 327 °C, and tin, which melts at 232 °C. Yet solder melts at just 185 °C. This makes it useful for joining electrical components without damaging them.

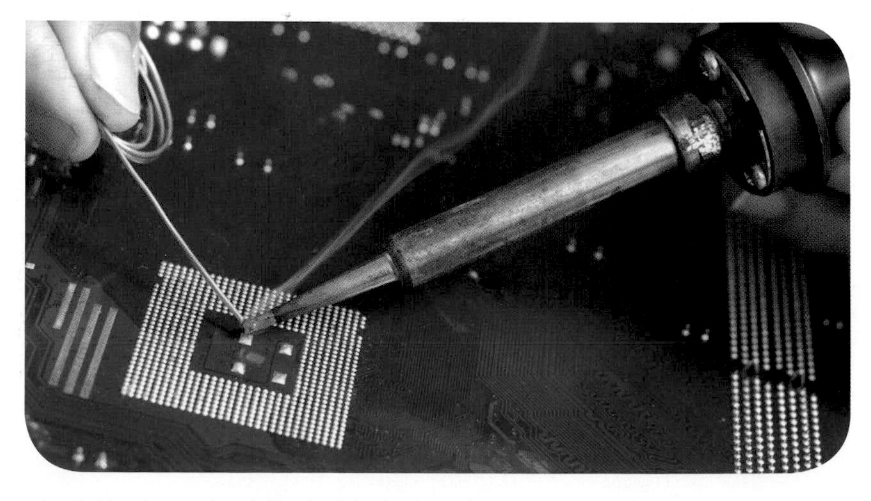

▲ Solder is used to join electrical wires

Alloys are often stronger and harder than the metals they contain. Brass is like this. Its copper and zinc atoms are different sizes. As a result, layers of atoms cannot slide over each other so easily if brass is stretched or bent.

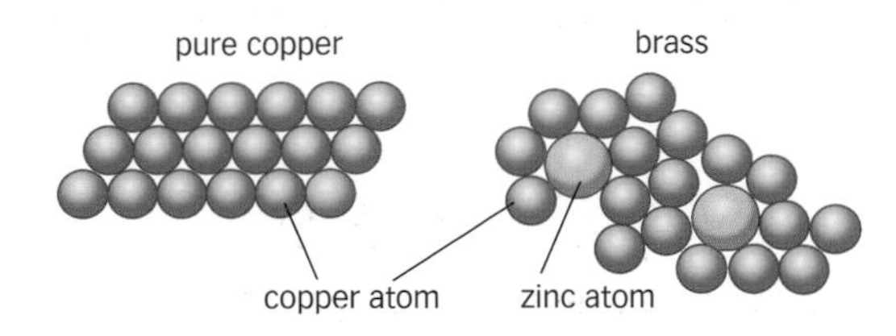


▲ Atoms in copper and brass

## Smart alloys

**Shape-memory alloys** are 'smart alloys'. If they are bent out of shape, they return to their original shape when they are warmed. This property has greatly increased the number of uses of alloys. For example, a shape-memory alloy called nitinol is used for spectacle frames. The composition of the alloy is adjusted during manufacture so that the frame returns to its original shape below room temperature. So if someone sits on their glasses, the frame returns to its proper shape when they stand up again.

### Did you know...?

Nitinol was discovered by accident in 1962. Its name comes from the metals it contains, nickel and titanium, and the place where it was discovered, the Naval Ordnance Laboratory in America.

### Questions

1 What is an alloy?

2 Why is brass, rather than copper, used for electrical plugs?

E

3 What is unusual about solder compared with the metals it contains?

4 Describe the properties needed for a tooth filling. Give some benefits and drawbacks of using amalgam for a filling.

C

5 A stent is a frame inserted into narrow arteries. It keeps the artery open for blood to flow normally. Shape-memory alloys may be used.

(a) Should the stent's normal shape be open or folded?

(b) Suggest why the surgeon would insert the stent in the other state.

(c) Explain at what temperature the stent should change to its normal shape.

A*

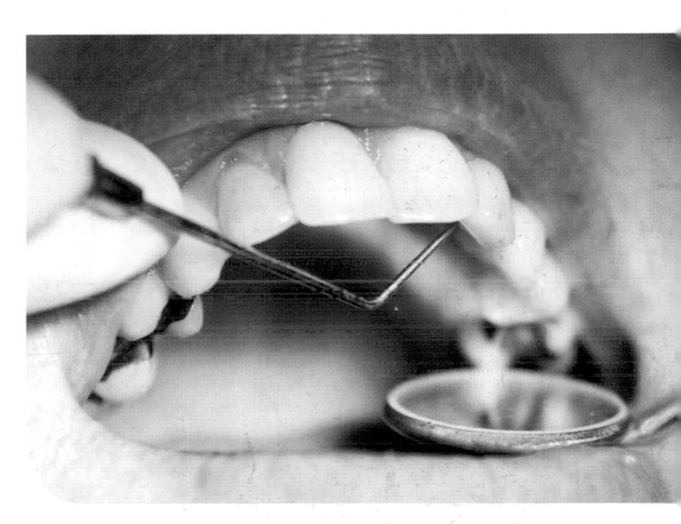

▲ Amalgam is used for tooth fillings

### Did you know...?

Amalgam for tooth fillings contains around 50% mercury. Mercury is a liquid at room temperature but amalgam is a tough solid. Some people with amalgam fillings get painful shocks if they chew aluminium foil. Their saliva acts as an electrolyte and a tiny electric current is produced, stimulating nerves in the tooth.

# 7: Corrosion

## Learning objectives

After studying this topic, you should be able to:

- explain why aluminium does not corrode
- describe the conditions needed for rusting to happen
- describe the similarities and differences between iron and aluminium

**A** Why does aluminium not corrode in moist conditions?

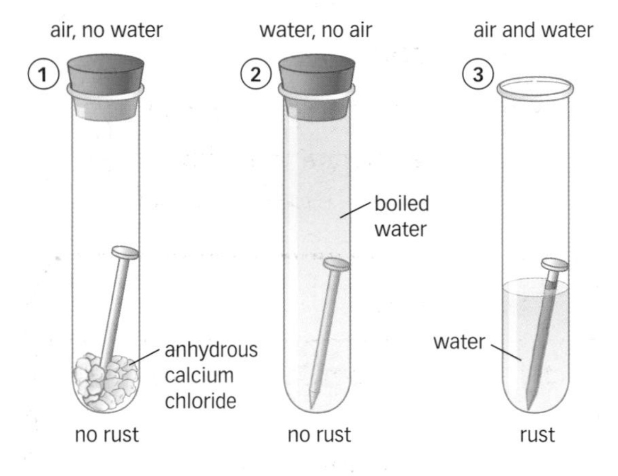


▲ An experiment to investigate rusting

### Exam tip OCR

- Remember that salt does not cause rusting, but it does speed up rusting.

**B** What is rusting?

## Corrosion

**Corrosion** happens when metals react with substances in the surroundings, such as air or water. Aluminium is a reactive metal but it does not corrode easily. Its surface is protected by a very thin natural layer of aluminium oxide. This layer is tightly bound to the metal and does not flake off. It stops air and water getting to the metal itself. This is why aluminium does not corrode in moist conditions. On the other hand, iron corrodes easily. This is called **rusting**. **Oxidation** is the addition of oxygen. Iron is oxidised when it rusts.

Iron is not protected by a layer of its oxide. Iron oxide flakes off easily, exposing fresh metal underneath. A piece of iron needs both oxygen *and* water for it to rust. It will not rust if only one of these substances is there. A layer of paint will stop oxygen and water getting to the metal underneath.

Iron rusts quickly at the seaside. This is because salt water speeds up the rate of corrosion. Acid rain in polluted air does this, too.

▲ Iron rusts quickly in salt water

## What is rust?

Iron is oxidised to hydrated iron(III) oxide when it reacts with oxygen and water:

iron + oxygen + water → hydrated iron(III) oxide

The familiar orange-brown rust is hydrated iron(III) oxide. Unlike the layer of aluminium oxide on the surface of aluminium, it flakes off easily. In time, an iron object can rust away completely.

## Iron and aluminium

Iron and aluminium are two commonly used metals. They have some properties in common. They are both

- **malleable** (they can be hammered or bent into shape)
- good electrical conductors.

However, iron corrodes easily but aluminium does not. Iron is also more **dense** than aluminium. Its **density** is 7.8 $g/cm^3$ while the density of aluminium is just 2.7 $g/cm^3$. This means, for example, that a 1 cm diameter iron ball would have a mass of 4.1 g, but the mass of an aluminium ball would only be 1.4 g.

Iron is a magnetic material. It is attracted to magnets. On the other hand, like most other metals, aluminium is not a magnetic material.

Scrap yards use electromagnets to lift scrap iron and steel

Aluminium is used for overhead electricity cables

### Did you know...?

Osmium is the densest metal. A 1 cm diameter ball of osmium would have a mass of 11.7 g, over eight times the mass of a similar aluminium ball.

### Key words

corrosion, rusting, oxidation, malleable, dense, density

### Questions

1 What is the role of salt in rusting?

2 Why does oiling a bicycle chain stop it rusting?

E

3 Make a table to compare the similarities and differences between the properties of iron and aluminium.

4 Explain why aluminium, rather than iron, is used to make overhead electricity cables.

C

5 Explain, with the help of a suitable equation, what happens when iron rusts.

A*

# 8: Making cars

## Learning objectives

After studying this topic, you should be able to:

- list the major materials used in a car and explain why they are used
- describe the advantages and disadvantages of building cars from steel or from aluminium
- explain the advantages and disadvantages of recycling cars

Robots help to build cars

A Name three metals used to make cars.

B Give three reasons why car manufacturers may look at alternative materials for their car parts.

## What's in a car?

A car may contain over 10 000 parts. The different parts are made from different materials, depending on the job of the part. For example, steel may be used for the car body, aluminium for the engine block, and copper for the electrical wiring. Glass is used for the windows, and plastic is used for the bumpers and dashboard. Fibres are used for the carpet and seat fabric. Car manufacturers are increasingly looking at alternative materials for their car parts. This may be to reduce costs and fuel consumption, to extend the useful life of the car, or to make recycling easier when the car is no longer fit for the road.

## Car bodies

Car bodies must be strong and tough. Modern cars often have plastic panels, for example the front wings. However, most of the body, including the frame and chassis, will be made from metal. Steel is an alloy of iron, harder and stronger than iron itself. It is usually used to make car bodies, but aluminium is being used, too.

Aluminium has a lower density than steel. A car body will weigh less if made from aluminium rather than steel. This should help the car to have a better fuel economy than the same car built from steel.

This car has an aluminium body

Aluminium does not corrode in moist conditions, but steel will rust unless it is protected with paint and other coatings. A car body will corrode less if it is made from aluminium rather than steel. This will help the car have a better useful lifetime. On the other hand, aluminium is more expensive than steel. A car body made from aluminium will cost more than the same one made from steel.

## Recycling cars

Recycling saves natural resources. It reduces the problem of how to dispose of parts and materials at the end of their useful lives. A European Union Directive requires car makers to improve their designs to make cars easier to recycle. They must also avoid certain toxic materials, and increase the percentage of a car that can be recycled. The target is a minimum of 95% of a car's weight by 2015. The Directive also allows owners to have their cars scrapped at no cost to them.

### Questions

1. What are the properties of glass that make it suitable for car windows?
2. State two advantages of recycling materials from cars.

E

3. What are the advantages and disadvantages of using aluminium instead of steel for car bodies?
4. Stainless steel resists corrosion, just as aluminium does. Explain why car manufacturers are unlikely to build car bodies from stainless steel.

C

5. Plastic parts are often used in modern cars where metal ones would have been used in the past. As a result, modern cars are often lighter than older ones. Different plastics are used for different purposes, depending on the properties required.
   (a) Explain two benefits of using plastics in cars.
   (b) To what extent does the use of plastics reduce the use of non-renewable resources?
   (c) Why may plastics be more difficult to recycle than steel and aluminium?

A*

▲ The steel in cars eventually rusts away

▲ There is profit to be made in selling second-hand car parts

### Did you know...?

Formula 1 racing cars are built for speed and agility. Their parts are made from the latest and most expensive materials. Yet, amazingly, they have a hard wooden strip fitted underneath. This runs down the middle from front to back. It is there to check that the car is not riding too close to the road.

# 9: Making ammonia

## Learning objectives

After studying this topic, you should be able to:

- describe factors affecting the cost of making a new substance
- describe some of the uses of ammonia
- describe how ammonia is made in the Haber process

**A** List five factors that affect the cost of making a new substance.

### Did you know...?

Fritz Haber was the German chemist who invented the process for making ammonia. He received the 1918 Nobel Prize in Chemistry for his work. Haber also helped to develop the use of chlorine and other poisonous gases during World War I.

### Key words

**reversible reaction, fertiliser, Haber process**

### Exam tip OCR

- The symbol ⇌ in an equation shows that the reaction is a **reversible reaction**.

## Making useful substances

Most of the substances you use in your everyday life will be manufactured. Very few are likely to be used in their natural state. The cost of making new and useful substances varies. The table describes several of the factors involved.

| Factor | How it affects the cost |
|---|---|
| Cost of starting materials | Different starting materials cost different amounts. The costs can be reduced if unreacted materials are recycled. |
| Cost of the equipment (plant) needed | Complex machinery and high pressures increase the cost of the plant. |
| Labour costs and wages | The more workers who are needed, the higher the costs. The costs may be reduced if a process is automated. |
| Price of the energy needed | Gas and electricity are expensive. The use of high temperatures increases the energy cost. |
| The speed at which the new substance is made | Catalysts increase the rate of chemical reactions, reducing costs. |

Ammonia is an important substance. Over 140 million tonnes of it are manufactured around the world each year. It provides a good example of how useful new substances are made.

▲ Chemical factories are very expensive to build and run

## Ammonia

Ammonia is a gas with a sharp smell. It is often used in cleaning fluids because it reacts with grease, making this easier to remove. It has other uses too, including making nitric acid and **fertilisers**. For example, ammonia and nitric acid react together to produce ammonium nitrate, which is an important source of nitrogen in artificial fertilisers.

Plants use up minerals in the soil as they grow. Fertilisers are substances applied to the soil to replace these minerals. Without fertilisers, crop yields would be much smaller. Fertilisers are so important to world food production that they use around 80% of all the ammonia manufactured.

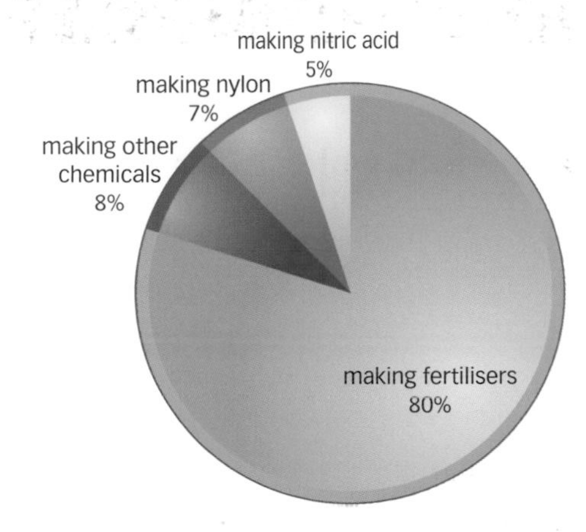

▲ The main uses of ammonia

**B** What are fertilisers and why are they important?

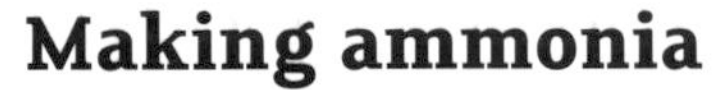

## Making ammonia

Ammonia is made from nitrogen and hydrogen using the **Haber process**:

$$\text{nitrogen} + \text{hydrogen} \rightleftharpoons \text{ammonia}$$

$$N_2 + 3H_2 \rightleftharpoons 2NH_3$$

The hydrogen often comes from cracking oil fractions or from natural gas. The nitrogen comes from air. Nitrogen is a relatively unreactive gas, so the conditions needed to get it to react with hydrogen must be chosen carefully. These include:

- a high pressure
- a temperature of around 450 °C
- adding an iron catalyst to speed up the reaction.

Costs are reduced by recycling unreacted nitrogen and hydrogen.

▲ Fritz Haber (1868–1934)

### Questions

1 What does the symbol $\rightleftharpoons$ mean?
2 Describe three uses of ammonia.
↓ E

3 Describe three factors that affect the cost of making a new substance.
4 Describe the conditions needed to make ammonia by the Haber process.
↓ C

5 Explain the importance of ammonia in world food production.
↓ A*

# 10: Reversible reactions

## Learning objectives

After studying this topic, you should be able to:

- interpret data about percentage yield in reversible reactions and changes in conditions
- explain the conditions used in the Haber process
- explain how economic considerations affect the choice of conditions used to manufacture new substances

## Key words

**yield**

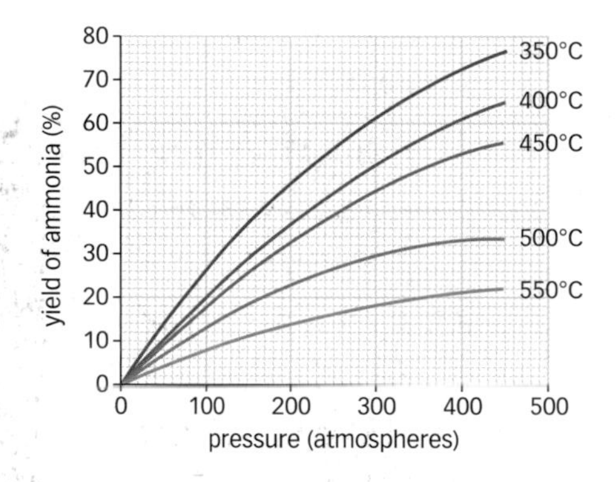


How the yield of ammonia varies if the temperature and pressure are changed

## Exam tip OCR

- You do not need to remember this graph, but you should be able to interpret a graph or table giving data about the percentage yield in a reversible reaction.
- You do not need to explain why reversible reactions behave as they do.

## Reversible reactions and yield

The signs of a chemical change include the production of a new substance and an irreversible change. Yet some chemical reactions *are* reversible. Under the right conditions, the products can reform the reactants.

The Haber process for making ammonia involves a reversible reaction:

$$\text{nitrogen} + \text{hydrogen} \rightleftharpoons \text{ammonia}$$

Two reactions are happening at once:

- a forward reaction, in which nitrogen and hydrogen react to make ammonia
- a backward reaction, in which ammonia breaks down to make nitrogen and hydrogen.

This is why the conditions used for the Haber process must be chosen carefully. The amount of a substance made in a reaction is called its **yield**. To get a high yield of ammonia, the conditions must favour the forward reaction rather than the backward reaction.

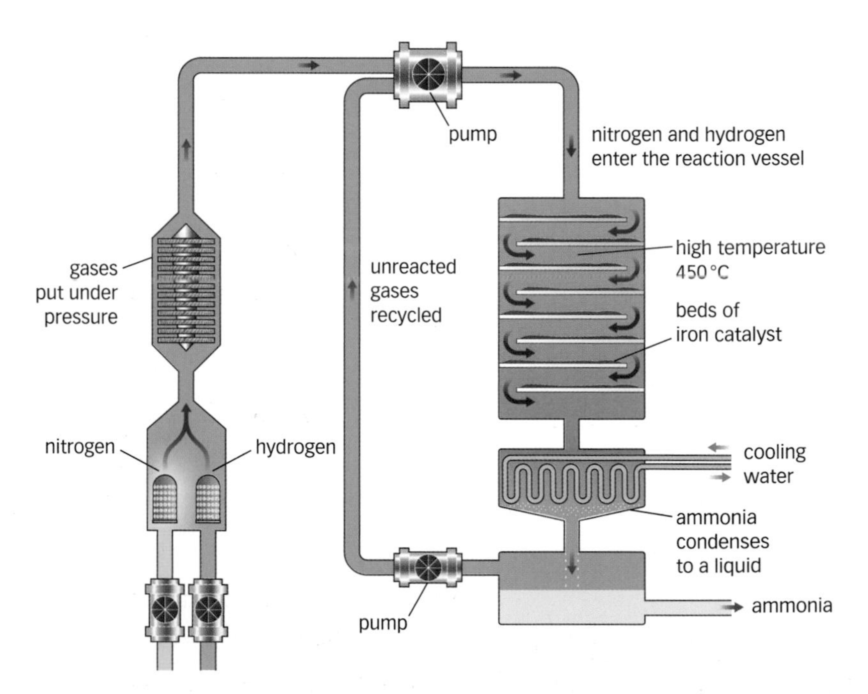


The Haber process

**A** What happens to the yield of ammonia if the pressure is increased, and if the temperature is increased?

You may remember that the conditions used for the Haber process include a high pressure and a temperature of around 450 °C. These conditions are chosen to obtain a reasonably high yield quickly; without using such a high pressure the costs would be too high.

## More about the Haber process

It is no good manufacturing a substance at such a high cost that no one wants to buy it. The chemical engineers who design a chemical process investigate many factors. The conditions they choose give the lowest cost, rather than the highest percentage yield or fastest rate of reaction.

The percentage yield must be high enough to give a sufficient daily yield of the product. This is achieved in the Haber process by using a high pressure. The higher the pressure, the greater the percentage yield of ammonia. However, there is a practical limit to how high this can be: if it is too high, the plant will be too expensive because very strong equipment will be needed. So a compromise pressure of around 200 atmospheres is often chosen.

The rate of reaction must be high enough to give a sufficient daily yield of the product. This is usually achieved by using a catalyst and a high temperature. Iron is used as the catalyst in the Haber process. It does not change the percentage yield, but it does increase the rate of reaction. However, in the Haber process the percentage yield decreases as the temperature increases. So a compromise must be made. The temperature must be high enough to get a fast reaction with a sufficiently high percentage yield. The optimum temperature chosen is about 450 °C.

Although the percentage yield is low, it is acceptable because the unreacted nitrogen and hydrogen are recycled in the process. This gives the unreacted gases additional chances to react to produce ammonia:

$$\text{nitrogen} + \text{hydrogen} \rightleftharpoons \text{ammonia}$$
$$N_2 + 3H_2 \rightleftharpoons 2NH_3$$

▲ Brown $NO_2$ reacts to produce colourless $N_2O_4$. The reaction is reversible. A high temperature decreases the percentage yield of $N_2O_4$. So the hot tube has less $N_2O_4$ and is dark brown. The cold tube has more $N_2O_4$ and is light brown.

## Questions

1. What does the term 'yield' mean?

E

2. From the graph on the previous page, what is the yield of ammonia at:
   (a) 100 atmospheres and 400 °C?
   (b) 400 atmospheres and 550 °C?
3. Why are a high pressure and 450 °C temperature chosen for the Haber process?

C

4. Explain why the pressure chosen for the Haber process is a compromise.
5. Explain the choice of 450 °C for the Haber process.
6. In the Haber process, why is iron used and unreacted gases recycled?

A*

## Learning objectives

After studying this topic, you should be able to:

- explain the use of the pH scale
- describe how universal indicator can be used to estimate pH
- explain that the pH of an acid is determined by its $H^+$ ion concentration
- describe neutralisation in terms of $H^+$ ions and $OH^-$ ions

▲ A selection of alkaline substances

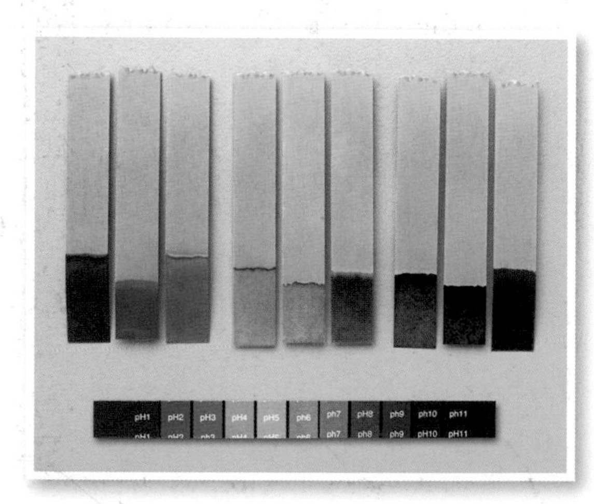


▲ Universal indicator paper and colour chart

## The pH scale

Acids dissolve in water to produce **acidic** solutions. These contain hydrogen ions, $H^+$. The concentration of these ions determines the pH of an acid. The higher the concentration of $H^+$ ions, the more strongly acidic the solution is, and the lower its pH. Acidic solutions have a pH less than 7. Pure water and other **neutral** substances have a pH of 7. Alkalis dissolve in water to produce **alkaline** solutions, with a pH more than 7.

▲ A selection of acidic substances

An **indicator** is a substance that changes colour depending on whether it is in an acidic solution or an alkaline solution. Litmus is an indicator that shows a sudden change in colour. It is red in acids and blue in alkalis. Litmus cannot show how acidic or alkaline a solution is, but **universal indicator** can.

Universal indicator is a mixture of different indicators. Each one changes colour over a different pH range. This means that the pH of a solution can be estimated using universal indicator paper or universal indicator solution.

**A** What is the pH scale?

**B** What is an indicator?

To estimate the pH of a solution, a spot of the solution is put onto a piece of universal indicator paper. After around 30 seconds, the colour of the spot is matched to the indicator chart. This shows the corresponding pH for each colour. It can be difficult to exactly match the colours, so the pH is just an estimate. A properly calibrated pH meter gives more accurate measurements of pH.

▲ Using a pH meter to measure the pH of a solution

## Neutralisation

A **base** is a substance that can neutralise an acid. The reaction between an acid and a base is called **neutralisation**. If a base is soluble in water, it is an alkali. The pH changes when acids and alkalis react together. For example, the pH decreases when an acid is added to an alkali. If enough acid is added, the pH decreases from above pH 7 to below pH 7. The opposite happens when an alkali is added to an acid. These changes can be sudden or gradual, depending on the acid and alkali used.

A salt and water are produced in neutralisation reactions. Here is the general word equation for the reaction between an acid and a base:

$$\text{acid} + \text{base} \rightarrow \text{a salt} + \text{water}$$

The particular salt produced in a neutralisation reaction depends upon the combination of acid and base used.

**C** Why are all alkalis bases, but not all bases are alkalis?

**D** Name the two products formed in the reaction between an acid and a base.

### Ions in solution

Acidic solutions contain hydrogen ions, $H^+$. Alkaline solutions contain hydroxide ions, $OH^-$. These react together in neutralisation reactions to produce water. This **ionic equation** describes neutralisation:

$$H^+ + OH^- \rightarrow H_2O$$

The particular salt produced in a neutralisation reaction depends upon the other ions present.

### Key words

**acidic, neutral, alkaline, indicator, universal indicator, base, neutralisation, ionic equation**

### Questions

1. Explain how you can tell from the pH of a liquid whether it is acidic, neutral, or alkaline.
2. State the general word equation for neutralisation.

↓ E

3. Describe how universal indicator is used to estimate the pH of a solution.
4. Explain how the concentration of $H^+$ ions determines the pH of an acid.

↓ C

5. Identify the ions present in acidic or alkaline solutions. Write an ionic equation for neutralisation.

↓ A*

# 12: Making salts – 2

## Learning objectives

After studying this topic, you should be able to:

- predict the name of the salt produced in a neutralisation reaction
- write word equations for neutralisation reactions
- write balanced symbol equations for neutralisation reactions

Carbon dioxide causes bubbling during the reaction between a carbonate and an acid

Copper sulfate is made from copper oxide and sulfuric acid

## Bases and carbonates

Metal oxides such as copper oxide are bases because they can neutralise acids. For the same reason, metal hydroxides such as sodium hydroxide are bases, too. Here are the general word equations for these reactions:

metal oxide + acid → a salt + water
metal hydroxide + acid → a salt + water

Notice that a salt and water are produced in both reactions.

Carbonates such as sodium carbonate can also neutralise acids. A third product, carbon dioxide, is also produced:

carbonate + acid → a salt + water + carbon dioxide

## Naming salts

You can predict the name of the salt produced in a neutralisation reaction if you know which base or carbonate is used, and the acid it neutralises. The names of salts have two parts. The first part comes from the name of the metal in the base or carbonate. The exception to this is when ammonia or ammonium carbonate is used. In these cases, the first part of the salt's name will be 'ammonium'. The second part comes from the acid used:

- hydrochloric acid produces chloride salts
- nitric acid produces nitrate salts
- sulfuric acid produces sulfate salts
- phosphoric acid produces phosphate salts.

### Worked example

What is the salt produced when copper oxide reacts with sulfuric acid? Write a word equation for the reaction.

The first part of the name will be copper (from copper oxide) and the second part of the name will be sulfate (from sulfuric acid). The salt produced is copper sulfate.

The equation for the neutralisation reaction is:

copper oxide + sulfuric acid → copper sulfate + water

## Balanced symbol equations

When writing the balanced symbol equation for a neutralisation reaction, it is a good idea to identify the salt formed first, and then work out its formula. The table shows the six bases or carbonates you need to know about.

| Bases | | | Carbonates | | |
|---|---|---|---|---|---|
| Name | Formula | Ion in salt | Name | Formula | Ion in salt |
| copper oxide | $CuO$ | $Cu^{2+}$ | calcium carbonate | $CaCO_3$ | $Ca^{2+}$ |
| sodium hydroxide | $NaOH$ | $Na^+$ | sodium carbonate | $Na_2CO_3$ | $Na^+$ |
| potassium hydroxide | $KOH$ | $K^+$ | | | |
| ammonia solution | $NH_4OH$ | $NH_4^+$ | | | |

| Acid | Formula | Ion in salt |
|---|---|---|
| hydrochloric acid | $HCl$ | $Cl^-$ |
| nitric acid | $HNO_3$ | $NO_3^-$ |
| sulfuric acid | $H_2SO_4$ | $SO_4^{2-}$ |
| phosphoric acid | $H_3PO_4$ | $PO_4^{3-}$ |

The formula for hydrochloric acid is HCl. Calcium carbonate reacts with hydrochloric acid to produce calcium chloride and water. The formula for calcium chloride is $CaCl_2$, so the balanced symbol equation is:

$$CaCO_3 + 2HCl \rightarrow CaCl_2 + H_2O + CO_2$$

Notice that, in this case, the equation is balanced by putting a 2 in front of HCl.

### Questions

1 When a carbonate reacts with an acid, what product is made that is not made when a base reacts with an acid?

E

2 Name the salt produced when sodium hydroxide reacts with hydrochloric acid.

3 Write a word equation for the reaction between sodium carbonate and sulfuric acid.

4 Write a word equation for the reaction between ammonia solution and nitric acid.

C

5 Write balanced symbol equations for the reactions in Questions 2, 3, and 4.

A*

### Exam tip OCR

✔ Remember that carbon dioxide makes limewater turn cloudy white, or milky.

# 13: Fertilisers and crop yields

## Learning objectives

After studying this topic, you should be able to:

- ✔ recall three essential elements needed for plant growth
- ✔ explain that fertilisers increase crop yields
- ✔ describe the process of eutrophication

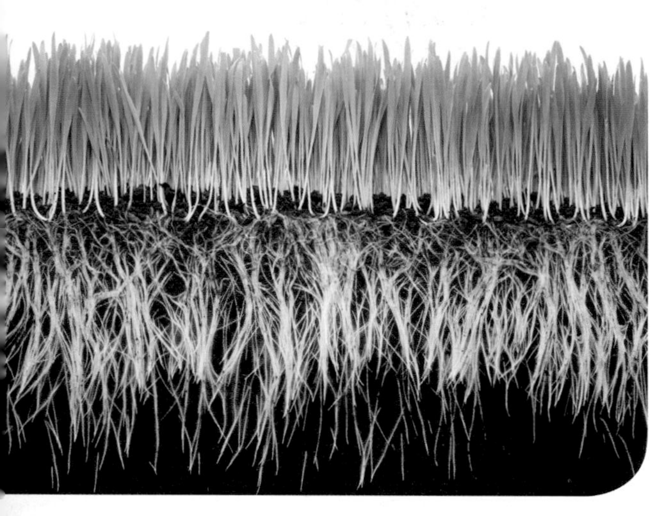

▲ Plants absorb minerals dissolved in water through their roots

## Exam tip OCR

- ✔ You might remember the three essential elements by their chemical symbols, NPK. You can then find their names on the periodic table.

## Key words

**essential element**, **eutrophication**, **algal bloom**, **aerobic bacteria**

## Essential elements

Plants need certain elements to make them grow properly. Three **essential elements** for plant growth are

- nitrogen
- phosphorus
- potassium.

These elements are found combined with other elements in the soil, forming compounds called minerals. For example, ammonium nitrate contains nitrogen, and potassium phosphate contains potassium and phosphorus. Plants absorb water and dissolved minerals through their roots. They cannot absorb insoluble minerals.

**A** Name three essential elements for plant growth.

**B** Name the essential minerals found in ammonium phosphate.

## Fertilisers

Fertilisers are chemicals that provide plants with essential elements. They make crop plants grow faster and bigger, and the crop yield is increased. Since plants can only absorb minerals dissolved in water, fertilisers must be soluble in water, too.

The world population is rising quickly, so we need to provide more food. Fertilisers can help achieve this. However, using too much fertiliser can pollute water supplies, leading to problems such as the death of living organisms in the water. This is known as **eutrophication**.

▲ Fertilisers must first dissolve in water so plants can use them

## More about fertilisers

As plants grow, they absorb dissolved minerals from the soil. If the plants are removed, for example by farmers at harvest time, these minerals are lost from the soil. Over time, minerals in the soil decrease. Fertilisers replace essential elements used by a previous crop. They may also provide extra essential elements. For example, plants need nitrogen to make their proteins. If extra nitrogen-containing minerals are applied to the soil, plants can make more protein, producing increased plant growth.

## Eutrophication

Overuse or careless use of fertilisers can cause eutrophication. The flow chart on the right describes how this process works. Excess fertiliser washes off the land into rivers and lakes. Depending on the fertiliser, this increases the concentration of nitrate or phosphate in the water. Tiny water plants called algae use these minerals to grow at a much faster rate. They produce a thick green layer on the water, called an **algal bloom**. This blocks off sunlight to other plants in the water, which die. **Aerobic bacteria** are bacteria that use oxygen. They feed on the dead plants, using up oxygen in the water.

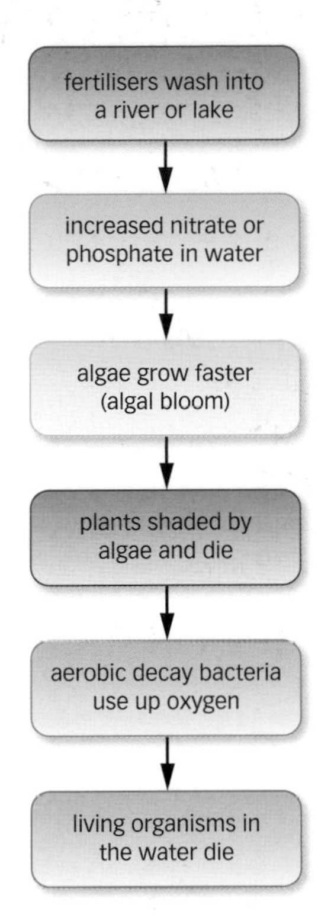

fertilisers wash into a river or lake
increased nitrate or phosphate in water
algae grow faster (algal bloom)
plants shaded by algae and die
aerobic decay bacteria use up oxygen
living organisms in the water die

This flow chart shows the process of eutrophication

An algal bloom in a lake, caused by over-use of fertilisers

### Questions

1. What is a fertiliser, and what does it do?
2. Name the essential elements in this fertiliser compound: $KNO_3$.

E

3. Explain why fertilisers must be water-soluble.

C

4. Explain how a fertiliser containing nitrogen can lead to increased plant growth.
5. Describe the process of eutrophication.

A*

### Did you know...?

About 7 billion more cattle would be needed to produce enough manure to do the same job as artificial fertilisers do now. Almost all the land currently covered by trees would be needed to grow the food they would need.

# 14: Making fertilisers

## Learning objectives

After studying this topic, you should be able to:

- name two nitrogenous fertilisers made from ammonia
- name the acid and alkali needed to make a particular fertiliser compound
- identify the apparatus needed to make a fertiliser compound in the lab
- describe how to make a fertiliser compound in the lab

## Key words

**nitrogenous fertiliser, burette, measuring cylinder, filter funnel**

A Name two nitrogenous fertilisers.

## Fertilisers from ammonia

Ammonia is made from nitrogen and hydrogen by the Haber process. It can be used on its own as a fertiliser to provide plants with nitrogen, as it is soluble in water. However, it is an alkali with a sharp smell, so it is usually used to manufacture other compounds for use as fertilisers. These **nitrogenous fertilisers** include

- urea, $(NH_2)_2CO$
- ammonium nitrate, $NH_4NO_3$
- ammonium sulfate, $(NH_4)_2SO_4$
- ammonium phosphate, $(NH_4)_3PO_4$.

Potassium nitrate, $KNO_3$, is also a nitrogenous fertiliser, but it is not manufactured directly from ammonia.

These crop scientists are investigating the effects of fertilisers on plant growth

Large bags of fertiliser stacked at a fertiliser factory

## Fertilisers from acids and alkalis

Fertilisers can be made by neutralising an acid with an alkali. For example, potassium nitrate can be made by neutralising nitric acid with potassium hydroxide. Ammonia solution is an alkali that produces ammonium salts when it neutralises acids:

- ammonium nitrate with nitric acid
- ammonium sulfate with sulfuric acid
- ammonium phosphate with phosphoric acid.

Fertiliser compounds like these are made in huge amounts at chemical factories, but you can easily make smaller amounts in the laboratory.

Just enough acid is mixed with an alkali to produce a neutral salt solution. A **burette** and **measuring cylinder** are used to measure the volumes of liquid needed. The salt solution is then heated to evaporate some of its water. This reduces the volume and makes it more concentrated. As it cools, crystals of salt form. They are separated from the solution by filtration using a **filter funnel** and filter paper.

**B** Name the acid and alkali needed to make potassium nitrate.

▲ This student is using a burette to measure the volume of acid needed to neutralise an alkali

## More about making fertiliser

There are four steps to making a fertiliser from an acid and an alkali in the laboratory.

1. Use a measuring cylinder to add a known volume of alkali to a conical flask. This is usually 25 $cm^3$. Add a few drops of indicator, such as phenolphthalein. Add acid from the burette until the indicator just changes colour. Note the volume of acid needed to do this.
2. Repeat Step 1 but without the indicator: just use the burette to add the correct volume of acid.
3. Heat the salt solution made in Step 2 to reduce its volume. Leave it to cool down so that crystals form.
4. Filter the mixture to separate the crystals, and leave them to dry.

## Questions

1 What type of fertiliser are urea and ammonium sulfate? ↓E

2 Name the acid needed to make ammonium nitrate.

3 Name the acid and alkali needed to make ammonium phosphate. ↓C

4 Describe how you could use neutralisation to prepare dry ammonium sulfate. ↓A*

## Exam tip OCR

- ✔ You should be able to label the apparatus needed to make a fertiliser from an acid and an alkali, including the burette, measuring cylinder, and filter funnel.
- ✔ You need to be able to name the acid and alkali needed to make a particular fertiliser. You also need to be able to describe the method needed to obtain solid fertiliser from them.

## Learning objectives

After studying this topic, you should be able to:

- describe how sodium chloride is mined
- recall the products formed by the electrolysis of brine
- describe some uses of these products

An underground salt mine

A What is brine?

B Suggest why salt for spreading on roads in the winter is mined using cutting machines, rather than by solution mining.

## Mining for salt

Common salt is sodium chloride, NaCl. It is not just useful as a flavouring or preservative in food. It is also an important raw material for the chemical industry.

Salt can be obtained from seawater by evaporating the water. It can also be obtained from salt deposits. There are huge underground deposits of salt in Cheshire and some other parts of the UK. These are the remains of ancient seas that have long since evaporated and been covered by rock. Salt for cooking and for applying to the roads in winter is usually mined in the normal way. Machines cut through the salt, forming large caverns as they work.

Sodium chloride is soluble in water. Since the chemical industry needs concentrated sodium chloride solution or **brine**, it is also mined by **solution mining**. Water is pumped from the surface into the deposit deep underground. The sodium chloride dissolves in the water, and the brine is pumped up to the surface. This is a continuous process that needs little labour, which keeps costs down. Mining salt may cause **subsidence**. This is where the land slumps downwards because of a mine below, leading to cracked buildings and holes appearing in the ground.

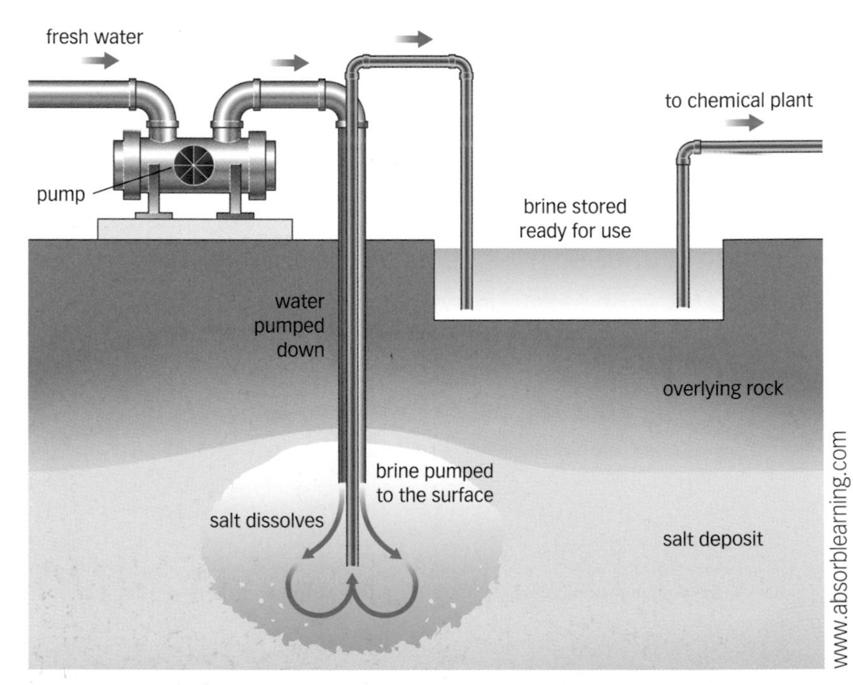


Solution mining

## Chemicals from salt

New substances are made from brine by passing electricity through it. This process is called electrolysis. There are more details about electrolysis on the next spread. The electrolysis of brine produces three substances:

- hydrogen
- chlorine
- sodium hydroxide solution.

**Key words**

**brine, solution mining, subsidence, bleach**

**Exam tip** OCR

✔ Make sure you can recall the three products made from the electrolysis of brine, and at least one use for each of them.

**C** List the three products from the electrolysis of brine.

Hydrogen is used in the manufacture of margarine. It is used to convert liquid vegetable oils into solid vegetable fats. These can be blended to make margarine that spreads from the fridge without running everywhere.

Chlorine is used to sterilise water. It is widely used to kill harmful microorganisms in swimming pool water and tap water. Chlorine is also important as a raw material for manufacturing other new substances. For example, the polymer PVC, polyvinylchloride, is made from chloroethene, which is itself made using chlorine. Chlorine is also used to make household **bleach** and some solvents.

Sodium hydroxide reacts with vegetable oils to make soap. It also reacts with chlorine to make household bleach (sodium chlorate solution):

sodium hydroxide + chlorine → sodium chloride + water + sodium chlorate (bleach)

### Questions

1 State two uses of sodium chloride in food.

2 State one use of hydrogen.

3 Describe two uses of chlorine, and two uses of sodium hydroxide.

E

4 Describe how sodium chloride can be mined.

5 Describe, with the help of a word equation, how household bleach is made.

C

▲ Household bleach is made from sodium hydroxide and chlorine

# 16: Electrolysis of brine

## Learning objectives

After studying this topic, you should be able to:

- describe the key features of the electrolysis of brine
- describe a laboratory test for chlorine
- explain the electrolysis of brine in terms of loss or gain of electrons

▲ A chlor-alkali plant

## Did you know...?

The European chlor-alkali industry produces around 10 million tonnes of chlorine every year. To do this, it uses 36 billion kilowatt hours of electricity. This is enough electricity for 7.5 million typical UK households.

## The chlor-alkali industry

Brine is concentrated sodium chloride solution. It is a raw material for the chlor-alkali industry. This is the chemical industry that produces useful new substances from the electrolysis of brine. These are chlorine, hydrogen, and sodium hydroxide. Large amounts of electricity are needed, and electricity accounts for about half of the production costs.

## Electrolysis

Electrolysis is the process in which a compound breaks down to form new substances when electricity is passed through it. Two electrodes are needed to pass the electricity into the salt solution. These are made from an inert material such as titanium. This stops the electrodes reacting with the salt or the products from it. Different products form at each electrode during electrolysis:

- chlorine forms at the positive electrode (the anode)
- hydrogen forms at the negative electrode (the cathode).

Sodium hydroxide is also made. This stays dissolved.

**A** Why are electrodes made from an inert material like titanium?

**B** What is the difference between an anode and a cathode?

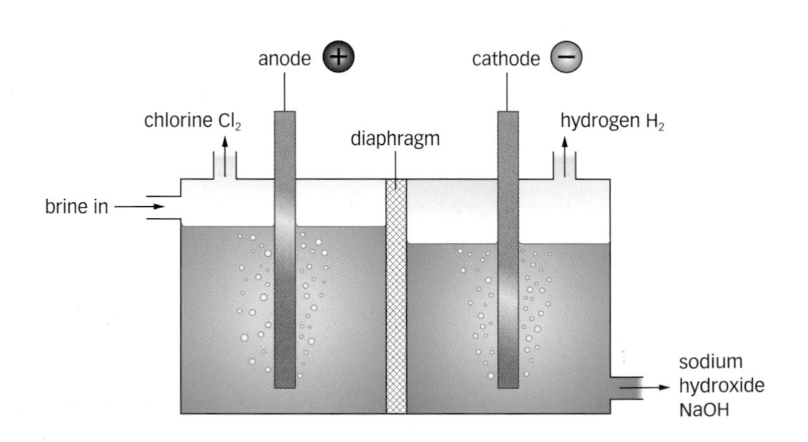


▲ The electrolysis of brine. The diaphragm helps to keep the chlorine away from the sodium hydroxide solution.

## Oxidation and reduction

Oxidation is the gain of oxygen by a substance, and reduction is loss of oxygen by a substance. However, oxidation and reduction can also be described in terms of loss or gain of electrons:

- oxidation is loss of electrons
- reduction is gain of electrons.

Sodium chloride solution (brine) contains a mixture of different ions. These are $Na^+$ ions and $Cl^-$ ions from the sodium chloride, and $H^+$ ions and $OH^-$ ions from the water.

At the anode (the positive electrode), chloride ions lose electrons. They are oxidised to chlorine gas:

$$2Cl^- \; - \; 2e^- \rightarrow Cl_2$$

At the cathode (the negative electrode), hydrogen ions gain electrons and are reduced to hydrogen gas:

$$2H^+ \; - \; 2e \rightarrow H_2$$

The remaining ions, the $Na^+$ ions and $OH^-$ ions, are not discharged. Instead, they make sodium hydroxide.

## Testing for chlorine

Chlorine is a gas with a faint yellow-green colour. It has a sharp smell. This is usually noticed in swimming pools or when bleach is used. Chlorine dissolves in water to form an acidic solution. It can also bleach dyes – change them from being coloured to colourless. These properties are the basis of a simple laboratory test for chlorine. When moist blue litmus paper is held in chlorine, it turns red and then white.

### Exam tip OCR

✔ Remember OIL RIG: oxidation is loss of electrons, and reduction is gain of electrons.

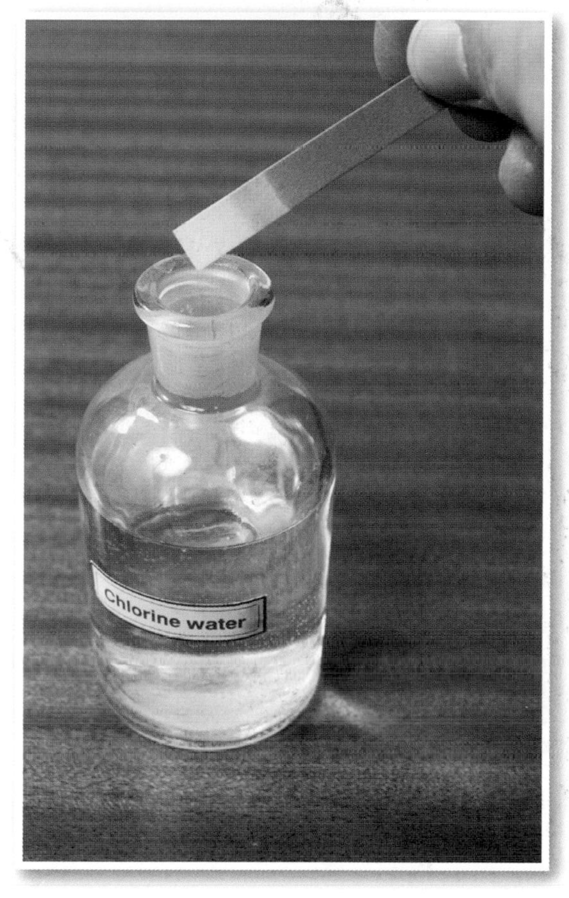


▲ Chlorine bleaches damp litmus paper

### Questions

1 Describe the laboratory test for chlorine.

2 Name the two gases produced by the electrolysis of brine.

↓ E

3 Describe what is produced at each electrode during the electrolysis of brine. Why are inert electrodes used?

4 Name the alkali produced by the electrolysis of brine.

↓ C

5 Describe, in terms of electrons and with the help of suitable equations, what happens during the electrolysis of brine.

↓ A*

# Module summary

## Revision checklist

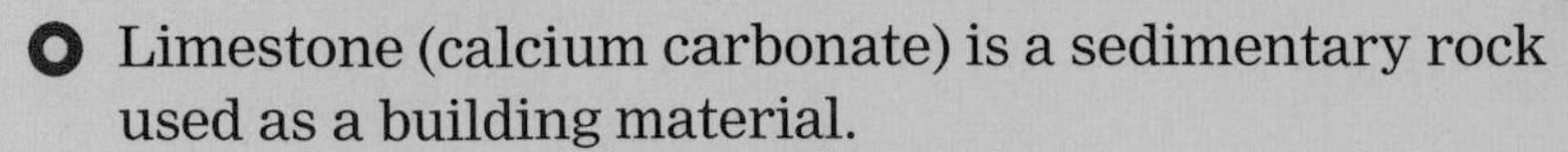

- Limestone (calcium carbonate) is a sedimentary rock used as a building material.
- Limestone is heated with clay to form cement; cement, water, and sand form concrete, which can be reinforced with steel.
- Igneous rocks (granite) and metamorphic rocks (marble) are harder than sedimentary rocks.
- Many metal carbonates are decomposed by heat to form metal oxides + carbon dioxide.
- The Earth has a layered structure. The thin rocky crust floats on the mantle and at the centre is the hot iron core.
- The crust is divided into tectonic plates. Convection currents in the mantle cause the plates to move. Earthquakes and volcanic eruptions occur at plate boundaries.
- Lava is liquid rock that cools to form igneous rocks with a variety of crystal sizes.
- Copper metal is extracted from copper ore by heating it with carbon. This is called reduction, because oxygen is removed. Impure copper can be purified by electrolysis.
- Recycling copper saves resources, energy, and reduces waste.
- Alloys are metal mixtures with properties making them more useful than pure metals. Some alloys are 'smart alloys'.
- Rusting is an example of metal corrosion and occurs when iron is in contact with air and water.
- Aluminium does not corrode easily and has a lower density than iron or steel.
- Iron has physical properties ideal for use in the manufacture of a car. Aluminium and plastic are easier to recycle and are used in cars to reduce weight and resist corrosion.
- Ammonia is manufactured in a reversible reaction using nitrogen (from the air) and hydrogen (from oil or natural gas).
- Manufacturing conditions are chosen to reduce the costs of energy and raw materials.
- Acid solutions contain $H^+$ ions and are neutralised by alkalis to form salts plus water.
- Neutralisation reactions are used to make fertilisers.
- Fertilisers increase crop yields by providing essential elements (N, P, K), but can cause eutrophication in lakes and rivers.
- Chlorine, sodium hydroxide, and hydrogen are all useful products of electrolysis of sodium chloride solution.

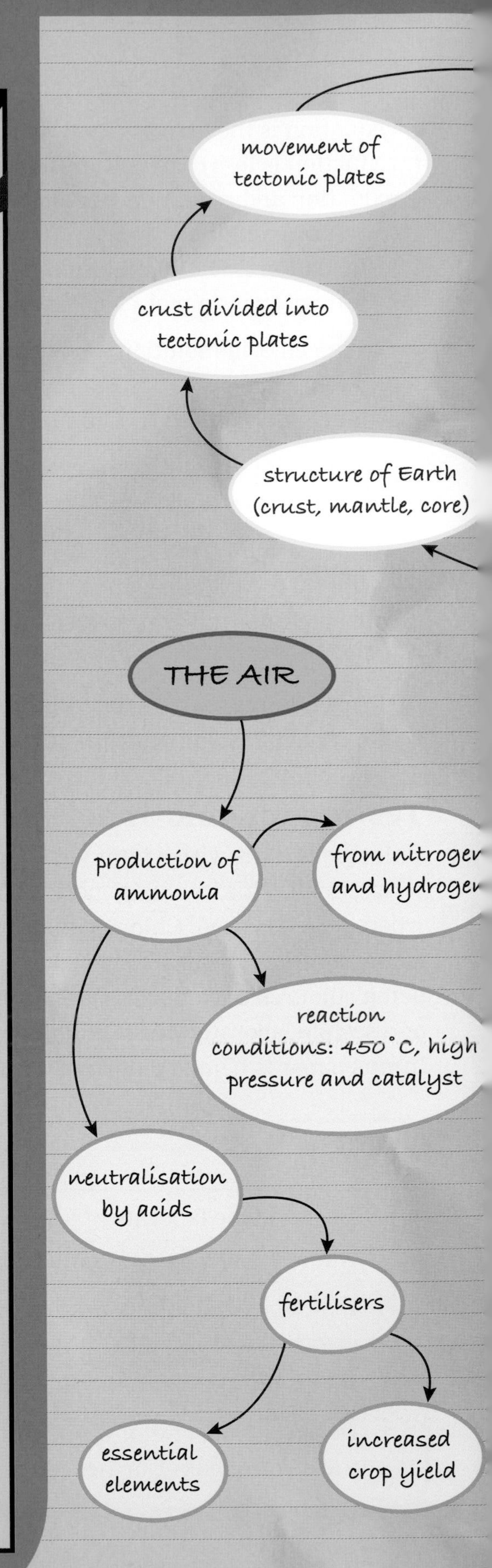



NOW USE THE C2 GRADE CHECKER ON PAGE 242

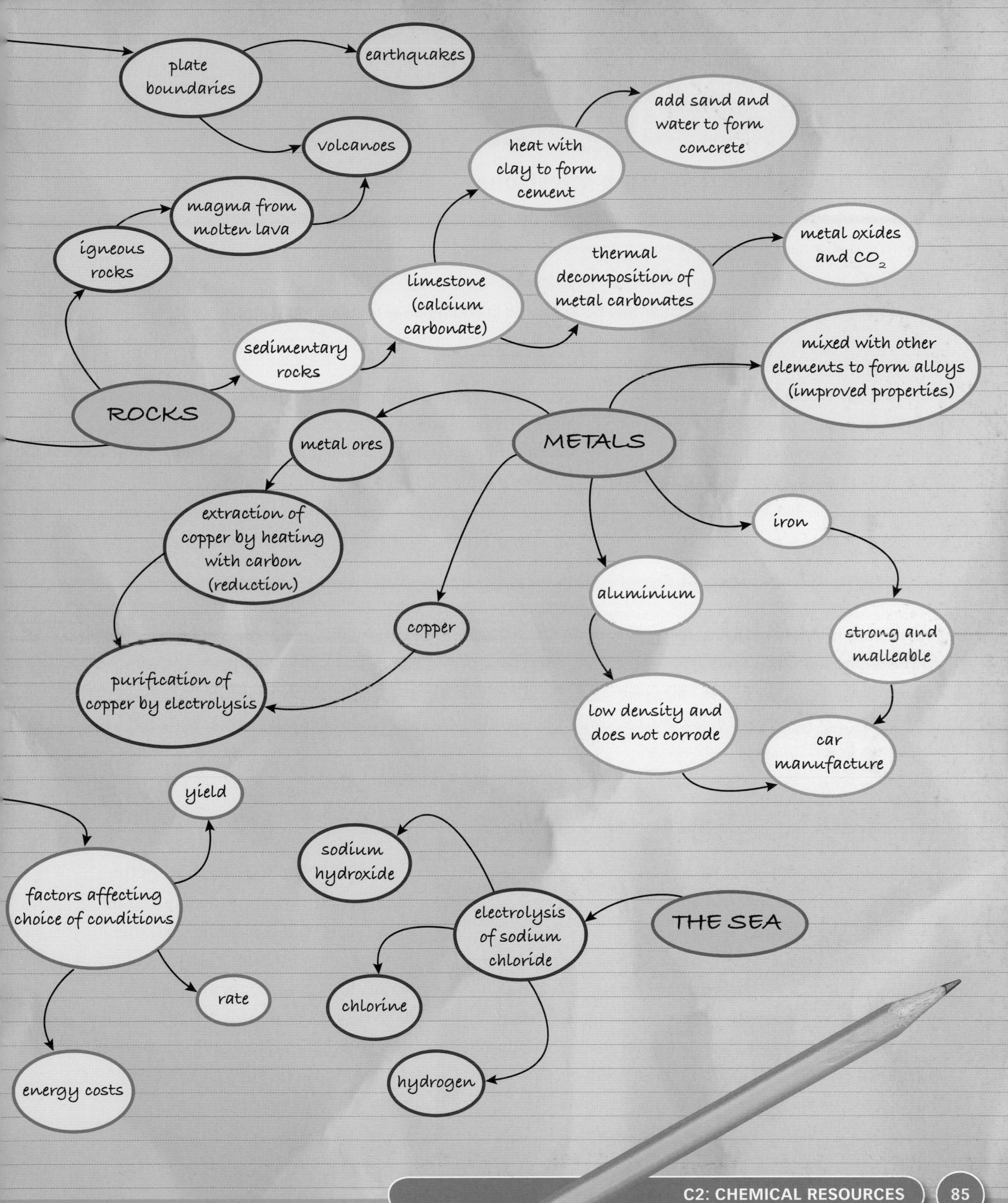
plate boundaries
earthquakes
volcanoes
magma from molten lava
igneous rocks
add sand and water to form concrete
heat with clay to form cement
metal oxides and $CO_2$
thermal decomposition of metal carbonates
limestone (calcium carbonate)
sedimentary rocks
ROCKS
mixed with other elements to form alloys (improved properties)
metal ores
METALS
extraction of copper by heating with carbon (reduction)
iron
aluminium
copper
strong and malleable
purification of copper by electrolysis
low density and does not corrode
car manufacture
yield
sodium hydroxide
factors affecting choice of conditions
electrolysis of sodium chloride
THE SEA
rate
chlorine
energy costs
hydrogen

## Answering Extended Writing questions

QUESTION

Outline the purposes of adding fertilisers to growing crops.

Describe arguments for and against the use of fertilisers.

**The quality of written communication will be assessed in your answer to this question.**

Crops grow faster with fertillisers but you mustn't youse them because I think they is not organic and so they is bad for the enviroment but there be hungry peeple in the world

E

*Examiner:* The candidate makes one relevant point, at the start of the answer. The disadvantage identified (being bad for the environment) is too vague. The candidate mentions hungry people, but does not link this to the use of fertilisers. There are several spelling, punctuation and grammar errors. The answer is poorly organised.

Fertilisers make crops grow fast. Minerals go up plant roots, so they must dissolve in water. They are good becoz we need more food becoz their are more peopel. They pollute water supplies, which is not grate.

C

*Examiner:* The candidate has given a reason for using fertilisers, and has explained one of their advantages. A disadvantage is also identified. The second sentence is not relevant. The answer is well organised, and the punctuation and grammar are accurate. There are several spelling mistakes.

Fertilisers provide plants with essential elements like nitrogen, which is incorporated into plant proteins. So plants grow bigger and faster.

The world population is increasing, so we must produce more food. This is an advantage of using fertilisers.

But fertilisers cause eutrophication in lakes. Extra nutrients cause algal bloom. This blocks off sunlight to plants beneath, which die. Then aerobic bacteria use up oxygen in the water, so living organisms die.

A*

*Examiner:* This answer covers the main scientific points clearly and in detail. It is well organised, and the spelling, punctuation and grammar are accurate. Scientific terms are used correctly. The candidate has not stated that eutrophication is a disadvantage, though this is implied by the word 'but' in the third paragraph.

# Exam-style questions

**1** Sam carries out an experiment to neutralise 10 $cm^3$ of hydrochloric acid by adding an alkali, sodium hydroxide.

| Volume of alkali added ($cm^3$) | pH |
|---|---|
| 0 | 2 |
| 3 | 5 |
| 6 | 8 |
| 9 | 10 |
| 12 | 11 |
| 15 | 12 |

A03 **a** Which result shows that the solution was barely alkaline?

A03 **b** Sam found it difficult to say how much alkali it took to neutralise the acid. Explain why.

E

**2** Ammonia is made in industry by reacting nitrogen with hydrogen. The graph shows how the yield of ammonia depends on the conditions used.

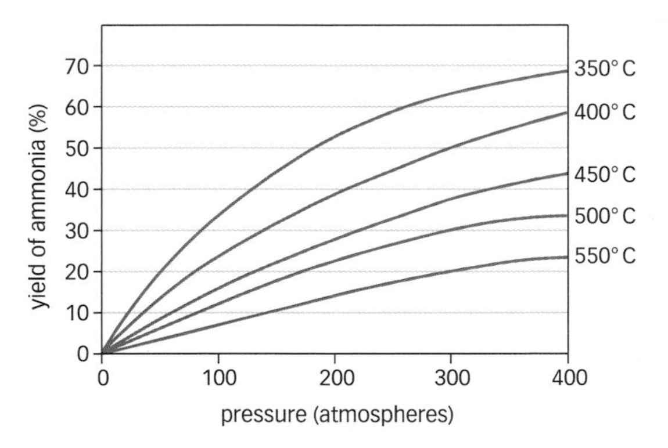


A03 **a** Describe what happens to the yield when:

**i** The temperature is increased.

**ii** The pressure is increased.

C

A03 **b** What is the yield at 450 °C and 200 atmospheres?

A02 **c** Why is production of ammonia an expensive process? State two reasons for this.

A02 **d** How could the cost of making ammonia be reduced?

C

**3** Synthetic fertilisers are used in modern farming. Many of the substances used in the fertilisers can be made easily in the laboratory.

A01 **a** Fertilisers increase crop yield. Explain why.

A01 **b** State one reason why the use of fertilisers on farmland needs to be carefully controlled.

A*

## Extended Writing

**4** A01 Cars are made using a number of different substances. Write about three of the different substances that are used, explaining what makes them suitable for use in a modern car.

E

**5** A01 Explain how chloralkali industry uses electrolysis of sodium chloride to make a number of useful products.

C

**6** A01 Alfred Wegener suggested the theory of continental drift in 1914. This theory was rejected by other scientists until the 1960s when it developed into the theory of plate tectonics which is widely accepted today. Describe what changed the views of other scientists.

A*

A01 Recall the science

A02 Apply your knowledge

A03 Evaluate and analyse the evidence

# Chemical economics

| OCR specification match | | |
|---|---|---|
| C3a | 3.1 | Reaction time |
| C3a | 3.2 | Measuring rates |
| C3b | 3.3 | Reaction rate and temperature |
| C3b | 3.4 | Pressure and concentration |
| C3c | 3.5 | Explosions |
| C3c | 3.6 | Catalysts |
| C3d | 3.7 | Relative formula mass |
| C3d | 3.8 | Conservation of mass |
| C3e | 3.9 | Percentage yield |
| C3e | 3.10 | Atom economy |
| C3f | 3.11 | Energy |
| C3f | 3.12 | Calorimetry |
| C3g | 3.13 | Batch or continuous? |
| C3g | 3.14 | Making medicines |
| C3h | 3.15 | Diamond and graphite |
| C3h | 3.16 | Bucky balls and nanotubes |
| Module summary | | |
| OCR Upgrade | | |

## Why study this module?

Explosions are very fast reactions, while rusting and other reactions are slow. In this module, you will investigate the factors affecting the rate of reactions, including temperature, concentration, pressure, and surface area. You will also investigate the effect of catalysts, and why they are useful in industrial chemical processes.

The chemical industry is working hard to be more economic and sustainable. You will discover some of the calculations chemical engineers use to make their processes as economic and 'green' as possible. These include predicting the mass of product that should be formed, and calculating the reaction's 'atom economy'. Some reactions release energy, while others absorb it. You will examine the reasons for this, and how the energy can be measured.

You will discover how pharmaceutical drugs are developed, tested, manufactured, and marketed. Finally, you will learn about the various forms of carbon including diamond, graphite, bucky balls, and nanotubes.

### You should remember

1. The particle model provides explanations for the different physical properties and behaviour of matter.
2. How to measure temperature, time, mass, and volume in experiments.
3. How to plot line graphs and calculate the gradient of the line.
4. How to represent chemical reactions using equations.
5. Carbon is one of around 100 different elements, and exists as diamond and graphite.

The British pharmaceutical industry spends over £10 million per day on research and development, and around 20% of the world's top medicines were discovered and developed in Britain. Many pharmaceutical drugs are based upon naturally-occurring substances with medicinal properties.

These flasks contain freeze-dried samples of active substances from mushrooms and bacteria. Freeze-drying involves drying the samples under a vacuum at low temperatures, so that they can be stored until needed by scientists.

# 1: Reaction time

## Learning objectives

After studying this topic, you should be able to:

- describe what is meant by reaction time
- recognise the relationship between the speed of a reaction and its reaction time
- interpret data involving reaction time

## Key words

**reaction time, reactant, excess, continuous variable**

**A** Give an example of a slow reaction and a fast one.

Magnesium ribbon reacting with hydrochloric acid

## How fast?

The **reaction time** is the time taken between a reaction starting and stopping. Slow reactions have long reaction times. For example, rusting is a slow reaction. It can take many years for a piece of iron or steel to completely rust away. On the other hand, burning and explosions are very fast reactions. They can take very little time to happen.

It is taking many years for this wrecked ship to rust away

A reaction stops when one of the **reactants** is all used up. For example, magnesium reacts with hydrochloric acid:

| magnesium | + | hydrochloric acid | → | magnesium chloride | + | hydrogen |
|---|---|---|---|---|---|---|
| $Mg$ | + | $2HCl$ | → | $MgCl_2$ | + | $H_2$ |

The reaction starts once the magnesium and acid are mixed together. The release of hydrogen causes bubbling. If there is enough hydrochloric acid to react with all the magnesium, we say that the acid is in **excess**. When the acid is in excess, the bubbling stops when the magnesium is all used up. The reaction time is found by timing how long it takes for the bubbling to stop after mixing the magnesium and acid.

**B** What is meant by reaction time in chemistry?

## Investigating reaction time

The table shows the results of an investigation into a reaction. Magnesium ribbon was cut into identical 1.5 cm pieces and 25 $cm^3$ of dilute hydrochloric acid was measured into a beaker. A piece of magnesium ribbon was added to the acid and the reaction time measured. Another piece was then added to the reaction mixture and the reaction time for this piece was measured. More pieces were added one at a time until a total of 12 pieces were added.

The total number of pieces of magnesium and the reaction time are both **continuous variables**, so a line graph is the best way to display them.

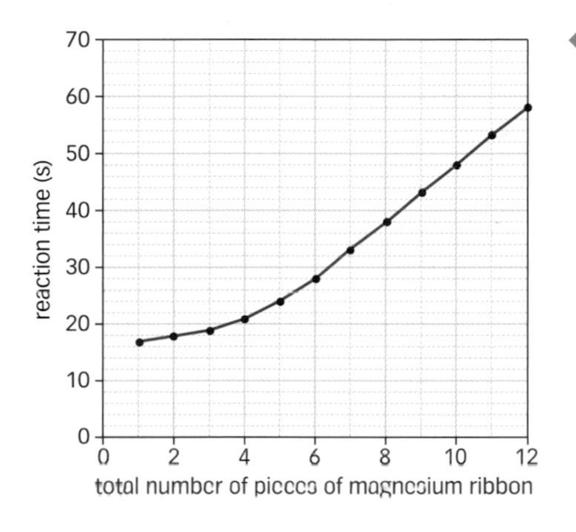


The reaction time between magnesium and hydrochloric acid changes as each extra piece of magnesium ribbon is added

| Total number of pieces of Mg | Reaction time (s) |
|---|---|
| 1 | 17 |
| 2 | 18 |
| 3 | 19 |
| 4 | 21 |
| 5 | 24 |
| 6 | 28 |
| 7 | 33 |
| 8 | 38 |
| 9 | 43 |
| 10 | 48 |
| 11 | 53 |
| 12 | 58 |

**C** What happens to the reaction time as more magnesium is added?

### Questions

1 Suggest why identical pieces of magnesium ribbon were used in the investigation.

2 Apart from noting when the bubbling stops, suggest another way to tell when the reaction stops in the investigation.

E

3 In the investigation, what was the reaction time for:

(a) the fastest reaction?

(b) the slowest reaction?

4 What is the relationship between the number of pieces of magnesium ribbon added and the reaction time?

C

5 Estimate the reaction time for adding 8.25 cm of magnesium ribbon.

A*

### Exam tip OCR

✔ Make sure you can read values from graphs.

## Learning objectives

After studying this topic, you should be able to:

- describe how to measure the volume of gas produced in a reaction
- interpret data involving the rates of reactions
- describe how the amount of product depends upon the amount of reactant
- calculate the rate of reaction

## Key words

**rate of reaction, product, gas syringe, limiting reactant, directly proportional**

## Rate of reaction

The **rate of reaction** measures the amount of **product** formed in a fixed period of time.

- In a slow reaction, a small amount of product will form in a long time.
- In a fast reaction, a large amount of product will form in a short time.

To measure the rate of a particular reaction, you need to decide which product you are going to measure, and how you are going to do this. For example, if a gas is produced in the reaction, it is often easier to measure its volume rather than its mass.

Explosions have a high rate of reaction

## Measuring the volume of a gas

A **gas syringe** measures the volume of a gas produced in a reaction, such as the one between calcium carbonate and hydrochloric acid:

calcium carbonate + hydrochloric acid → calcium chloride + water + carbon dioxide

$CaCO_3 + 2HCl \rightarrow CaCl_2 + H_2O + CO_2$

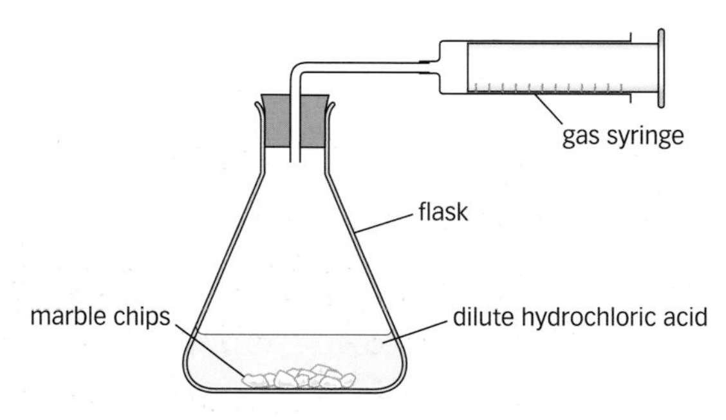


This apparatus can be used to measure the rate of a reaction producing a gas

## Exam tip OCR

- Make sure you can label the gas syringe and flask in a diagram of laboratory apparatus.
- You need to be able to choose the most appropriate format for presenting data. The data for the experiments described on this spread are shown on a line graph rather than a bar chart or a pie chart. This is because time and rate of reaction are both continuous variables.

The reaction starts once the calcium carbonate and acid are mixed together. Carbon dioxide pushes the plunger out, and its volume is read at regular intervals from the graduations on the gas syringe. The graph shows the results from an investigation involving this reaction. The line on the graph is steepest at the beginning, showing that the rate of reaction is greatest then. The line eventually becomes horizontal, showing that the reaction has stopped.

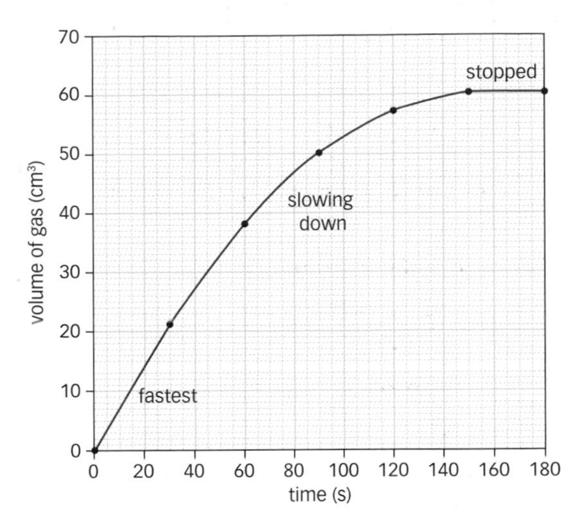


▲ The volume of gas produced during the reaction between calcium carbonate and hydrochloric acid

The reactant that is all used up first is the **limiting reactant**. For example, in this reaction between calcium carbonate and hydrochloric acid, calcium carbonate will be the limiting reactant if the acid is in excess. The reaction will stop when the calcium carbonate is all used up, and some acid will be left over afterwards.

The amount of product formed in a reaction is **directly proportional** to the amount of limiting reactant. For example, if the mass of calcium carbonate is doubled, the amount of carbon dioxide produced doubles, too. A graph of carbon dioxide produced against mass of calcium carbonate would give a straight line with a positive gradient that passes through the origin.

If the volume of a gas is being measured, the units for rate of reaction will be $cm^3/s$ or $cm^3/min$. If the mass of a product is being measured, the units will be g/s or g/min.

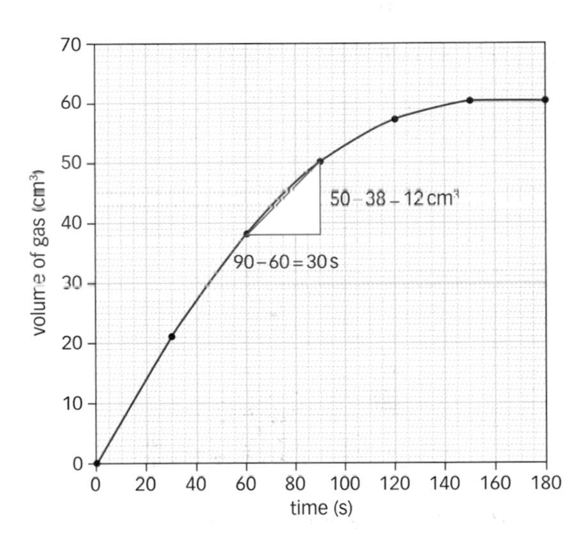


▲ The rate of reaction between 60 s and 90 s is 0.4 $cm^3/s$ (12 ÷ 30)

## Calculating rates of reaction

The rate of reaction is the gradient or slope of the line on a graph of amount of product against time.

## Questions

1. What is meant by rate of reaction?
2. Name the apparatus used to measure gas volume.

E

3. Explain how you can tell from the graph that the reaction has stopped by 150 s.
4. Explain how you know that the rate of reaction decreases as time goes by.

C

5. Use the graph to calculate the rate of reaction between 0 s and 30 s.

A*

## Did you know...?

Scientists use lasers to study how chemical bonds are broken and made in chemical reactions. Their lasers are capable of incredibly fast flashes, lasting just one millionth of a billionth of a second.

# 3: Reaction rate and temperature

## Learning objectives

After studying this topic, you should be able to:

- recall that a chemical reaction happens when particles collide
- describe and explain the effect of changing the temperature on the rate of reaction

Raised temperatures are often used to get a reasonable rate of reaction

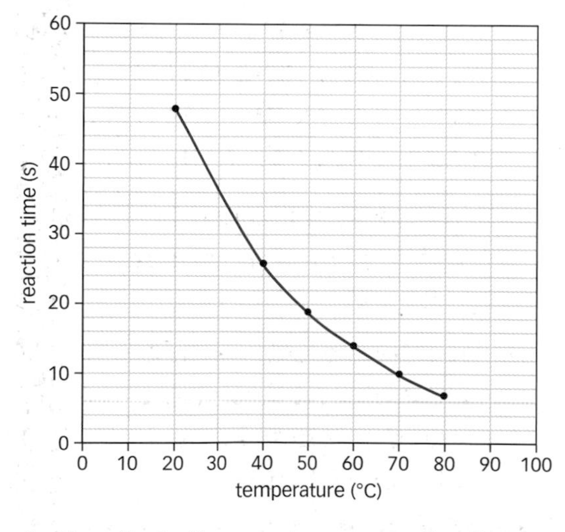


The effect of temperature on reaction time

## Changing the temperature

The rate of reaction for chemical reactions depends on the temperature of the reaction mixture. As the temperature increases, the rate of reaction increases, too. Industrial processes often use high temperatures to achieve a high rate of reaction.

## Investigating the effect of temperature

The reaction between sodium thiosulfate solution and hydrochloric acid is often used to investigate the rate of reaction. The equations look complicated, but it is a simple task to measure the reaction time:

sodium thiosulfate + hydrochloric acid → sodium chloride + water + sulfur dioxide + sulfur

$$Na_2S_2O_3 + 2HCl \rightarrow 2NaCl + H_2O + SO_2 + S$$

You might expect the sulfur dioxide gas to cause bubbling, but it dissolves too quickly for this to happen. Instead, the production of sulfur is used to measure the reaction time. Sulfur is insoluble in water and causes the reaction mixture to turn cloudy. The reaction time is measured by seeing how long it takes for the mixture to become too cloudy to see a cross drawn on paper placed beneath the reaction vessel.

Before and after the reaction between sodium thiosulfate solution and hydrochloric acid

The graph on the left shows the results of an investigation into this reaction. The same concentrations and volumes of sodium thiosulfate solution and hydrochloric acid were used each time. Only the temperature of the reaction mixture was changed.

> **A** What happens to the reaction time as the temperature is increased?

## When particles collide

For a reaction to happen, reactant particles must collide with each other. The more frequent the collisions are, the faster the reaction. The particles have more energy at higher temperatures. They move faster, causing an increased rate of reaction.

> **B** What must reactant particles do for a reaction to happen?

### More about particles and rates

The rate of reaction depends upon

- the amount of energy transferred during the collisions; and
- the frequency of collisions.

At low temperatures the amount of energy transferred in a collision may not be enough to cause a reaction. Increasing the temperature increases the chance of a **successful collision** (one that causes a reaction) because the particles have more energy. It also increases the collision frequency. Not only do the reactant particles collide more often, the collisions themselves have more energy and so are more likely to be successful.

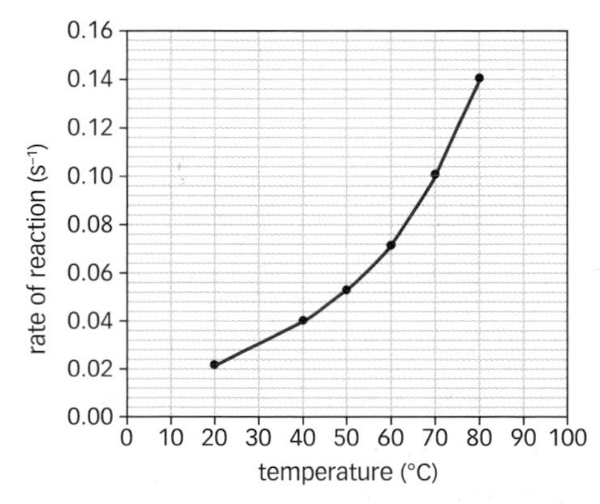


▲ Increasing the temperature has a large effect on the rate of reaction between sodium thiosulfate solution and hydrochloric acid

**Key words**

**successful collision**

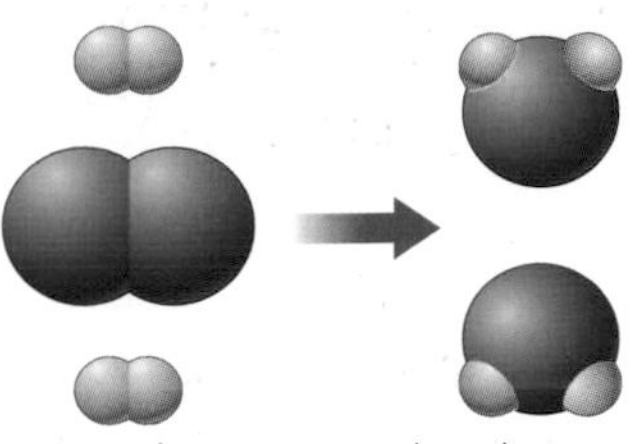

▲ Hydrogen and oxygen molecules must collide so that they can react to produce water

**Exam tip** OCR

✔ You need to be able to explain a scientific process using ideas or models. Explaining why substances react together using the idea of colliding particles is an example of doing this.

### Questions

1. What happens to the rate of reaction if the temperature is decreased?
2. Suggest why the rate of production of water was not measured in the investigation.

3. Explain, in terms of particles, what happens to the rate of reaction when the temperature is increased.

4. What is an unsuccessful collision between reactant particles?
5. Use the graph to explain, in terms of particles, why the rate of reaction doubles between 60 °C and 80 °C.

# 4: Pressure and concentration

## Learning objectives

After studying this topic, you should be able to:

- describe and explain the effect of changing the pressure on the rate of reactions involving gases
- describe and explain the effect of changing the concentration on the rate of reactions

## Key words

**concentration**, **solution**, **solute**, **solvent**

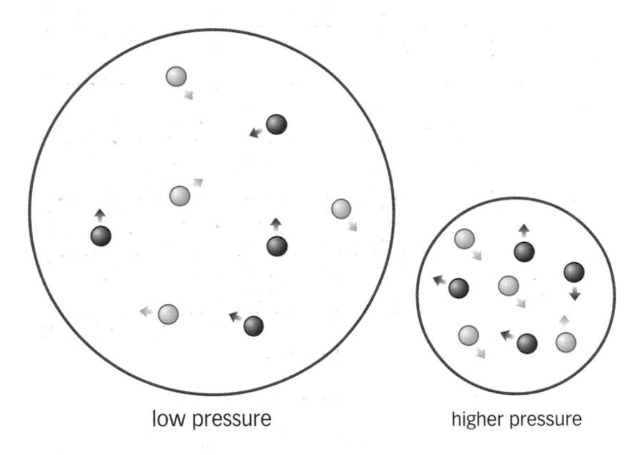


There is a greater chance of gas particles colliding at higher pressures

## Exam tip OCR

- Increasing the pressure only increases the rate of a reaction if at least one of the reactants is a gas.
- You need to be able to explain a scientific process using ideas or models. Using the idea of colliding particles to explain the factors that change the rate of a reaction is an example of doing this.

## Under pressure

Unlike normal boiling where the steam escapes, the steam in a pressure cooker cannot escape. As the pressure inside increases, the boiling point of the water increases, too. The food cooks faster at the higher temperature.

Food cooks more quickly in a pressure cooker

## Particles and pressure

The rate of reaction for chemical reactions involving gases depends on the pressure of the reaction mixture. As the pressure increases, the rate of reaction increases. Industrial processes often use high pressures to achieve a high rate of reaction. When gases are put under pressure, their particles become more crowded. They do not gain energy, but they do have more frequent collisions.

## Particles and concentration

The **concentration** of a **solution** is a measure of how much **solute** is dissolved in the **solvent**. The more concentrated a solution is, the more solute is dissolved in the solvent. If a reaction involves one or more reactants in solution, the rate of reaction increases as the concentration increases. When solutions are more concentrated, their particles become more crowded. They do not gain energy, but they do have more collisions.

The graphs show the results of an investigation into the effect of concentration on reaction rate. The reaction involved was between magnesium and hydrochloric acid:

magnesium + hydrochloric acid → magnesium chloride + hydrogen

$$Mg + 2HCl \rightarrow MgCl_2 + H_2$$

The length of magnesium ribbon used was kept the same, as were the temperature and volume of acid. Only the concentration of the acid was changed, by diluting it with water.

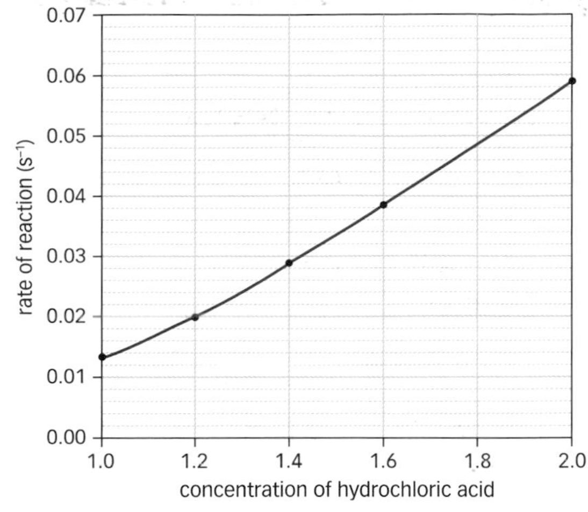


The effect of changing the concentration of hydrochloric acid on the rate of reaction with magnesium

## More about particles and rates

The rate of a reaction involving gases increases as the pressure increases. This is because the collision frequency between reactant particles increases. The rate of a reaction involving solutions increases as the concentration increases, for the same reason.

In chemical reactions, the amount of product formed is directly proportional to the amount of limiting reactant used. The greater the amount of limiting reactant there is, the more reacting particles there are, so the more product is formed.

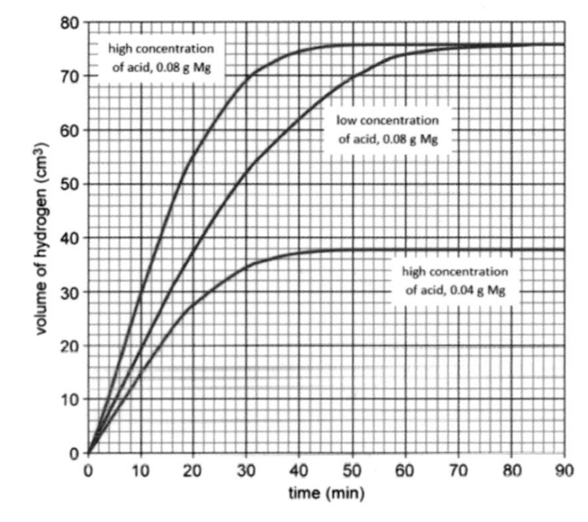


The rate of reaction is greater at a high concentration of acid rather than a low one. If the same mass of magnesium is used, the volume of hydrogen produced is the same, provided the acid is not the limiting reagent. If the mass of magnesium is halved, the total volume of hydrogen is also halved. The graphs would have similar shapes if the temperature was increased instead of the acid concentration.

### Questions

1. In a reaction involving gases, what happens when the pressure is decreased?
2. In a reaction involving solutions, what happens when the concentration is reduced?

E

3. In the investigation, what was the relationship between reaction time and concentration of acid?
4. Explain, in terms of particles, why the rate of reaction increases if the pressure or concentration of certain reactants is increased.

C

5. Use the graph at the top of the page to estimate the reaction time, and rate of reaction, for an acid concentration of 1.8 units.

A*

**A** Give two ways in which reactant particles can be made more crowded.

**B** What happens to the reaction time as the concentration of acid is increased?

## Learning objectives

After studying this topic, you should be able to:

- recall examples of explosions
- explain why powders react faster than lumps
- explain the dangers of powders in factories

## Key words

**explosion, explosive, surface area, combustible**

**A** What is an explosion?

## Exam tip OCR

- A simple laboratory test for hydrogen involves a lighted splint. The flame ignites hydrogen in a test tube with a 'pop' sound.

## Did you know...?

Dynamite was invented by a Swedish chemist called Alfred Nobel. The Nobel family business was the manufacture of nitroglycerine. After his younger brother was killed in a nitroglycerine explosion in 1864, Alfred set to work to invent a safer explosive. He patented dynamite in 1867.

## Bang!

An **explosion** is a very fast reaction in which a large volume of hot gases is released in a short time. The expanding gases create a shock wave that travels at high speed, damaging objects in its path. Hydrogen, for example, explodes when it is ignited:

$$\text{hydrogen} + \text{oxygen} \rightarrow \text{water}$$
$$2H_2 + O_2 \rightarrow 2H_2O$$

Hydrogen and oxygen mix completely because they are gases, and the chance of collisions between their molecules is very high. The reaction produces a lot of heat and the water vapour expands rapidly. Solids can cause explosions, too, even custard powder.

## Explosives

TNT (trinitrotoluene) is a high **explosive**. The shock waves that exploding TNT produces travel faster than the speed of sound, giving it the potential to cause a lot of damage. TNT is a yellow solid at room temperature. When it explodes, it produces solid carbon and a large volume of gases:

$$\text{trinitrotoluene} \rightarrow \text{carbon} + \text{nitrogen} + \text{water} + \text{carbon monoxide}$$
$$2C_7H_5N_3O_6 \rightarrow 7C + 3N_2 + 5H_2O + 7CO$$

Dynamite is another high explosive. It consists of an explosive liquid, nitroglycerine, absorbed in a mineral powder. This makes it safer to handle than the liquid alone.

▲ Explosives are used to break the rock face in opencast mines

## Increasing the surface area

The rate of reaction can be increased if a powdered reactant is used, rather than a lump of reactant. A powder has a larger **surface area** than a lump of the same mass. This is because particles that were inside the lump become exposed on the surface when it is crushed.

Remember that reactant particles must collide for a reaction to happen. The larger the surface area, the greater the frequency of collisions and the faster the reaction. The graph shows what happens when the same mass of calcium carbonate reacts with excess hydrochloric acid as a powder and as a lump.

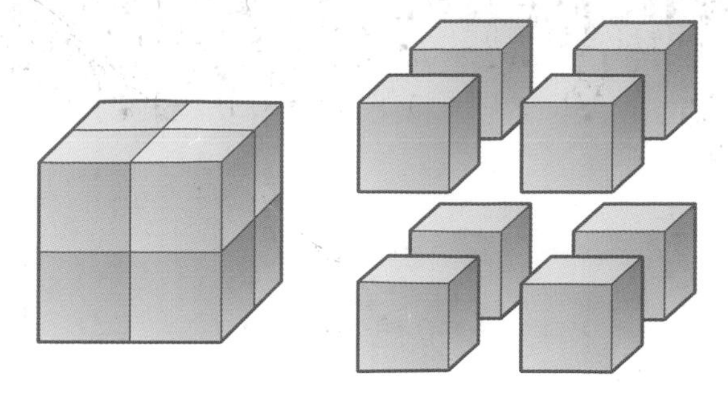

Eight 1 cm cubes have twice the surface area of one 2 cm cube, but the same volume

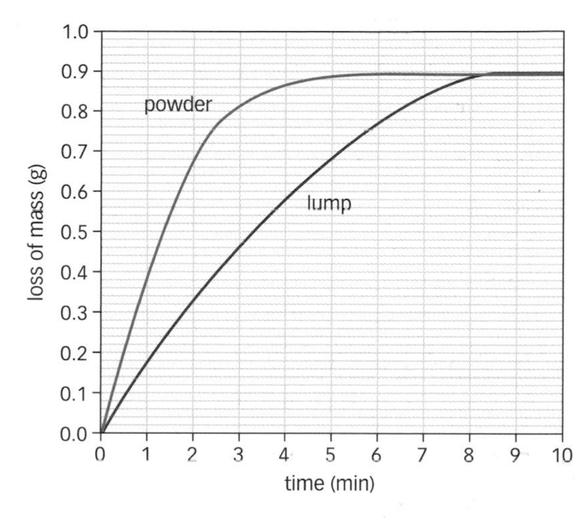


The loss in mass during the reaction between calcium carbonate and hydrochloric acid

**B** Explain how you know from the graph that the powder reacts faster than the lump.

## Powders in factories

**Combustible** substances burn easily. Combustible powders such as flour, custard powder and sulfur burn explosively in air. This makes them very dangerous in factories where they are manufactured or used. Care must be taken to stop dust escaping into the air inside buildings, and to prevent sparks or naked flames.

Powdered milk burns explosively

### Questions

1. Give two examples of explosions.
2. In general, which react faster, lumps or powders?

↓ E

3. Explain the effect of increasing the surface area on the rate of reaction.
4. Use the graph to answer these questions:
   (a) When did the two reactions finish?
   (b) Which reaction was the faster in the first two minutes?

↓ C

5. Calculate the mean rate of reaction for each reaction shown in the graph.
6. Explain why powders such as flour are dangerous in factories.

↓ A*

# 6: Catalysts

## Learning objectives

After studying this topic, you should be able to:

- describe the effect of catalysts on reactions

## Key words

**catalyst**, **catalyse**

## Did you know...?

Catalytic converters are fitted to the exhaust systems of cars and other vehicles. They contain expensive metal catalysts to convert harmful waste gases into less harmful ones. Around half the world's production of platinum and palladium, and almost all its production of rhodium, go into catalytic converters.

## Speeding up reactions

Depending on the reactants involved, the rate of a reaction can be increased by increasing the

- temperature
- concentration, if solutions are involved
- pressure, if gases are involved
- surface area of solids, by powdering them.

The rate of a reaction can also be increased by adding a **catalyst**.

## Catalysts

A catalyst is a substance which changes the rate of reaction but is unchanged at the end of the reaction. For example, if 1 g of a catalyst is added to a reaction mixture, there is still 1 g of it left after the reaction has finished. A small amount of a catalyst will **catalyse** the reaction between large amounts of reactants.

Catalysts are very useful in the chemical industry. Without them, many industrial processes would be too slow to be economical. For example, iron is used in the Haber process, which makes ammonia for fertilisers and explosives. The iron catalyst greatly increases the rate of reaction between nitrogen and hydrogen.

A catalytic converter

This fine rhodium–platinum gauze is the catalyst used in the manufacture of nitric acid from ammonia

## Specific catalysts

Catalysts are specific to particular reactions. A substance that acts as a catalyst for one reaction may not be able to act as a catalyst for a different reaction.

## Catalysts in the lab

Hydrogen peroxide is a liquid used to bleach hair and, in industry, to bleach wood pulp for making paper. It is usually diluted in water, where it breaks down very slowly to form water and oxygen:

hydrogen peroxide → water + oxygen

$$2H_2O_2 \rightarrow 2H_2O + O_2$$

Powdered manganese(IV) oxide, $MnO_2$, can catalyse the breakdown of hydrogen peroxide.

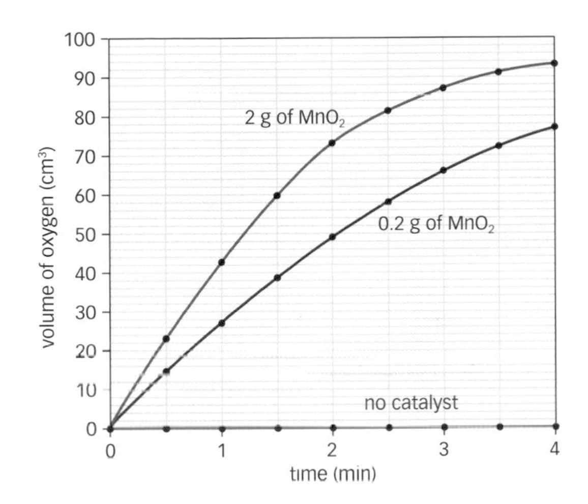


Adding a lot more catalyst makes little difference to the rate at which hydrogen peroxide breaks down

Large amounts of oxygen are produced rapidly when manganese(IV) oxide is added to hydrogen peroxide solution

**Exam tip** OCR

✔ Remember that a catalyst does not get used up in the reaction it catalyses.

### Questions

1 Give an example of a catalyst and the reaction it catalyses. (E)

2 What is a catalyst?

3 Use the graph to answer these questions:

(a) What was the effect on the rate of reaction of adding manganese(IV) oxide?

(b) Explain which reaction was the fastest in the first two minutes.

(c) How can you tell that the reactions were not complete after four minutes? (C)

4 Calculate the mean rate of reaction over four minutes for each reaction shown in the graph.

5 Suggest why adding ten times as much manganese(IV) oxide did not increase the rate of reaction by ten times. (A*)

# 7: Relative formula mass

## Learning objectives

After studying this topic, you should be able to:

- ✔ calculate the relative formula mass of a given substance

## Key words

**relative atomic mass, formula, relative formula mass**

## Did you know...?

A gold atom has a mass of $3 \times 10^{-21}$ kg. That's just 0.000 000 000 000 000 000 003 kg.

▲ Atoms of gold piled on top of a layer of carbon atoms

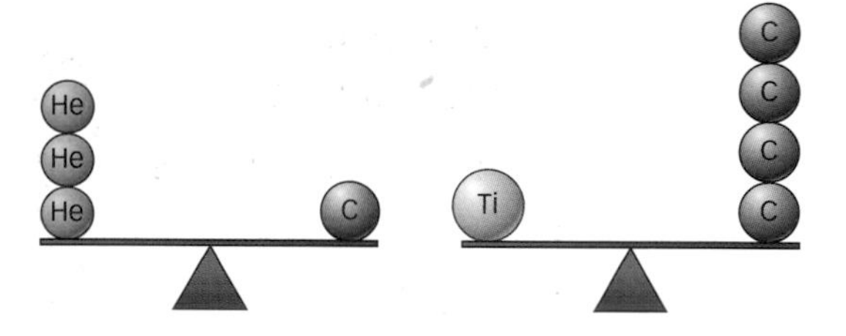


▲ Three helium atoms have the same mass as one carbon atom, and one titanium atom has the same mass as four carbon atoms

## Relative atomic mass

Individual atoms are incredibly small and they have very little mass. The gold atoms in the photograph are less than a millionth of a millimetre in diameter. Each one has such a small mass that it is more useful to use their **relative atomic mass**, rather than their actual mass in kg. Carbon atoms are the standard atom against which all the others are compared. The relative atomic mass, $A_r$, of the most common type of carbon atoms is exactly 12. Atoms with an $A_r$ of less than 12 have less mass than a carbon atom, and those with an $A_r$ greater than 12 have more mass than a carbon atom.

**A** How many helium atoms does it take to equal the mass of one titanium atom?

The periodic table shows the relative atomic masses of the elements. The top number in each box is the relative atomic mass for that element's atoms.

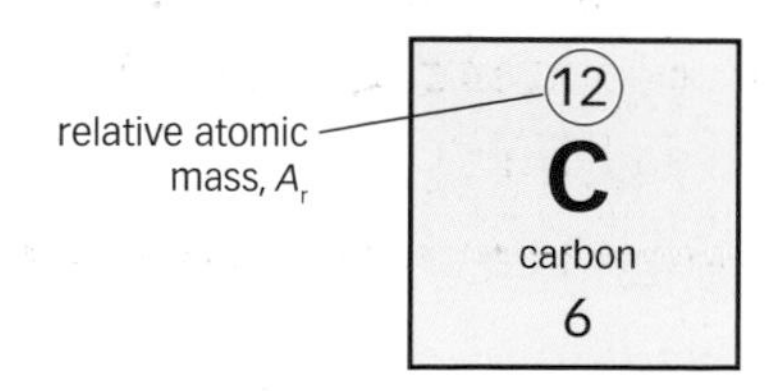


▲ Information about carbon from the periodic table

## Relative formula mass

The chemical **formula** of a substance tells you the number of each type of atom in a unit of that substance. For example, the formula for water is $H_2O$. It shows that each water molecule contains two hydrogen atoms and one oxygen atom, joined together. The **relative formula mass** or $M_r$ of a substance is the mass of a unit of that substance compared to the mass of a carbon atom. It is worked out by adding together all the $A_r$ values for the atoms in the formula.

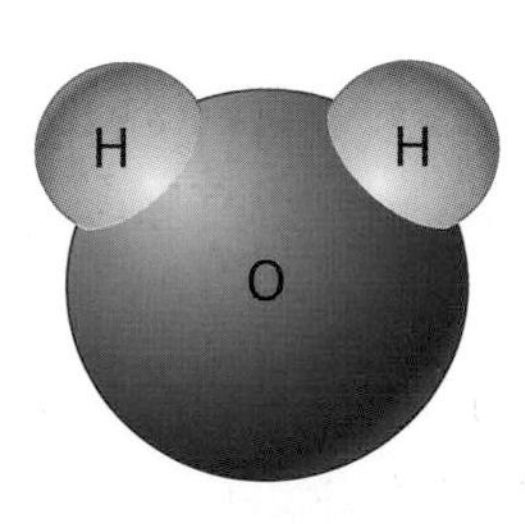


▲ The $M_r$ of water is 18. Each water molecule has 1.5 times the mass of a carbon atom.

### Worked example 1

What is the $M_r$ of water, $H_2O$?

$A_r$ values: H = 1, O = 16

$M_r$ of $H_2O$ = 1 + 1 + 16 = 18

### Worked example 2

What is the $M_r$ of magnesium hydroxide, $Mg(OH)_2$?

$A_r$ values: Mg = 24, O = 16, H = 1

$M_r$ of $Mg(OH)_2$ = 24 + [2 x (16 + 1)]

= 24 + 34 = 58

(Notice that the 2 outside the brackets in $Mg(OH)_2$ means that there are two oxygen atoms and two hydrogen atoms.)

### Worked example 3

What is the mass of one mole of magnesium hydroxide?

$M_r$ of $Mg(OH)_2$ = 58, so the mass of one mole = 58 g

**B** What is the relative formula mass of magnesium oxide, MgO?

### Exam tip OCR

✔ You will given a copy of the periodic table in the exam.

### Questions

Use the relative atomic masses from the periodic table to help you answer the questions.

1. What are the relative atomic masses of nitrogen, chlorine, and sodium?
2. What is the relative formula mass of:
   (a) oxygen, $O_2$?
   (b) carbon dioxide, $CO_2$?
   (c) ammonia, $NH_3$?
   (d) sodium chloride, NaCl?

E

3. What is the relative formula mass of aluminium hydroxide, $Al(OH)_3$?

C

4. What is the relative formula mass of ammonium sulfate, $(NH_4)_2SO_4$?

A*

## Learning objectives

After studying this topic, you should be able to:

- recall and describe how mass is conserved in chemical reactions
- calculate reacting masses
- explain why mass is conserved in chemical reactions

## Key words

**conservation of mass**

Green copper carbonate decomposes to form black copper oxide and carbon dioxide

## A precipitation reaction

The photograph shows the result of an investigation using silver nitrate solution and sodium chloride solution. They react together to form white solid silver chloride suspended in sodium nitrate solution. Notice that the total mass is the same before *and* after the reaction. This is an example of the principle of **conservation of mass**.

The total mass stays the same when silver nitrate solution reacts with sodium chloride solution

## Conservation of mass

Here is the balanced equation for the reaction between silver nitrate solution and sodium chloride solution, with the $M_r$ values underneath:

$$\begin{array}{ccccccc} AgNO_3 & + & NaCl & \rightarrow & AgCl & + & NaNO_3 \\ 170 & & 58.5 & & 143.5 & & 85 \end{array}$$

The total $M_r$ of the reactants is 228.5 (170 + 58.5), and so is the total $M_r$ of the products (143.5 + 85). Mass is conserved in the reaction, even though a solid is formed from two solutions. All chemical reactions show conservation of mass.

Copper carbonate breaks down when it is heated, forming copper oxide and carbon dioxide:

| copper carbonate | → | copper oxide | + | carbon dioxide |
|---|---|---|---|---|
| $CuCO_3$ | → | $CuO$ | + | $CO_2$ |
| 123.5 | | 79.5 | | |

It is possible to work out the $M_r$ of carbon dioxide from this information using the principle of conservation of mass. Its $M_r$ must be (123.5 – 79.5) = 44. A quick check using the $A_r$ values for carbon (12) and oxygen (16) shows that it really is 44.

## Why is mass conserved?

Mass is conserved because no atoms are created or destroyed in chemical reactions. This also explains why symbol equations are balanced when there are the same number, and type, of atom on each side of the equation.

### Reacting masses

The principle of conservation of mass can be used to calculate the masses of reactant and product in a reaction. In the copper carbonate example, 123.5 g of copper carbonate would decompose to form 79.5 g of copper oxide and 44 g of carbon dioxide. Looking again at the symbol equation, you will see that all the substances are in a 1:1 ratio. What if they are not?

Magnesium reacts with oxygen to make magnesium oxide:

| $2Mg$ | + | $O_2$ | → | $2MgO$ |
|---|---|---|---|---|
| 24 | | 32 | | 40 |

The $M_r$ values do not add up correctly at first sight. This is because the ratio of the substances must be taken into account in any calculations:

| $2Mg$ | + | $O_2$ | → | $2MgO$ |
|---|---|---|---|---|
| 2 × 24 | | 32 | | 2 × 40 |

For example, what mass of oxygen, $O_2$, will react completely with 12 g of magnesium, Mg?

$$\text{mass of oxygen} = \frac{12}{2 \times 24} \times 32 = \frac{12}{48} \times 32 = 8\text{ g}$$

▲ Magnesium reacts with oxygen to form magnesium oxide

### Questions

1. If 8 g of oxygen reacts with 12 g of magnesium, what mass of magnesium will 4 g of oxygen react with?

E

2. What does conservation of mass mean?
3. Carbon reacts with oxygen to produce carbon dioxide: $C + O_2 \rightarrow CO_2$. $A_r$ of C = 12, $M_r$ of $O_2$ = 32, $M_r$ of $CO_2$ = 44.
   (a) Use this information to show that mass is conserved in the reaction.

C

   (b) Calculate the mass of carbon dioxide formed from 3 g of carbon.
4. Calculate the mass of magnesium needed to form 8 kg (8000 g) of magnesium oxide.

A*

# 9: Percentage yield

## Learning objectives

After studying this topic, you should be able to:

- explain what is meant by percentage yield
- calculate percentage yield using given data
- recognise reasons why the percentage yield may be less than 100%
- explain why a high percentage yield is desirable in industrial processes

**A** If the percentage yield is 25%, what percentage of product is lost?

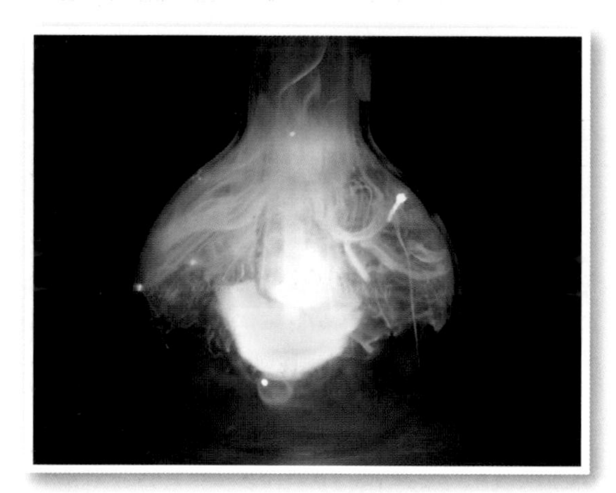

Sodium reacts violently with chlorine to make sodium chloride

## Exam tip OCR

- If you are asked to calculate the percentage yield in the exam, you will be given the actual yield and predicted yield.

## Yields

The **yield** of a substance is how much of it there is. Usually there is a difference between the **predicted yield** and the **actual yield**.

- The predicted yield is the expected mass calculated using reacting masses.
- The actual yield is the mass of product made.

The **percentage yield** is a way to compare these two amounts. Here is the formula needed to calculate it:

$$\text{percentage yield} = \frac{\text{actual yield}}{\text{predicted yield}} \times 100$$

For example, in the reaction between sodium and chlorine, the actual yield of sodium chloride was 8 g but the predicted yield was 10 g. This means that the percentage yield was 80%:

$$\text{percentage yield} = \frac{8}{10} \times 100 = 80\%$$

The percentage yield varies according to the reaction being carried out, and the way in which the product is obtained. For example:

- A yield of 0% means that no product has been obtained at all.
- A yield of 100% means that no product has been lost.

## Calculating predicted yields

The predicted yield for a product is calculated using reacting masses and the principle of conservation of mass. Think about the reaction between sodium and chlorine. The expected mass of sodium chloride depends upon the masses of the reactants:

$$\text{sodium} + \text{chlorine} \rightarrow \text{sodium chloride}$$
$$2Na + Cl_2 \rightarrow 2NaCl$$

What mass of sodium chloride could be made from 2.3 g of sodium reacting with excess chlorine? $A_r$ of Na = 23 and $M_r$ of NaCl = 58.5.

$$\text{percentage yield} = \frac{2.3}{2 \times 23} \times (2 \times 58.5) = 0.05 \times 117 = 5.85 \text{ g}$$

## Losing product

There are several reasons why the percentage yield of a product is less than 100%. Very often, product is lost during handling and purification. It may be lost in

- filtration
- evaporation
- the transfer of liquids from one container to another
- heating.

Ammonium sulfate, for example, is a useful fertiliser. It is a soluble salt, so plants can easily absorb it through their roots. It provides nitrogen, needed by plants to make proteins.

Ammonium sulfate is made from ammonia solution and sulfuric acid. In a typical laboratory preparation of the salt, the correct volumes of ammonia solution and sulfuric acid are determined using titration. They are then mixed and the solution of ammonium sulfate is filtered. The water is evaporated from the solution by gentle heating, leaving dry crystals of ammonium sulfate. The percentage yield of ammonium sulfate is likely to be less than 100%. For example, some ammonium sulfate solution will stay behind, soaked into the filter paper.

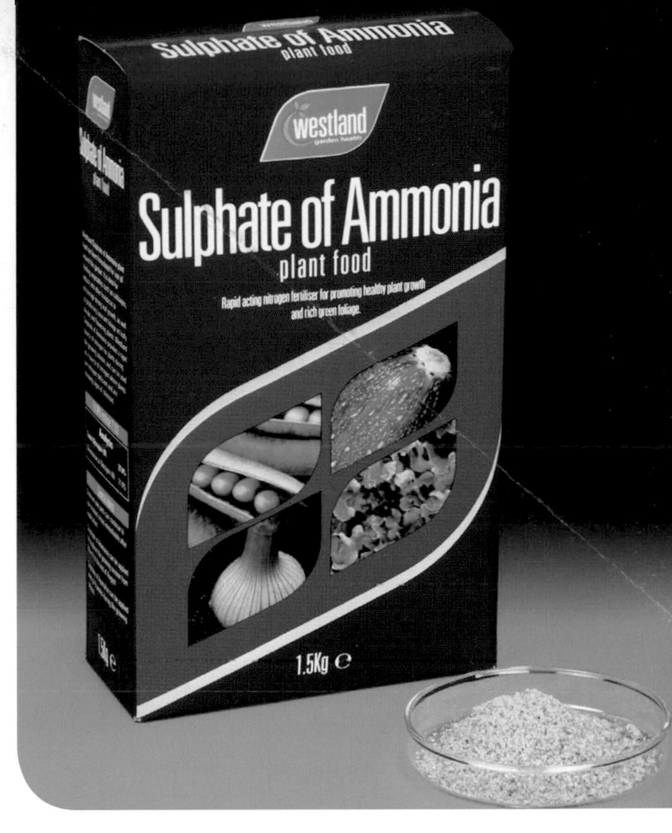


▲ Ammonium sulfate is useful as a fertiliser

**B** Why do plants easily absorb ammonium sulfate?

### Questions

1. What is meant by percentage yield?
2. What is the percentage yield if none of the product is lost, and if all of the product is lost?

↓ E

3. Describe three ways in which ammonium sulfate could be lost in the experiment described above.
4. The predicted yield of ammonium sulfate is 4 g, but the actual yield is 2.8 g. Calculate the percentage yield.

↓ C

5. A certain industrial chemical may be made in two ways. The first method gives a percentage yield of 75%, and the second method a percentage yield of 80%. Explain why the second method should be preferred.

↓ A*

### Industrial processes

It is important in industry to avoid wasting reactants. As much product as possible must be obtained to reduce costs. So industrial processes are designed to achieve high percentage yields.

### Key words

**yield, predicted yield, actual yield, percentage yield**

# 10: Atom economy

## Learning objectives

After studying this topic, you should be able to:

- ✔ explain what is meant by atom economy
- ✔ recognise that processes with high atom economy are 'greener'
- ✔ calculate atom economy using given data
- ✔ explain why a high atom economy is desirable

## Key words

**atom economy**

**A** Give one way in which industrial chemical processes may be wasteful.

**B** How is the atom economy of a reaction calculated?

## Reducing waste

Industrial chemical processes are wasteful if they use more energy or raw materials than they should. Some may produce hazardous chemical waste. One way to reduce waste is to use processes with a high **atom economy**.

▲ Hazardous chemical waste is difficult to dispose of

## Atom economy

Atom economy is a way to measure the amount of atoms that are wasted when a certain chemical is made. Here is the formula needed to calculate it:

$$\text{atom economy} = \frac{M_r \text{ of desired product}}{\text{total } M_r \text{ of all products}} \times 100$$

The atom economy varies according to the reaction being carried out. For example:

- A low atom economy means that few atoms in the reactants have been converted into the desired product.
- An atom economy of 100% means that all the atoms in the reactants have been converted into the desired product.

In general, the higher the atom economy of a process, the 'greener' it is – fewer resources are wasted to make the desired product.

## Exam tip OCR

- ✔ If you are asked to calculate an atom economy in the exam, you will be given the balanced symbol equation and relative formula masses.

## Industrial processes and atom economy

The higher the percentage yield and atom economy of a process, the more sustainable it is. The desired product will be made using fewer natural resources and with less waste. A process has an atom economy of 100% if there is only one product. For example, the Haber process has an atom economy of 100%:

$$N_2 + 3H_2 \rightleftharpoons 2NH_3$$

Any process with a waste product will have an atom economy of less than 100%.

## Calculating atom economy

Hydrogen is used in the manufacture of ammonia by the Haber process. It can be made by reacting steam with coke, which is mostly carbon:

| carbon | + | steam | → | carbon monoxide | + | hydrogen |
|---|---|---|---|---|---|---|
| C | + | $H_2O$ | → | CO | + | $H_2$ |
| 12 | | 18 | | 28 | | 2 |

The total $M_r$ of the products is 30 (28 + 2), and the $M_r$ of the desired product is 2.

$$\text{atom economy} = \frac{2}{30} \times 100 = 6.7\%$$

This is a process with a low atom economy. Most of the atoms in the reactants are converted into an unwanted waste product, carbon monoxide.

## More difficult calculations

Hydrogen can also be made by reacting steam with natural gas, which is mostly methane:

| methane | + | steam | → | carbon monoxide | + | hydrogen |
|---|---|---|---|---|---|---|
| $CH_4$ | + | $H_2O$ | → | CO | + | $3H_2$ |
| 16 | | 18 | | 28 | | 3 × 2 |

The total $M_r$ of the products is 34 (28 + 6). The $M_r$ of the desired product is 2, but its total $M_r$ is 6 because of the ratio in which it appears in the equation.

$$\text{atom economy} = \frac{3 \times 2}{34} \times 100 = 17.6\%$$

This process has a higher atom economy than the one which uses coke. More of the atoms in the reactants are converted into the desirable product.

▲ Older, less efficient chemical processes waste resources

### Questions

1 What is meant by atom economy?

2 What is the atom economy if all of the atoms in the reactants are converted into the desired product?

↓ E

3 Calculate the atom economy for making hydrogen using this process:

$CO + H_2O \rightarrow CO_2 + H_2$

$M_r$ of $CO_2 = 44$

$M_r$ of $H_2 = 2$

↓ C

4 Hydrogen can be made by reacting methane with steam:

$CH_4 + 2H_2O \rightarrow CO_2 + 4H_2$

(a) Use the equation, and the $M_r$ values in Question 3, to calculate the atom economy.

(b) Explain why a high atom economy is desirable.

↓ A*

## Learning objectives

After studying this topic, you should be able to:

- recognise and describe reactions as exothermic or endothermic
- explain energy changes in reactions in terms of making and breaking bonds

## Did you know...?

Silver fulminate is a contact explosive. Very little energy is needed to detonate it. Slight pressure, even from its own weight, is enough to set if off.

## Key words

**exothermic reaction, endothermic reaction**

## Exam tip OCR

- Think of a fire exit sign when trying to remember which way round exothermic and endothermic reactions go. A fire is hot and you go out of an exit. Exothermic reactions get hot as they give energy out.

## Exothermic reactions

Combustion is an **exothermic reaction** because energy is released during the reaction. Exothermic reactions transfer energy to the surroundings, usually as heat. This can make exothermic reactions easy to spot. If the temperature of the reaction mixture and its container goes up, the reaction is an exothermic one.

Fireworks use exothermic reactions

Exothermic reactions can also transfer other forms of energy to the surroundings. For example, explosions release light and sound as well as heat. The reactions in batteries release electrical energy when the battery is connected in a circuit.

The reactions in a glow worm give out light energy

**A** State four forms of energy that can be given out by chemical reactions.

## Endothermic reactions

**Endothermic reactions** take in energy during the reaction. They absorb energy from the surroundings, usually as heat. If the temperature of the reaction mixture and its container goes down, the reaction is an endothermic one. Sherbet sweets feel cold on your tongue because of an endothermic reaction between sodium hydrogencarbonate and a weak acid such as citric acid.

Sherbet sweets cause an endothermic reaction in your mouth

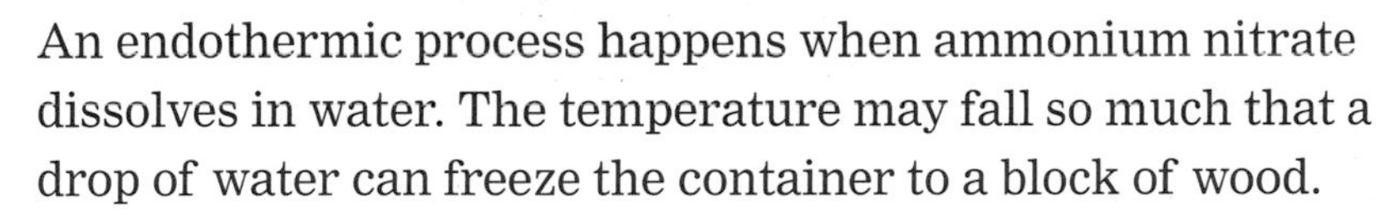

An endothermic process happens when ammonium nitrate dissolves in water. The temperature may fall so much that a drop of water can freeze the container to a block of wood.

Endothermic reactions can also take in other forms of energy from the surroundings. For example, electrolysis is an endothermic process. Electrical energy is absorbed to decompose a compound.

Chemical bonds in the reactants break during a chemical reaction, and new bonds are made as the products form.

- Bond breaking is an endothermic process (it takes in energy).
- Bond making is an exothermic process (it releases energy).

Whether a reaction is exothermic or endothermic is determined by the difference between the energy needed to break bonds, and the energy given out when bonds form.

For an exothermic reaction, the amount of energy taken in to break bonds is *less* than the amount of energy released to make new bonds.

For an endothermic reaction, the amount of energy taken in to break bonds is *more* than the amount of energy released to make new bonds.

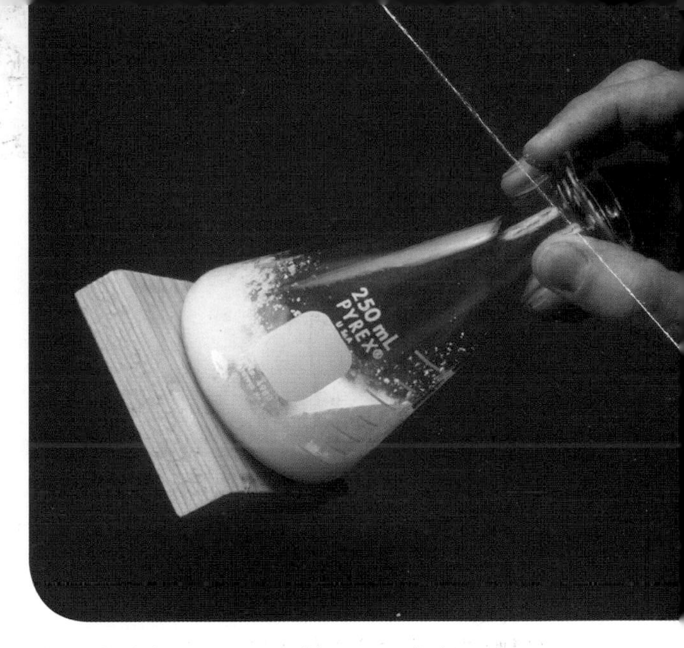


A flask frozen to wood by an endothermic reaction

### Questions

1 Give an example of an exothermic reaction and an example of an endothermic reaction.

2 When an acid is mixed with an alkali, the temperature of the mixture goes up. Explain whether this is an exothermic or an endothermic reaction.

E

3 In terms of energy transfer to or from the surroundings, describe the difference between an exothermic reaction and an endothermic reaction.

4 State whether bond making is an endothermic process or an exothermic one.

C

5 In terms of energy changes and bonds, explain the difference between an exothermic and an endothermic reaction.

## Learning objectives

After studying this topic, you should be able to:

- state the units used to measure energy and temperature
- describe how the energy released in a combustion reaction can be measured
- use data from calorimetry experiments to compare fuels
- calculate the energy transferred when a fuel burns
- calculate the energy output of a fuel in J/g

## Key words

**calorimetry, calorimeter**

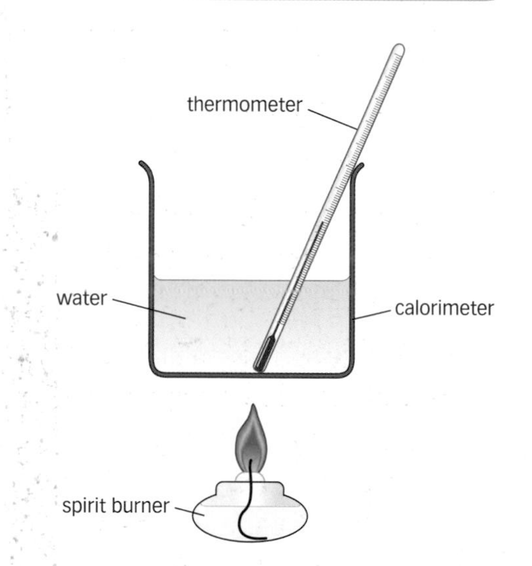

A calorimetry experiment

## Measuring energy changes

**Calorimetry** is a method used to measure the energy transferred in chemical reactions. It is often used to investigate fuels.

The fuel is put into a spirit burner. This is a container with a lid, and a wick so that the fuel can be lit. The whole spirit burner is weighed. A copper container called a **calorimeter** is filled with a known mass of water. It is then clamped above the spirit burner. The temperature of the water is measured. The wick is lit so the flame warms up the water. The lid is put back on at the end of the experiment. This puts the flame out. The spirit burner is weighed again, and the temperature of the water is also measured.

A spirit burner

The table shows some typical results in which the same mass of fuel was burnt.

| Fuel | Temperature of water at start (°C) | Temperature of water at end (°C) | Change in water temperature (°C) |
|---|---|---|---|
| ethanol | 18 | 38 | |
| propanol | 18 | 41 | |

Spirit burners work well if the fuel is a liquid. If it is a gas such as propane or butane, a bottled gas burner is used.

## Fair testing

Several variables can affect the energy transferred in a calorimetry experiment, not just the type of fuel used. For example, the volume of water used, the starting temperature and temperature change, and the height of the calorimeter over the flame can all affect it. They must be controlled to ensure fair testing.

It is assumed that all the energy from the burning fuel goes up to heat the water. However, a lot of energy will also be transferred to the surroundings. The experiments should be repeated to obtain reliable results. Any anomalous readings can then be identified and ignored when calculating the mean energy transferred for a particular fuel.

## Calculating the energy transferred

The amount of heat energy transferred to the water in a calorimetry experiment can be calculated:

$$\text{energy transferred (J)} = \text{mass of water heated (g)} \times 4.2 \times \text{temperature change (°C)}$$

The 4.2 in the equation is the specific heat capacity of water. It means that 1 g of water needs 4.2 J of energy to increase its temperature by 1 °C.

### Worked example 1

What is the energy transferred to heat 100 g of water by 20 °C?

energy transferred = 100 x 4.2 × 20
= 8400 J

The energy output of the fuel in J/g can be calculated:

$$\text{energy per gram} = \frac{\text{energy supplied (J)}}{\text{mass of fuel burnt (g)}}$$

### Worked example 2

In a calorimetry experiment, 7400 J of energy is transferred to water when 0.5 g of ethanol burns. What was the energy output of the ethanol?

$$\text{energy per gram} = \frac{7400}{0.5} = 14\,800 \text{ J/g}$$

**A** Work out the increase in water temperature for each fuel in the table on the previous page.

**B** The same mass of water was used in each experiment. Which fuel released the greater amount of energy?

**C** How do you know?

### Did you know...?

The unit of energy is the joule, J, named after James Joule. In 1843, he showed that heat was a form of energy. He was so keen to investigate heat, he even got his wife to help him whilst on their honeymoon.

### Questions

1 What is calorimetry?

2 Explain why a lid might be used on a calorimeter.

3 Describe three factors to control in a calorimetry experiment to make it a fair test.

4 250 g of water is warmed from 20 °C to 44 °C using butane. The gas burner weighed 645.4 g at the start and 644.2 g at the end.

(a) Calculate the energy transferred in J.

(b) Calculate the energy output in J/g.

(c) What assumption is made in the calculation?

# 13: Batch or continuous?

## Learning objectives

After studying this topic, you should be able to:

- distinguish between a batch process and a continuous process
- explain why batch processes or continuous processes may be used
- describe how chemicals are extracted from plant sources
- evaluate the advantages and disadvantages of each type of process

**A** What is a bulk chemical?

**B** What is a speciality chemical?

## Making chemicals

You will probably have made a chemical in a science practical. However you did this, you are likely to have made a small amount using a **batch process**. Some industrial chemicals are made using batch processes, while others are made using **continuous processes**.

Liquid chlorine, a bulk chemical, being transported by a freight train

## Continuous processes

**Bulk chemicals** are needed in large amounts. They are usually made in continuous processes in which the product is made all the time. For example, ammonia is made in the Haber process. The reactants needed are continually fed into a reaction vessel, where they react together. The ammonia produced is collected all the time. Sulfuric acid and chlorine are also made in continuous processes.

## Batch processes

A product made in a batch process is not made all the time. For example, wine is made in a batch process. Grapes are pressed to release the grape juice, which is then fermented to produce the wine. Other steps may be needed before it is ready for sale. In the chemical industry, **speciality chemicals** are high value chemicals needed in small amounts. They are made on demand when a customer needs them. **Pharmaceuticals** and other speciality chemicals are made in batch processes. The raw materials needed may be made synthetically using chemical reactions, or they may be extracted from plants (see flow diagram on next page).

Wine is made in vats in a batch process

## Batch vs continuous

Bulk chemicals are usually made by continuous processes. Some could be made by batch processes instead. For example, ethanol can be made by reacting ethene with steam in a continuous process. It can also be made in a batch process similar to wine-making. The table summarises some of the advantages and disadvantages of each type of process.

| | Continuous process | Batch process |
|---|---|---|
| Relative cost of factory equipment | high | low |
| Rate of production | high | low |
| Shut-down times | rare | frequent |
| Workforce needed | small | large |
| Ease of automating the process | high | low |

The choice of process may involve balancing the advantages with the disadvantages. Many speciality chemicals, like pharmaceuticals, must be made in several steps. They can only be made by a batch process.

### Questions

1 Name two chemicals made in continuous processes.

2 What sort of process is used to make pharmaceuticals?

E

3 Suggest why speciality chemicals are more expensive than bulk chemicals.

C

4 Refer to the information in the table to help you answer these questions:

(a) give three reasons why ethanol should be made in a continuous process

(b) give three reasons why ethanol should not be made in a batch process.

A*

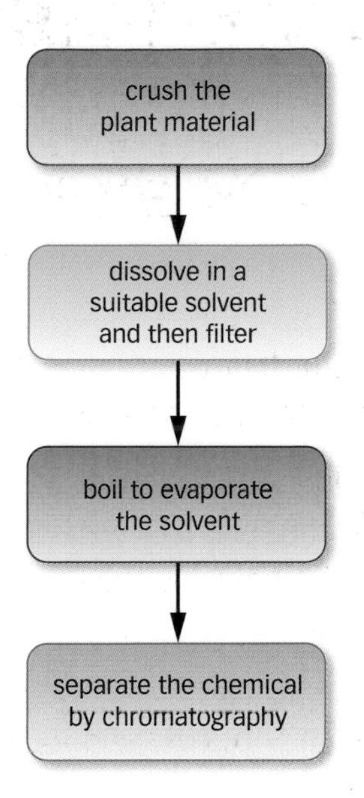


▲ Several steps are need to extract chemicals from plants

### Key words

**batch process, continuous process, bulk chemical, speciality chemical, pharmaceutical**

### Exam tip OCR

✔ Make sure you can distinguish between a batch process and a continuous process.

### Did you know...?

The world's chemical industry makes around 130 million tonnes of ammonia, and around 165 million tonnes of sulfuric acid, each year. Aspirin is the world's most widely used pharmaceutical drug, yet only around 40 thousand tonnes are made each year.

# 14: Making medicines

## Learning objectives

After studying this topic, you should be able to:

- describe the factors that contribute to the cost of new pharmaceutical drugs
- interpret melting point, boiling point, and chromatographic data to do with a substance's purity
- explain why it is difficult to develop and test safe new pharmaceutical drugs

▲ These 'bioreactors' contain bacteria that produce proteins for use in pharmaceuticals

**A** Give one example of a stage in drug development that is expensive and takes a long time.

### Did you know...?

On average, it takes around £550 million, and 10 to 12 years, to develop a new pharmaceutical drug. Even then, there is no guarantee that it will be a commercial success.

## Development and production

It is often expensive to develop and make new pharmaceutical drugs. Several factors are involved:

| Factor | Description |
|---|---|
| Research and testing | Suitable new substances must be identified, then tested to make sure they are safe and effective. |
| Labour costs | Many skilled people are needed. |
| Energy costs | Electricity and fuel are needed, and these are expensive. |
| Raw materials | The raw materials may be rare or expensive, and complex chemical reactions may be needed to make the drug from them. |
| Time taken for development | Research and testing take a long time, and a new drug must be licenced for use. |
| **Marketing** | Healthcare professionals have to be told about the drug and how to use it. |

Thousands of new substances may be made and tested in the development of a new drug. The first tests involve computer simulations and tests on cells grown in the laboratory. The most promising substances are then tested on laboratory animals. If it passes these first stages, a substance is checked for side effects in healthy human volunteers. It is then tested on a small group of patients to see if it works as expected, and then on a larger group of patients to gather more information about it. All these stages in development are expensive and time consuming.

## Payback time

A pharmaceutical company that develops a new drug will apply for a patent. This prevents other companies making and selling the drug for up to 20 years. Once the patent expires, these companies are free to make and sell the drug. The original company must make enough money from its sales to pay for the development costs. The payback time is the time taken for this to happen. If the patent expires earlier than this, the original company may lose a lot of money on the drug.

## Testing for purity

The presence of harmful impurities could make people ill, so it is important to make pharmaceutical drugs as pure as possible. The purity of a drug can be tested by measuring its melting or boiling point. Impurities alter the temperature at which a drug melts or boils. The further the temperature is away from the correct one, the less pure the drug is.

**Thin layer chromatography** (TLC) can also be used to see how pure a drug is. TLC is similar to the paper chromatography used to separate coloured substances like ink. Instead of paper, the different substances move through a thin layer of powder coated onto a glass or plastic plate. Colourless substances show up as spots on the plates when reacted with certain chemicals. These may be fluorescent under ultraviolet light, or they may become coloured when heated.

## The cost of production

The raw materials for pharmaceuticals may be made synthetically using chemical reactions, or they may be extracted from plants. The flow chart shows the main steps in doing this.

### Questions

1 List the factors that affect the cost of making and developing a drug.

2 Describe how the purity of a pharmaceutical drug can be tested using melting and boiling points.

E

3 Study the diagram of the TLC plate. Explain which sample, **A** or **B**, is the least pure.

4 Explain why a new substance must be tested before it can be licenced for use.

5 Explain why it is often expensive to make and develop new pharmaceutical drugs.

C

6 Explain why it is difficult to test and develop new pharmaceutical drugs that are safe to use.

A*

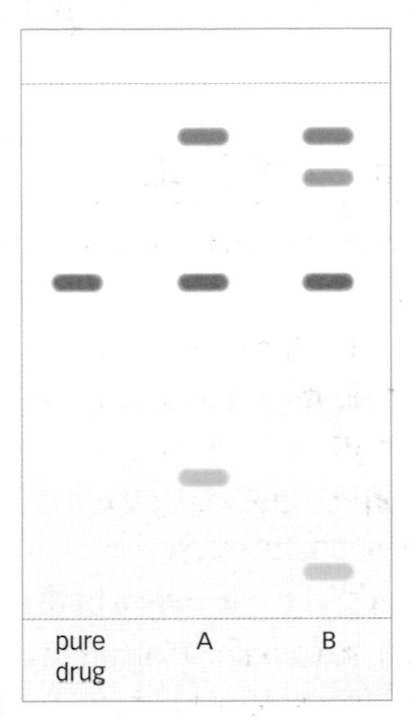


▲ Three samples of a pharmaceutical drug analysed by thin layer chromatography

**B** What is meant by payback time?

### Did you know...?

The world market for pharmaceuticals is worth at least £500 billion a year.

### Key words

**marketing, thin layer chromotography**

# 15: Diamond and graphite

## Learning objectives

After studying this topic, you should be able to:

- recognise the structures of diamond and graphite, and describe their properties
- explain some of the uses of diamond and graphite
- explain the properties of diamond and graphite in terms of their structure
- predict and explain the properties of substances that have a giant molecular structure

## Key words

**lustrous, lubricant, electrode, electrolysis, allotrope, covalent bond, delocalised electron**

This diamond-tipped drill bit is used to drill for oil

## Did you know...?

In ancient Rome, people wrote with a thin metal rod made of lead. Graphite pencils became widely used after a large deposit of graphite was discovered in 1564 in Borrowdale, England. The first pencils consisted of a graphite stick wrapped in string.

## Diamond

Diamonds are insoluble in water and they do not conduct electricity. Some diamonds are used in cutting tools, while others may be used in jewellery. Diamonds are hard and they have a high melting point. This is why they are used in drill bits, glass cutters, and some dental drills.

Diamonds are colourless and transparent (clear). They can be cut into shapes that allow light to pass through in such a way that they seem to sparkle. This **lustrous** appearance makes them highly valued gemstones for jewellery.

## Graphite

Graphite is black, lustrous, and opaque (not clear). It is also soft and slippery. Pencil 'lead' is not lead at all, but graphite. The black slippery graphite easily wears away on paper, leaving a black line. Graphite's properties also make it useful in **lubricants** that work at high temperature. Oil is commonly used as a lubricant to help moving parts in machinery slide past each other easily, but it begins to break down at high temperatures. Graphite, however, is slippery and it has a high melting point.

Graphite is used in pencils

Like diamond, graphite is insoluble in water and it has a high melting point. However, unlike diamond, graphite conducts electricity. This makes it useful as **electrodes** in **electrolysis**. For example, graphite is used in the production of aluminium by the electrolysis of a mixture of molten aluminium oxide and cryolite.

## The structures of diamond and graphite

Diamond and graphite are **allotropes** of carbon – they are different forms of the same element in the same physical state. Although diamond and graphite both consist of carbon atoms, their structures are different.

### Structure and properties

The carbon atoms in diamond and graphite are joined to each other by **covalent bonds**. This is why the two substances have high melting points. A lot of energy is needed to break the many strong covalent bonds.

Graphite has a layered structure in which each carbon atom is joined to three others by covalent bonds. There are only weak forces between the layers, so the layers can slide over each other easily. This is why graphite is slippery.

In diamond, each carbon atom is joined to four other carbon atoms. There are no free electrons, so diamond cannot conduct electricity. However, graphite does have electrons that are free to move. These **delocalised electrons** let graphite conduct electricity.

### Questions

1 Make a table to compare diamond and graphite.

2 Explain why diamond and graphite are allotropes of carbon.

3 Explain why diamond is used in cutting tools and jewellery.

4 Explain why graphite is used in pencils, lubricants, and electrodes.

5 In terms of their structure, explain why:

(a) graphite conducts electricity but diamond does not

(b) graphite is slippery but diamond is hard

(c) graphite and diamond both have high melting points.

6 Silicon has a structure similar to diamond. Predict and explain two physical properties of silicon.

E ↓ C ↓ A*

### Exam tip OCR

✔ Make sure you can recognise the structures of diamond and graphite.

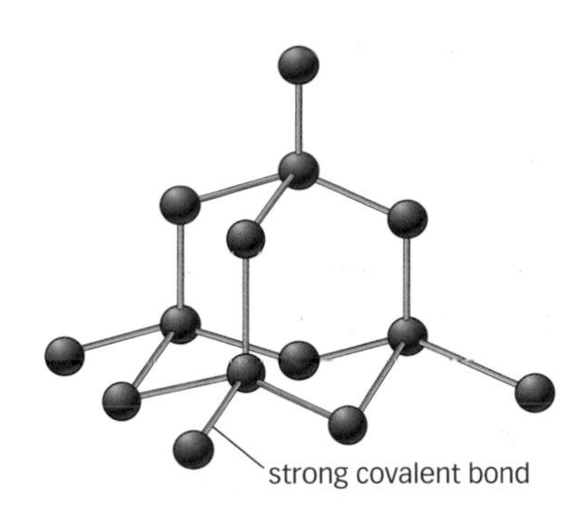


▲ The structure of diamond

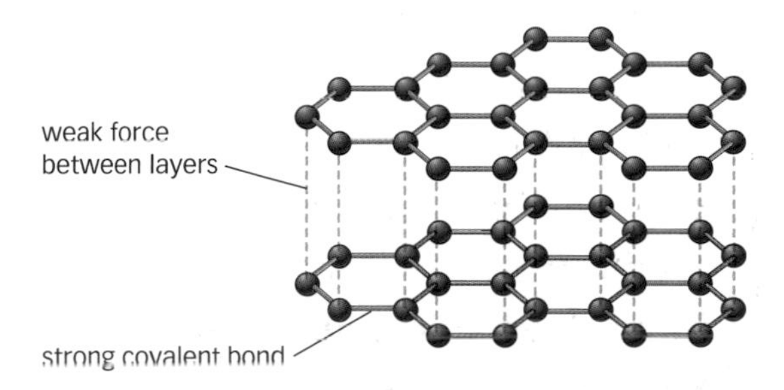


▲ The structure of graphite

### Giant molecules

Diamond and graphite have a giant molecular structure. Each molecule contains very many atoms joined together by covalent bonds. As covalent bonds are very strong, and there are many of them in each molecule, the melting and boiling points of substances with giant molecular structures are high. For example, sand contains silica. This has a giant molecular structure containing silicon and oxygen atoms. Its melting point is around 1610 °C.

# 16: Bucky balls and nanotubes

## Learning objectives

After studying this topic, you should be able to:

- recognise the structure of Buckminster fullerene
- recall the use of nanotubes in electrical circuits and describe the use of fullerenes in drug delivery systems
- explain how the structure of nanotubes allows them to be used as catalysts

## Key words

**Buckminster fullerene, nanotube**

## Did you know...?

Buckminster fullerene is named after an American architect called Richard Buckminster Fuller. He designed 'geodesic domes' which consist of the same sort of arrangement of hexagons and pentagons seen in the molecule.

Large amounts of Buckminster fullerene are made using an electric arc between carbon electrodes

## Bucky balls

**Buckminster fullerene**, $C_{60}$, was discovered in 1985. Each molecule has 60 carbon atoms, forming a hollow sphere. Since its discovery, many other different-sized fullerenes or 'Bucky balls' have been found. The fullerenes, diamond, and graphite are allotropes of carbon. They are in the same state at room temperature, but their carbon atoms are arranged differently.

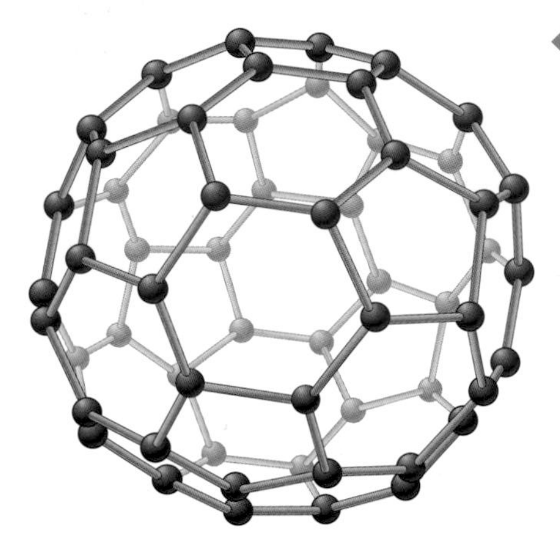

The structure of Buckminster fullerene

The Eden Project in Cornwall, England, is made using giant geodesic domes

## Fullerene 'cages'

Buckminster fullerene and other fullerenes are hollow. The space inside is large enough to contain atoms or other molecules. Scientists have discovered how to 'cage' radioactive metal atoms and drug molecules inside fullerenes. These fullerenes have potential use as new drug delivery systems. For example, they can be coated with chemicals that cause them to gather next to cancer cells after being injected into the body. In this way, the drug can be delivered to its target without damaging normal cells.

## Nanotubes

Fullerene molecules can be joined together to make **nanotubes**. These resemble a layer of graphite rolled into a tube. Nanotubes are the strongest and stiffest materials yet discovered. They are used to reinforce graphite in tennis racquets because they are so strong.

Nanotubes have electrical properties. Depending on their structure, nanotubes can conduct electricity like tiny wires, or act as semiconductors for electrical circuits. It is difficult to manufacture nanotubes for electrical uses at the moment.

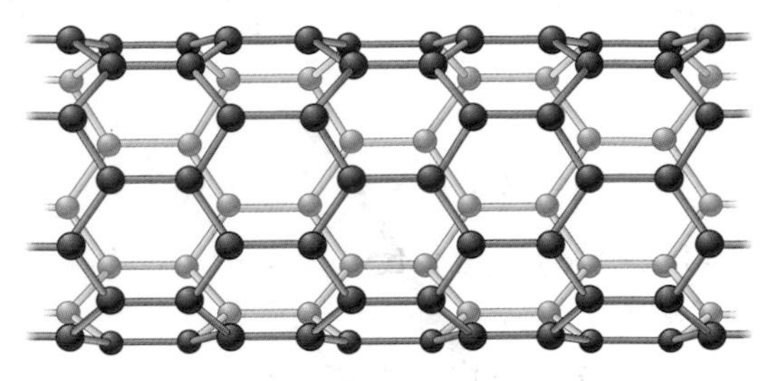

▲ The structure of a nanotube

### Did you know...?

Carbon nanotubes are very thin. Their diameter is about 10 000 times less than the diameter of a human hair.

### Nanotube catalysts

Nanotubes can act as catalysts, especially when they are stacked side by side. They have a huge surface area compared to their volume, allowing a high collision frequency with reactant molecules. Their properties are modified to make them effective catalysts by attaching other substances, such as nitrogen or iron, to their surface.

▲ Carbon nanotubes seen through an electron microscope. Each little bump is a carbon atom.

### Questions

E

1 What are nanotubes?

2 Describe and explain two uses of nanotubes.

3 Explain why fullerenes can be used in new drug delivery systems.

4 Explain why the fullerenes, diamond, and graphite are allotropes of carbon.

C

5 Explain how the structure of nanotubes enables them to be used as catalysts.

A*

# Module summary

## Revision checklist

- Reactions can be exothermic (release energy) or endothermic (absorb energy).
- Calorimetry experiments use reactions to heat water and are used to compare fuels.
- Some reactions are fast (short reaction time). Others are slow (long reaction time).
- Rate of reaction = $\frac{\text{mass of product formed}}{\text{time taken}}$
- Reactions are faster at higher temperatures, high concentrations and high pressures, and when solids are crushed into a powder.
- Catalysts increase the rate of a reaction but are unchanged at the end.
- The relative formula mass of a substance is calculated by adding up all the relative atomic masses of its atoms.
- The mass of the substances in a chemical reaction remains the same before and after the reaction.
- The predicted mass of products (yield) can be predicted from the mass of a reactant using relative formula masses and a balanced equation.
- Percentage yield = $\frac{\text{actual yield}}{\text{predicted yield}} \times 100$
- Percentage yields are usually less than 100% because product is lost in some practical techniques.
- Atom economy = $\frac{M_r \text{ of desired product}}{\text{total } M_r \text{ of all products}} \times 100$
- Industrial processes ideally need to use reactions that have high atom economy and high percentage yield.
- Some chemicals (eg ammonia) are made in large amounts in continuous processes.
- Developing new medicines is expensive because of the time and labour needed for research, development, and trials.
- The purity of compounds is tested by using thin-layer chromatography and measuring melting or boiling points.
- Diamond and graphite are two forms of carbon (allotropes) with a giant molecular structure. They have different properties and uses that can be explained using ideas about their structure.
- Fullerenes are a third form of carbon and can form nanotubes that are used as semiconductors and catalysts.

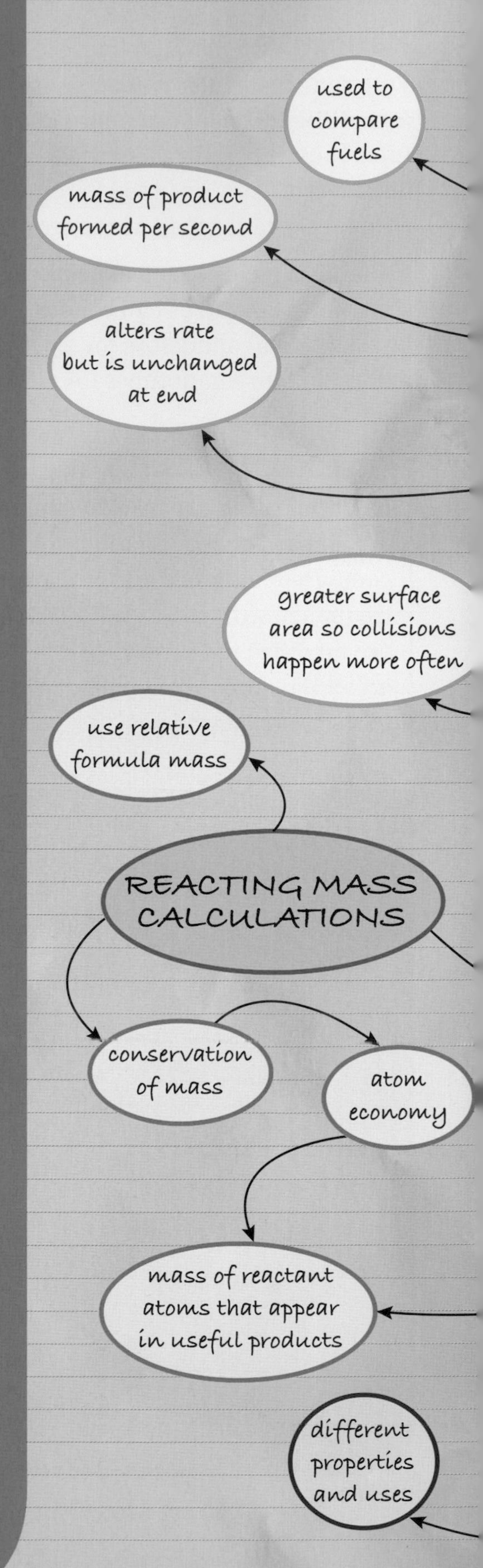




NOW USE THE C3 GRADE CHECKER ON PAGE 244

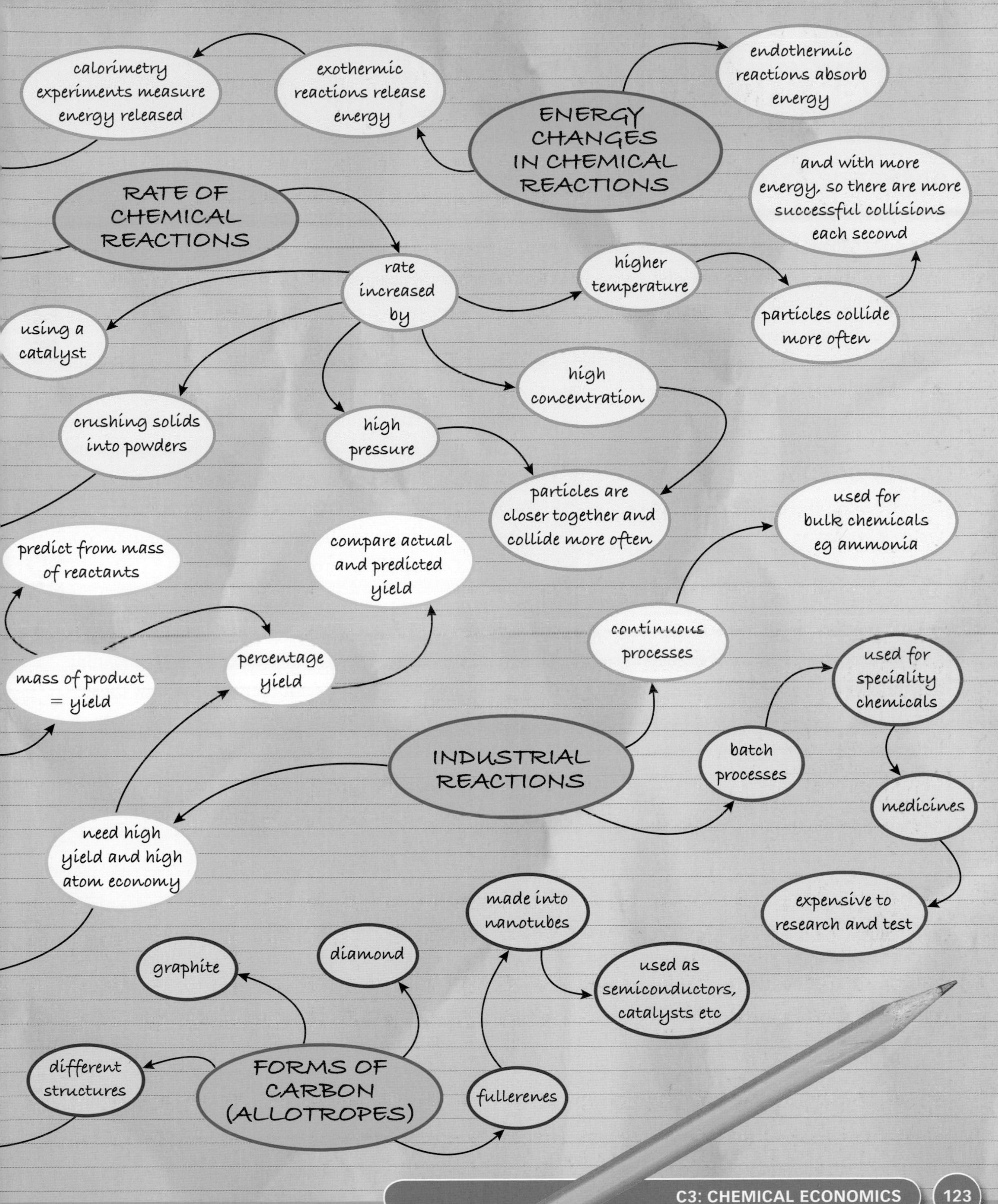
calorimetry experiments measure energy released
exothermic reactions release energy
endothermic reactions absorb energy
ENERGY CHANGES IN CHEMICAL REACTIONS
and with more energy, so there are more successful collisions each second
RATE OF CHEMICAL REACTIONS
rate increased by
higher temperature
particles collide more often
using a catalyst
high concentration
crushing solids into powders
high pressure
particles are closer together and collide more often
used for bulk chemicals eg ammonia
predict from mass of reactants
compare actual and predicted yield
continuous processes
mass of product = yield
percentage yield
used for speciality chemicals
INDUSTRIAL REACTIONS
batch processes
medicines
need high yield and high atom economy
made into nanotubes
expensive to research and test
graphite
diamond
used as semiconductors, catalysts etc
different structures
FORMS OF CARBON (ALLOTROPES)
fullerenes

## Answering Extended Writing questions

**QUESTION**

Energy changes happen during chemical reactions. Describe the kinds of changes that can happen.

In your answer you should describe how you can tell that an energy change has happened, and explain why energy changes happen.

**The quality of written communication will be assessed in your answer to this question.**

Reactions make things heat up or cool down because energy is given. Sometimes they light up as well. I would use a thermometre to find this out, it would go up.

**E**

*Examiner:* The candidate knows that reactions cause temperature changes, although the term 'temperature' is not used. The answer doesn't fully explain why temperature changes during a chemical reaction (energy is given out to heat things and taken in to cool things). One spelling error.

In an exothermic reaction energy is given out as heat so the temperature rises. The heat ends up all around us and can be used to heat things. Endothermic reactions is reactions where energy is taken in. The temperature falls. You can measure the temperature change on a thermometer.

**C**

*Examiner:* The candidate explains the meaning of exothermic and endothermic and how these are recognised in an experiment. The answer explains what happens to the energy, but does not describe transfer of energy to the surroundings. One instance of poor grammar, but the spelling is good.

In an exothermic reaction energy is transferred to the suroundings as heat so the temperature rises (you can measure this on a thermometer). This happens because bonds are formed. Endothermic reactions are reactions where energy is taken in from the suroundings. The temperature goes down, this is because bonds are broken.

**A***

*Examiner:* A good answer, explaining energy changes well. The candidate doesn't quite cope with the difficult idea that in most reactions bonds are both broken and formed. For example, in exothermic reactions more energy is released in forming bonds than is needed to break bonds. One spelling error.

# Exam-style questions

**1** Olivia heats 200 g of water using 1 g of two fuels, ethanol and hexane, in an experiment to help her decide which one is the best fuel.

| Fuel | ethanol | hexane |
|---|---|---|
| Start temperature (°C) | 21 | 21 |
| End temperature (°C) | 43 | |
| Temperature rise (°C) | | 16 |

A02 **a** Complete the missing boxes.

A02 **b** Choose the correct description of the energy changes.

- **i** Both reactions release energy.
- **ii** The hexane reaction releases energy and the ethanol reaction absorbs energy.
- **iii** The hexane reaction absorbs energy and the ethanol reaction releases energy.

A03 **c** Which fuel releases the most energy? Explain your answer.

A03 **d** Give two ways in which Olivia can make her experiment a fair test.

**2** Jack investigates the reaction between magnesium and hydrochloric acid. He performs the reaction with a low concentration of hydrochloric acid and then with a high concentration, and measures the volume of gas produced over time.

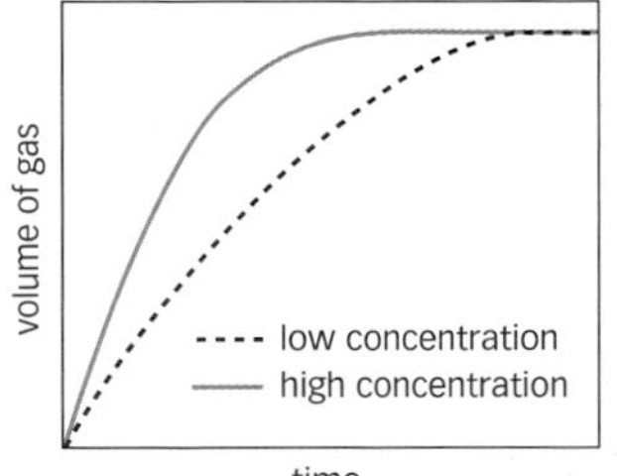


A02 **a** Which reaction has the fastest rate? Explain your answer.

A02 **b** What can Jack conclude about the effect of increasing concentration on rate of reaction?

A01 **c** Jack finds that increasing the temperature speeds up the reaction. Explain why using ideas about particles.

**3** Ammonia, $NH_3$, is manufactured in industry by reacting nitrogen and hydrogen. The equation for the reaction is $N_2 + 3H_2 \rightleftharpoons 2NH_3$.

A02 **a** The predicted yield of ammonia in a reaction was 15 tonnes. However, only 6 tonnes were produced. Calculate the percentage yield.

A01 **b** Industrial processes require the yield of a reaction to be as high as possible. Give two reasons why.

## Extended Writing

**4** A01 New medicines are often very expensive. Explain why.

**5** A01 Explain the difference between a batch process and a continuous process, and give examples of industrial reactions that use each type.

**6** A01 A02 Look at the diagram showing the structure and bonding in silicon dioxide.

● – silicon atoms
● – oxygen atoms

Use information from the diagram to predict and explain the properties of silicon dioxide.

A01 Recall the science
A02 Apply your knowledge
A03 Evaluate and analyse the evidence

# The periodic table

| OCR specification match | | |
|---|---|---|
| C4a | 4.1 | Atomic structure |
| C4a | 4.2 | Electronic structure |
| C4b | 4.3 | Ionic bonding |
| C4b | 4.4 | Ionic compounds |
| C4c | 4.5 | Electrons and the periodic table |
| C4c | 4.6 | Covalent bonding |
| C4d | 4.7 | The Group 1 elements |
| C4d | 4.8 | Flame tests |
| C4e | 4.9 | The Group 7 elements |
| C4e | 4.10 | Displacement reactions |
| C4f | 4.11 | The transition elements |
| C4f | 4.12 | Precipitation reactions |
| C4g | 4.13 | Metallic structure |
| C4g | 4.14 | Superconductors |
| C4h | 4.15 | Purifying water |
| C4h | 4.16 | Testing water |
| Module summary | | |
| OCR Upgrade | | |

## Why study this module?

In this module you will find out about the structure of the atom, and how this is linked to the properties of the elements and their positions in the periodic table. You will discover how ionic bonding, covalent bonding, and metallic bonding work. In addition, you will investigate the properties of substances with these different types of chemical bond.

You will examine the reactions of the Group 1 elements, which are very reactive metals, and the reactions of the Group 7 elements, which are very reactive non-metals. The transition elements include very important metals like iron and copper, so you will investigate their chemical and physical properties, too. You will discover why metals behave as they do, and the strange world of superconductors.

Water is vital to your survival. You will learn how water is treated to make it safe to drink, how it is tested, and what can go wrong if it becomes polluted.

### You should remember

1. Elements and compounds are made from atoms.
2. Substances may contain ionic bonds or covalent bonds.
3. The periodic table is a chart showing information about the elements.
4. Some elements are more reactive than others.
5. Some compounds break down when heated, forming other compounds.
6. Some dissolved compounds may react together to produce an insoluble compound.
7. Metals are good conductors of electricity.
8. Simple laboratory tests may be used to detect carbon dioxide and other substances.
9. Filtration can be used to remove insoluble particles from water.

This is part of the Large Hadron Collider at CERN, near Geneva, under the border between France and Switzerland. It weighs over 38 000 tonnes and runs through a circular tunnel almost 27 km long.

Powerful superconducting magnets accelerate subatomic particles to almost 99.9998% of the speed of light. The magnets are cooled by liquid helium to a very chilly −271.25 °C. Enough superconducting cable was used in its construction to go around the equator almost seven times. Beams of particles smash into each other with enough energy to recreate conditions believed to have existed billionths of a second after the Big Bang.

# 1: Atomic structure

## Learning objectives

After studying this topic, you should be able to:

- describe the structure of the atom
- recall the relative charge and relative mass of protons, neutrons, and electrons
- describe what is meant by an isotope
- work out the number of protons, neutrons, and electrons in an atom or ion

## Exam tip OCR

- You should be able to explain how a scientific idea has changed as new evidence has been found. For example, the results of lots of experiments have gradually led to changes in ideas about the structure of atoms.

| Subatomic particle | Relative charge | Relative mass |
|---|---|---|
| proton | +1 | 1 |
| neutron | 0 | 1 |
| electron | –1 | 0.0005 |

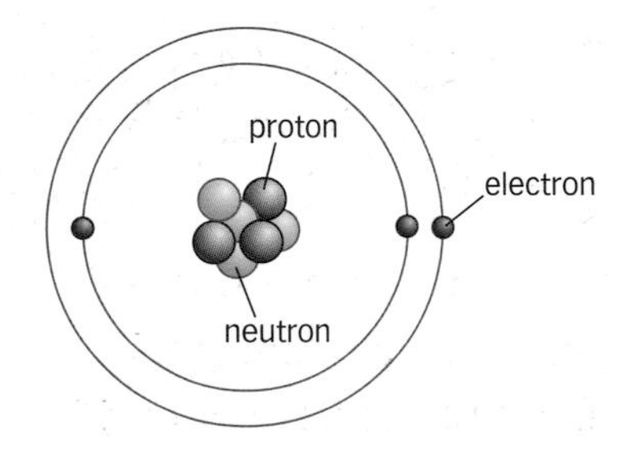


The structure of an atom

**A** Which subatomic particle has the least mass, and which one has no electrical charge?

## What's in an atom?

In 1803, John Dalton suggested that everything was made from tiny objects called **atoms**. Other scientists had suggested this before, but Dalton's ideas were more detailed. He suggested that:

- all matter is made from atoms
- atoms cannot be made or destroyed
- all atoms of an **element** are identical
- different elements contain different types of atoms.

Individual atoms are far too small to see. Scientists imagined them as tiny solid balls, but this simple idea was to change. As new evidence was found, Dalton's ideas were developed into a detailed theory of atomic structure.

In 1897, J.J. Thomson discovered that atoms contain even smaller **subatomic particles** called **electrons**. In 1911, Ernest Rutherford discovered that an atom is mostly empty space, with its electrons arranged around a central object called the **nucleus**. Two years later, Niels Bohr improved on Rutherford's theory, calculating that electrons move in fixed orbits or shells around the nucleus.

The Atomium in Brussels, Belgium, is 102 m high. It represents the arrangement of atoms in iron. You can even go inside.

Scientists now know that the nucleus is made from two subatomic particles called **protons** and **neutrons**. Protons are positively charged, neutrons have no electrical charge, and electrons are negatively charged. An atom is electrically neutral overall because it contains equal numbers of protons and electrons. Most of the mass of an atom is found in the nucleus. The radius of an atom is about $10^{-10}$ metres. Atoms have a mass of about $10^{-23}$ grams.

## Isotopes

The **atomic number** of an atom is the number of protons it contains. The atoms of different elements have different atomic numbers. The **periodic table** contains information about each element, including the atomic number of its atoms.

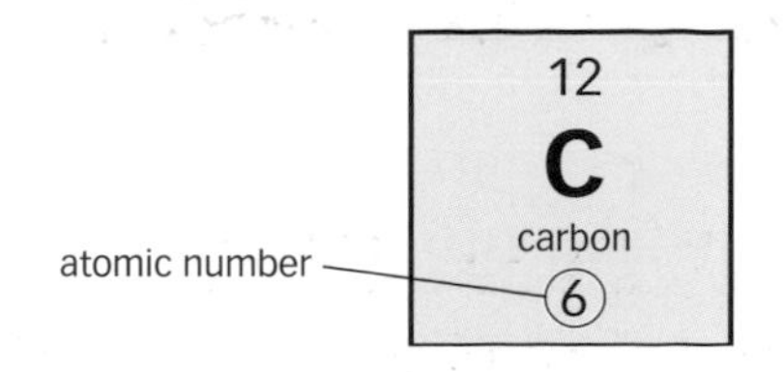


The atomic number is the bottom number for each element in the periodic table

> **B** Use the periodic table to find the atomic numbers of oxygen, O, and sodium, Na.

**Key words**

**atom, element, subatomic particle, electron, nucleus, proton, neutron, atomic number, periodic table, mass number, isotope**

The **mass number** of an atom is the total number of protons and neutrons it contains. **Isotopes** are varieties of an element that have the same atomic number, but different mass numbers.

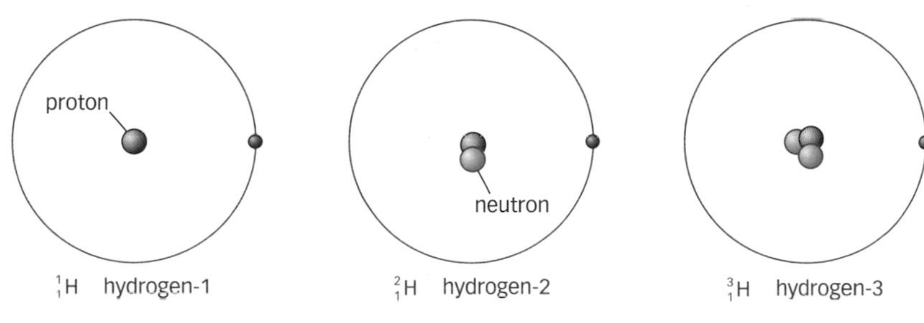


Three different isotopes of hydrogen. They contain the same number of protons and electrons, but different numbers of neutrons.

## More about isotopes

Isotopes are shown by a full chemical symbol, for example $^{12}_{6}C$. The bottom number is the atomic number: it is the number of protons, and also the number of electrons. The top number is the mass number. The number of neutrons is the mass number minus the atomic number.

For example, $^{12}_{6}C$ contains 6 protons and 6 electrons. It contains 6 neutrons (12 – 6). An isotope is named after the element and its mass number. For example, $^{12}_{6}C$ is carbon-12.

> When an atom becomes an ion, its nucleus stays the same and so do the numbers of protons and neutrons. An ion with a single negative charge has one more electron than the original atom, and an ion with a single positive charge has one less electron.

### Questions

Use the periodic table to help you answer Questions 1 and 5.

1. What are the names and symbols of the elements with the following atomic numbers?
   (a) 3
   (b) 13
   (c) 18.
2. Describe the structure of an atom.

E

3. Explain what is meant by atomic number and mass number.

C

4. Calculate the number of protons, neutrons, and electrons in $^{19}_{9}F$ and in $^{19}_{9}F^{-1}$.
5. Explain why $^{40}_{20}X$ and $^{40}_{18}Y$ are not isotopes of the same element.

A*

## Learning objectives

After studying this topic, you should be able to:

- explain the difference between an element and a compound
- describe how the elements are arranged in the periodic table
- recall that electrons occupy shells around the nucleus
- work out the electronic structure of the first 20 elements

## Exam tip OCR

- You will be given a copy of the periodic table in the exam.

## Elements and the periodic table

An element is a substance that cannot be broken down chemically, and it contains only one type of atom. For example, a lump of pure carbon cannot be broken down into any other substance, as it contains only carbon atoms.

An element contains only one type of atom, just as this train sculpture in Darlington is made from only one type of brick

There are just over 100 different elements. The periodic table contains information about them, including their name, **chemical symbol**, and atomic number. No two elements have the same atomic number. The elements are arranged in order of increasing atomic number in the periodic table.

**A** What are the chemical symbols for arsenic and tin?

## Elements and compounds

A **compound** is a substance that contains two or more different elements, chemically combined. Chemical bonds join the elements together in a compound, so the elements are not just mixed together. The chemical formula of a compound shows the symbols of each element it contains, and the number of atoms of each element. For example, the chemical formula of copper sulfate is $CuSO_4$. It shows that this compound contains copper, sulfur, and oxygen.

Glycine molecules contain four different elements

**B** Which elements are contained in glycine, $C_2H_5NO_2$?

## Electron shells

In an atom, the electrons occupy the space around the nucleus. They are not just randomly arranged. Instead, they occupy **shells** around the nucleus. Different shells can hold different numbers of electrons. The way that the electrons are arranged in an atom is called its **electronic structure**.

### Electronic structure

The first shell can only hold up to two electrons, but the second and third can hold up to eight electrons. The electronic structure of sulfur, for example, is 2.8.6 – this shows that each atom has two electrons in the first shell, eight in the second and six in its outer shell. You can work this out by counting towards the right across the periodic table from hydrogen to sulfur, adding a full stop each time you have to drop down to start on the left of a new row.

### Evidence for the nucleus

Rutherford developed his 1911 theory using results obtained by two of his students, Hans Geiger and Ernest Marsden, in 1909. At the time, scientists thought that an atom was a cloud of positive charge with electrons dotted about inside. In Geiger and Marsden's experiments, positively charged particles were 'fired' at gold atoms. Instead of going straight through as expected, some of these particles were deflected. Some even came straight back. These unexpected results led to Rutherford's theory of a nuclear atom.

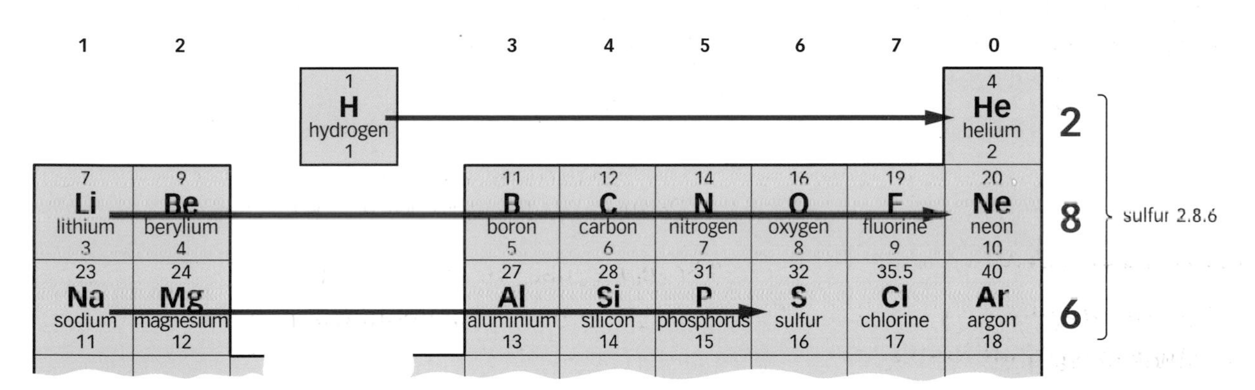


▲ Finding the electronic structure of sulfur

### Questions

Use the periodic table to help you answer Questions 1 and 5.

1. About how many different elements are there?
2. What is meant by element and compound?

E

3. Describe the arrangement of the elements in the periodic table.
4. Where in an atom are the electrons found?

C

5. Write down the electronic structures for lithium, fluorine, argon, and calcium.

A*

### Exam tip OCR

✔ Make sure you can work out the electronic structure of the first 20 elements in the periodic table.

### Key words

**chemical symbol, compound, shell, electronic structure**

# 3: Ionic bonding

## Learning objectives

After studying this topic, you should be able to:

- explain what ions are
- describe the formation of positive and negative ions
- explain what happens in ionic bonding
- use 'dot and cross' diagrams to describe ionic bonding

## Key words

**ion**, **ionic bond**, **dot and cross diagram**

## Exam tip OCR

- Dot and cross diagrams are an example of a scientific model. Electrons are not really dots and crosses, and the shells are not really circles, but the diagrams help you to understand and explain how bonding happens.

**A** How do positively charged ions form?

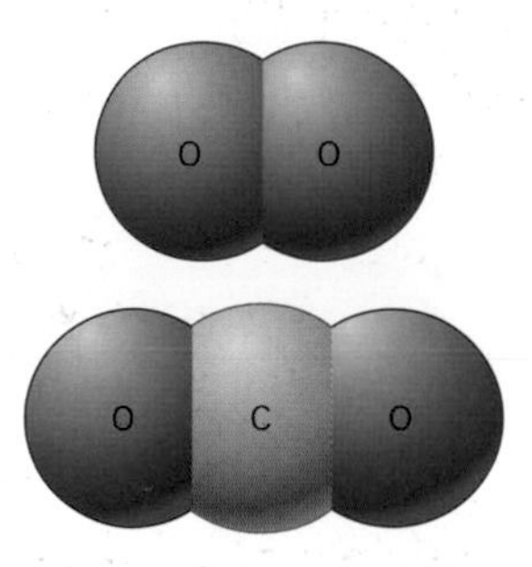


Oxygen and carbon dioxide molecules

## Ions

Electrically charged particles from the Sun stream towards the Earth and interact with its magnetic field, especially near the poles. Different substances in the upper atmosphere glow in different colours, providing a spectacular natural light show in the sky.

The Northern Lights or aurora borealis

An **ion** is an electrically charged atom or group of atoms. Electrons are negatively charged.

- If an atom loses one or more electrons, it becomes a positively charged ion.
- If an atom gains one or more electrons, it becomes a negatively charged ion.

Hydrogen atoms and metal atoms usually form positively charged ions, and non-metal atoms usually form negatively charged ions.

## Ions and formulae

Chemical formulae give information about the type of particle, for example whether it is an atom, molecule, or ion. A single atom has a symbol made from one or two letters, starting with a capital letter. For example, carbon atoms are shown as C and oxygen atoms as O. Molecules contain two or more atoms. For example, oxygen is shown as $O_2$ and carbon dioxide as $CO_2$.

The formulae for ions are easy to spot. They have a + or – sign at the top right, showing that the ion is positively charged or negatively charged. For example:

- A sodium atom, Na, loses one electron to form an $Na^+$ ion.
- A magnesium atom, Mg, loses two electrons to form an $Mg^{2+}$ ion.
- A chlorine atom, Cl, gains one electron to form a $Cl^-$ ion.
- An oxygen atom, O, gains two electrons to form an $O^{2-}$ ion.

## Ionic bonding

When a metal reacts with a non-metal, electrons are transferred from the metal atoms to the non-metal atoms. The positive metal ions and the negative non-metal ions attract one another, forming **ionic bonds**.

### Dot and cross diagrams

A **dot and cross diagram** is used to show ionic bonding. The electrons in the metal are shown as spots or dots and the electrons in the non-metal as crosses. The diagrams show how sodium ions and chloride ions form when sodium reacts with chlorine. The outer electron from a sodium atom transfers to the outer shell of a chlorine atom. In this way the outer shells become complete, making the atoms more stable.

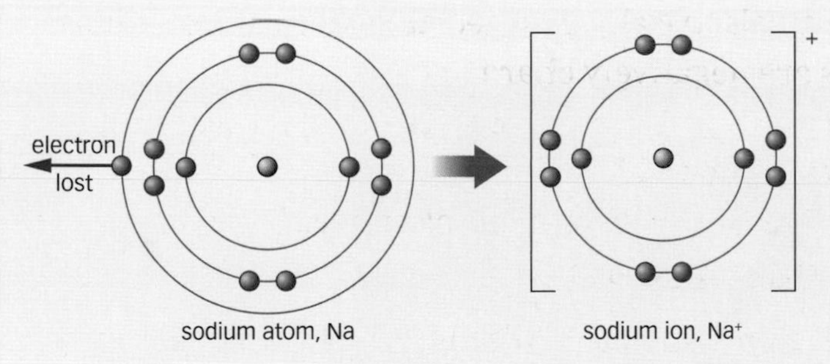


▲ A dot and cross diagram for the formation of a sodium ion

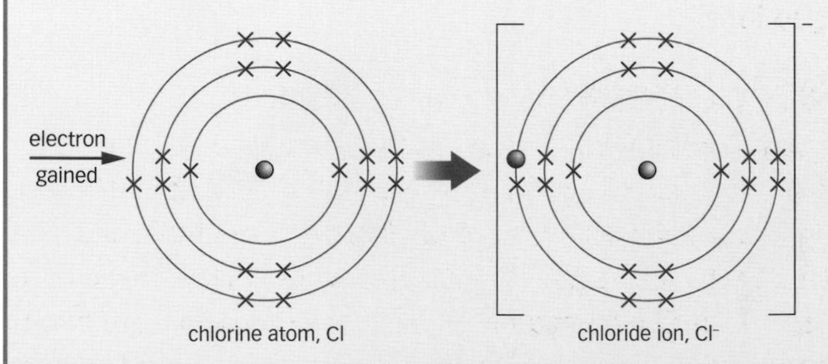


▲ A dot and cross diagram for the formation of a chloride ion

**B** An aluminium atom, Al, loses three electrons. What is the formula of the aluminium ion?

### Exam tips OCR

✔ The number next to the + or – sign shows the number of charges. If there is no number, there will only be one charge.

✔ Higher Tier students must be able to show the ionic bonding in simple compounds such as sodium chloride, magnesium oxide, sodium oxide, or magnesium chloride.

### Questions

Use the periodic table to help you answer Questions 1 and 5.

1 What is an ion?

2 Which of the following formulae represents an ion, and which represents a molecule? Ar, C, $NH_4^+$, $SO_2$.

3 Explain how 2+ ions and 2– ions form.

4 Explain what happens in ionic bonding involving a metal and non-metal.

5 Draw a dot and cross diagram to show the formation of:

(a) a magnesium ion, $Mg^{2+}$, from a magnesium atom, Mg

(b) an oxide ion, $O^{2-}$, from an oxygen atom, O.

## Learning objectives

After studying this topic, you should be able to:

- recall properties of sodium chloride and magnesium oxide
- describe the structure of ionic compounds
- explain physical properties of sodium chloride and magnesium oxide

**A** Suggest why sodium chloride is not used in wall boards.

## Key words

**ionic compound**, **giant ionic lattice**

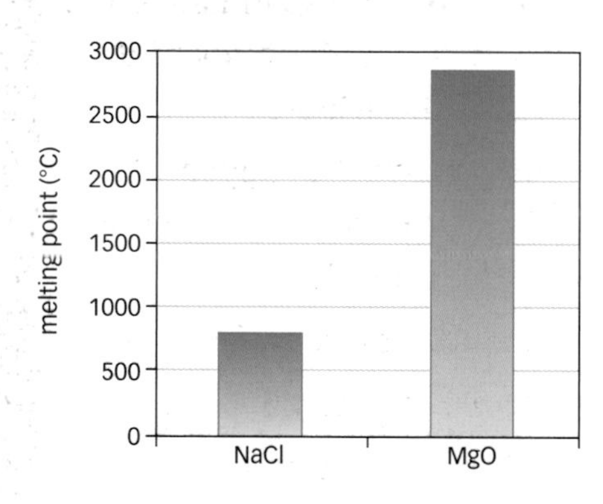


The melting points of sodium chloride and magnesium oxide

## Exam tip OCR

- Sodium chloride and magnesium oxide have similar physical properties, except that sodium chloride is soluble in water but magnesium oxide is insoluble.

## Ionic compounds

Magnesium oxide has a very high melting point. It is used in fire-resistant building materials such as wall boards and protection for metal beams. It is an **ionic compound**, just like sodium chloride, common table salt. Sodium chloride has a high melting point, too. However, it dissolves in water whereas magnesium oxide does not.

Sodium chloride and magnesium oxide do not conduct electricity when they are solid. They do conduct electricity when they are molten, but their high melting points make this difficult to show in the laboratory. However, when sodium chloride dissolves in water, its solution conducts electricity.

The Taipei 101 building in Taiwan is over half a kilometre high. Its 101 storeys contain magnesium oxide sheeting.

## Giant ionic lattices

The ions in ionic compounds like sodium chloride and magnesium oxide are arranged in a regular way. This arrangement is called a **giant ionic lattice**. The word giant tells you that the structure is repeated very many times, not that it is huge. The positive ions are strongly attracted to the negative ions, forming ionic bonds.

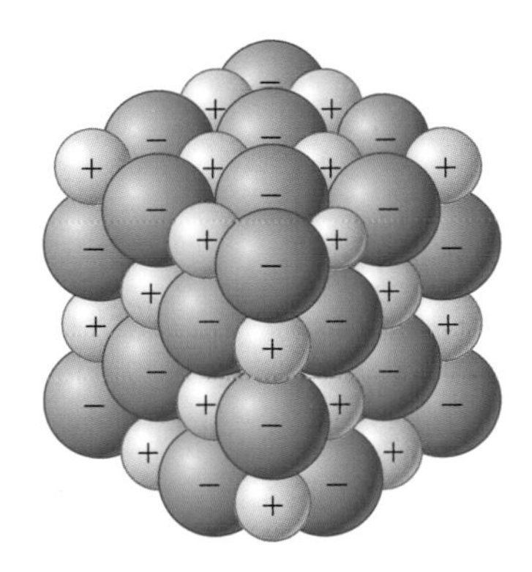

▲ The lattice structure of sodium chloride

### Explaining physical properties

Ionic bonds are strong chemical bonds. A lot of energy is needed to break them. This is why sodium chloride and magnesium oxide have high melting points. For an ionic compound to conduct electricity, its ions must be free to move from place to place. This can happen in a molten liquid or in a solution, but not in a solid.

## Formulae of ionic compounds

You can work out the formula of an ionic compound if you know the formulae of its ions. The number of positive charges has to be equal to the number of negative charges. For example, sodium chloride contains $Na^+$ ions and $Cl^-$ ions. These each carry one charge, so the formula for sodium chloride is NaCl.

If the positive ions carry a different number of charges from the negative ions, the number of ions in the formula has to be adjusted. For example, sodium oxide contains $Na^+$ ions and $O^{2-}$ ions. Two $Na^+$ ions are needed to balance the two charges on an $O^{2-}$ ion, so the formula for sodium oxide is $Na_2O$.

**B** Magnesium oxide contains $Mg^{2+}$ ions and $O^{2-}$ ions. What is its formula?

**C** Magnesium chloride contains $Mg^{2+}$ ions and $Cl^-$ ions. What is its formula?

### Questions

1. Under what conditions does magnesium oxide conduct electricity?
2. Under what conditions does sodium chloride conduct electricity?

E

3. Suggest why magnesium oxide bricks are used to line the inside of furnaces.
4. Describe the structure of magnesium oxide.

C

5. Work out the formulae of the following ionic compounds, using these ions:
   potassium $K^+$, copper(II) $Cu^{2+}$, chloride $Cl^-$, oxide $O^{2-}$, hydroxide $OH^-$
   (a) potassium hydroxide
   (b) copper(II) chloride
   (c) potassium oxide
   (d) copper oxide
   (e) copper(II) hydroxide.

A*

### Exam tip OCR

✔ Remember to write the formula of an ionic compound without showing any charges.

# 5: Electrons and the periodic table

## Learning objectives

After studying this topic, you should be able to:

- ✔ describe the main stages in the development of the classification of elements
- ✔ identify groups and periods
- ✔ describe the link between group number and electronic structure
- ✔ work out an element's group and period from its electronic structure

## Key words

**period, group**

**A** Which period is phosphorus in?

**B** Name two elements in Group 1 and two elements in Group 7.

| 1 | 2 | 3 | 4 | 5 | 6 | 7 | 0 |
|---|---|---|---|---|---|---|---|
| metals | | non-metals | | | | | 4 **He** helium 2 |
| 7 **Li** lithium 3 | 9 **Be** berylium 4 | 11 **B** boron 5 | 12 **C** carbon 6 | 14 **N** nitrogen 7 | 16 **O** oxygen 8 | 19 **F** fluorine 9 | 20 **Ne** neon 10 |
| 23 **Na** sodium 11 | 24 **Mg** magnesium 12 | 27 **Al** Aluminium 13 | 28 **Si** silicon 14 | 31 **P** phosphrous 15 | 32 **S** sulfur 16 | 35.5 **Cl** chlorine 17 | 40 **Ar** argon 18 |
| 39 **K** potassium 19 | 40 **Ca** calcium 20 | 70 **Ga** galium 31 | 73 **Ge** germanium 32 | 75 **As** arsenic 33 | 79 **Se** selenium 34 | 80 **Br** bromine 35 | 84 **Kr** krypton 36 |
| 85 **Rb** rubidium 37 | 88 **Sr** strontium 38 | 115 **In** indium 49 | 119 **Sn** tin 50 | 122 **Sb** antimony 51 | 128 **Te** tellurium 52 | 127 **I** iodine 53 | 131 **Xe** xenon 54 |
| 133 **Cs** caesium 55 | 137 **Ba** barium 56 | 204 **Ti** thalium 81 | 207 **Pb** lead 82 | 209 **Bi** bismuth 83 | [209] **Po** polonium 84 | [210] **At** astatine 85 | [222] **Rn** radon 86 |
| [223] **Fr** francium 87 | [226] **Ra** radium 88 | elements with atomic numbers 112–116 have been reported but not fully authenticated | | | | | |

▲ A shortened version of the periodic table. The black line separates the metals from the non-metals. Hydrogen is not shown here.

## Developing the periodic table

During the 1820s, Johann Döbereiner noticed that certain elements with similar properties could be organised into groups of three. For example, lithium, sodium, and potassium formed one such 'triad'. When listed in order of reactivity, the relative atomic mass of the middle element was roughly the mean of the other two masses. Unfortunately, Döbereiner could not explain his observations.

John Newlands classified the elements according to his 'law of octaves' in 1865. In Newlands' table, elements with similar properties (including triads) appeared in the same row. Newlands found patterns in the properties of elements when they were arranged in order of relative atomic mass, but this only worked as far as calcium. His table was not a success.

Dmitri Mendeleev developed a much more successful classification system for the elements. His first table, published in 1869, contained the elements in order of increasing relative atomic mass. This was the same as Newlands' table, but unlike Newlands, Mendeleev swapped some elements round if that worked better. He also left gaps for elements that had not yet been discovered. These changes helped to make Mendeleev's table the direct ancestor of the modern periodic table.

## Groups and periods

The elements are arranged in order of increasing atomic number in the modern periodic table. A horizontal row of elements is called a **period**.

- Hydrogen and helium occupy the first period.
- The elements lithium to neon occupy the second period.
- The elements sodium to argon occupy the third period.

A vertical column of elements is called a **group**. The elements in a group have similar chemical properties. For example, the elements in Group 1 are all reactive metals and the elements in Group 7 are all reactive non-metals.

## Electronic structure

The electronic structure of an atom is the way that its electrons are arranged. There are links between the position of an element in the periodic table and the electronic structure of its atoms.

The period and group that an element belongs to can be worked out from the electronic structure of its atoms:

- the period is the number of numbers;
- the group is the last number.

For example, silicon atoms have the electronic structure 2.8.4. Silicon is in the third period because there are three numbers. It is in Group 4 because the last number is 4.

**C** How many electrons are there in the outer shells of the atoms of Group 7 elements?

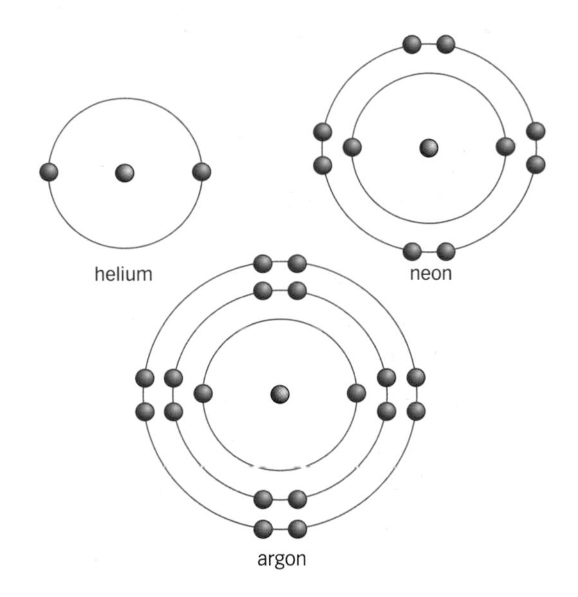


▲ The elements on the far right of the periodic table all have complete outer shells. Their group number can be shown as 0 or 8.

## Filling the gaps

One of the gaps that Mendeleev left in his periodic table was for 'eka-aluminium'. This element was not yet discovered, but Mendeleev predicted its likely properties by studying the properties of the elements around the gap. Gallium, an element with similar properties to 'eka-aluminium', was later discovered in 1875.

| Property | Eka-aluminium, Ea | Gallium, Ga |
|---|---|---|
| Atomic mass | About 68 | 70 |
| Melting point | low | 30 °C |
| Density (g/cm$^3$) | 6.0 | 5.9 |
| Formula of oxide | $Ea_2O_3$ | $Ga_2O_3$ |

These properties of gallium closely matched Mendeleev's predicted properties for eka-aluminium. Gallium was the element that filled the gap. Other elements were later discovered that filled Mendeleev's remaining gaps. The last one, 'eka-manganese' or technetium, was discovered in 1937.

## Iodine and tellurium

Mendeleev had a problem with iodine and tellurium. Iodine had the lower relative atomic mass, so should have gone first in his table. However, he swapped their positions so that iodine was in a group with chlorine, bromine, and iodine. Some scientists criticised Mendeleev for this. However, in 1913 Henry Moseley discovered how to measure atomic number. It turned out that iodine has a greater atomic number than tellurium. Mendeleev was right to put iodine after tellurium, even if he could not explain why at the time.

## Questions

1. What is a group in the periodic table?
2. What is a period in the periodic table?

E

3. Phosphorus is in Group 5. How many electrons do its atoms have in their outer shell?
4. How many occupied shells of electrons do the elements in the fourth period have
5. For each electronic configuration below, name the element, and its group and period:
   (a) 2 (b) 2.5
   (c) 2.8.7 (d) 2.8.8.2.

C

6. Explain how new discoveries confirmed Mendeleev's ideas about the periodic table.

A*

# 6: Covalent bonding

## Learning objectives

After studying this topic, you should be able to:

- describe covalent bonding
- work out the number of atoms of each element in a molecule
- describe the properties of carbon dioxide and water
- draw dot and cross diagrams for simple molecules
- relate the properties of carbon dioxide and water to their structure

**A** How many carbon atoms, C, and oxygen atoms, O, does a molecule of carbon dioxide contain? What is its molecular formula?

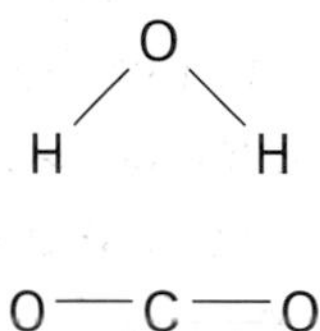


The displayed formulae for water and carbon dioxide

This is dry ice or solid carbon dioxide. It turns directly from a solid to a gas at –78.5 °C.

## A shared pair of electrons

Metals and non-metals combine to form ionic compounds, which have ionic bonding. On the other hand, non-metals combine with each other to form **covalent compounds**. These have covalent bonding rather than ionic bonding. A **covalent bond** is a shared pair of electrons.

Carbon dioxide causes the bubbles in this sparkling mineral water

## The formulae of molecules

A **molecule** is a particle containing two or more atoms chemically bonded together. The **molecular formula** shows the number of atoms of each element a molecule contains. For example, the molecular formula of water is $H_2O$. It shows that each water molecule contains two hydrogen atoms and one oxygen atom.

The **displayed formula** shows each atom in a molecule and the covalent bonds between them. Each covalent bond is shown as a straight line in a displayed formula.

## Carbon dioxide and water

Carbon dioxide and water both exist as **simple molecules**. Each molecule only contains a few atoms, joined together by covalent bonds. There are weak **intermolecular forces** between these molecules. Carbon dioxide has a low melting point. It is a gas at room temperature. Water also has a low melting point, 0 °C, but it is a liquid at room temperature. Neither compound conducts electricity.

## Explaining properties

Although the covalent bonds between the atoms in a molecule are strong, the intermolecular forces between molecules are weak. Little energy is needed to overcome them, which is why carbon dioxide and water have such low melting points.

Carbon dioxide and water molecules have no overall electrical charge. Their electrons are not free to move from place to place, so these compounds do not conduct electricity.

## Drawing molecules

Dot and cross diagrams are used to show covalent bonding in molecules. Unlike the diagrams used to show ionic bonding, only the outer shells need be shown. Each covalent bond is shown in a shared area as a dot and a cross. The electrons in the bonded atom are shown as dots, and the electrons from the other atom as crosses.

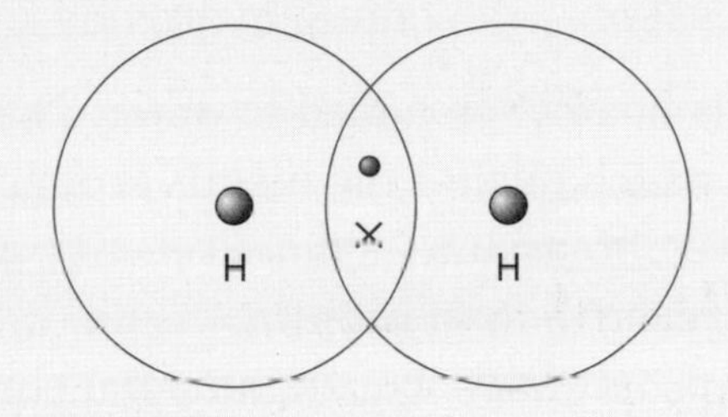


▲ A dot and cross diagram for hydrogen, $H_2$

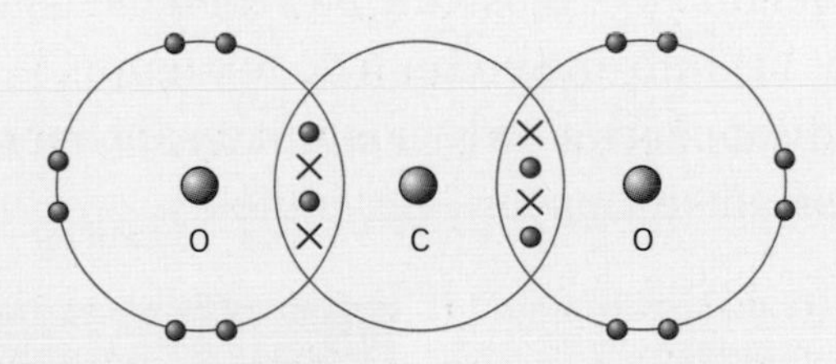


▲ A dot and cross diagram for carbon dioxide. In this example, electrons from the oxygen atoms are shown as dots.

**B** The dot and cross model can be used to explain ionic bonding and covalent bonding. Describe a difference in how the model is used when explaining these two types of bond.

### Key words

**covalent compound, covalent bond, molecule, molecular formula, displayed formula, simple molecule, intermolecular force**

### Questions

1. What information do a molecular formula and a displayed formula give?
2. How many atoms of each element are there in:
   (a) methane, $CH_4$?
   (b) ethanol, $C_2H_5OH$?

E

C

3. What is a covalent bond?
4. Explain, in terms of structure and bonding, why carbon dioxide is a gas at room temperature, and why it does not conduct electricity.

A*

5. Use the dot and cross model to describe the bonding in chlorine $Cl_2$, methane $CH_4$, and water $H_2O$.

# C4 7: The Group 1 elements

## Learning objectives

After studying this topic, you should be able to:

- ✔ describe the reactions of Group 1 elements with water
- ✔ explain why the elements in Group 1 have similar properties
- ✔ predict the reactions of rubidium and caesium with water
- ✔ predict the properties of rubidium and caesium
- ✔ explain the trend in reactions with water

| 1 |
|---|
| 7<br>**Li**<br>lithium<br>3 |
| 23<br>**Na**<br>sodium<br>11 |
| 39<br>**K**<br>potassium<br>19 |
| 85<br>**Rb**<br>rubidium<br>37 |
| 133<br>**Cs**<br>caesium<br>55 |
| [223]<br>**Fr**<br>francium<br>87 |

◀ Group 1 is on the far left of the periodic table

▲ Lithium, sodium, and potassium are stored in oil

**A** Write the word equation for the reaction of lithium with water.

**B** What colour flame does potassium give during its reaction with water?

## Group 1

The **Group 1** elements are placed in the vertical column on the far left of the periodic table. They are all metals and they react vigorously with water to form **alkaline** solutions, so they are called the **alkali metals**.

Lithium, sodium, and potassium are Group 1 elements. They are so reactive with air and water that they must be stored under oil. This stops air and water reaching the metals during storage.

## Reactions with water

Lithium, sodium, and potassium react with water in a similar way. Here is a general word equation for their reaction with water:

metal + water → metal hydroxide + hydrogen

For example, sodium reacts with water to form sodium hydroxide and hydrogen:

sodium + water → sodium hydroxide + hydrogen

All three metals float when added to water. The hydrogen causes fizzing, which pushes the metals around on the surface. However, potassium is more reactive than sodium, which is more reactive than lithium.

- Lithium keeps its shape, but sodium and potassium melt to form silvery balls.
- Potassium ignites explosively and then burns with a lilac flame, but lithium and sodium do not ignite.
- Lithium disappears slowly, sodium disappears quickly, and potassium disappears very quickly.

▲ Sodium reacting with water. Universal indicator in the water turns purple, showing the presence of an alkali (sodium hydroxide).

▲ Potassium gives a lilac flame during its reaction with water

Notice that the reactivity of the metals with water increases as you go down Group 1. The atoms of the Group 1 elements each have one electron in their outer shell. This is why they have similar chemical properties.

**Key words**

**Group 1, alkaline, alkali metal**

## Losing an electron

Rubidium and caesium are below potassium in Group 1, so they are even more reactive than potassium. Rubidium and caesium both react explosively on contact with water. Here is a general balanced symbol equation for the reaction of Group 1 metals with water, where M stands for the symbol of the metal:

$$2M + 2H_2O \rightarrow 2MOH + H_2$$

The Group 1 elements have similar properties because, when they react, each atom loses its outer electron to become an ion with a single positive charge. This ion has the stable electronic structure of a noble gas.

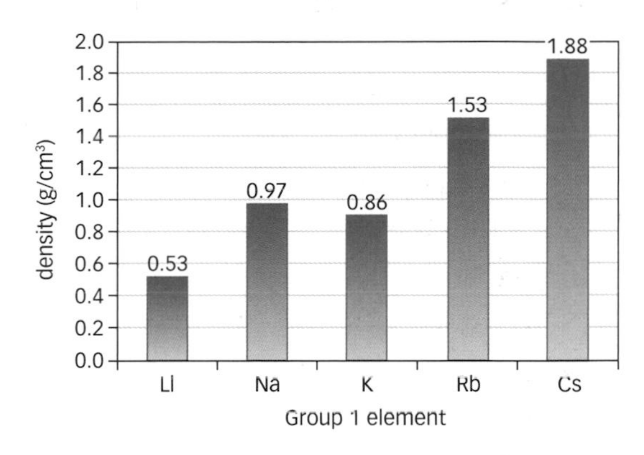


▲ The densities of the Group 1 elements

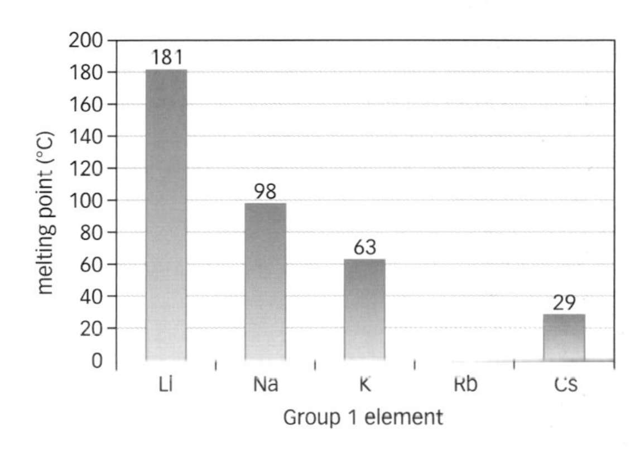


▲ The melting points of the Group 1 elements

### More about trends

The easier it is for an atom of a Group 1 element to lose one electron, the more reactive the element is. For example, potassium is more reactive than lithium because its atoms lose electrons more easily than lithium ions do. The loss of electrons is oxidation. Here is a general ionic equation for such oxidation, where M stands for the symbol of the metal:

$$M \rightarrow M^+ + e^-$$

There are trends in the physical properties of the Group 1 elements, not just in their chemical properties. Their densities generally increase going down the group, and their melting points decrease.

**C** Write the balanced symbol equation for the reaction of caesium, Cs, with water.

### Questions

1 Why are Group 1 elements stored in oil?

2 Describe the reactions of lithium, sodium, and potassium with water.

3 Write a balanced symbol equation for the reaction of sodium, Na, with water.

E

4 Explain why Group 1 elements have similar chemical properties.

C

5 Predict, with reasons, the melting point of rubidium. Explain why it is more reactive than potassium, and describe how it would react with water. Include a balanced equation and an ionic equation in your answer.

A*

# 8: Flame tests

## Learning objectives

After studying this topic, you should be able to:

- describe how to carry out flame tests
- interpret information from flame tests

## Key words

**flame test**

## Flame colours

Have you ever wondered what gives fireworks their different colours? Various metal compounds are added to the explosive mixture in the firework. The very high temperatures produced when it burns or explodes vaporises the metal ions in the metal compounds. They absorb heat energy, causing their electrons to jump into higher shells. When the electrons drop back into their normal shells, light energy is released.

Different metal ions release different coloured light. For example, sodium ions release orange light. This is why street lamps produce an orange glow, although they use electrical energy instead of heat energy.

Sodium vapour street lamps produce an orange light

## Flame tests

Different metal ions can be identified by the colours they produce in a **flame test**. The table shows the colours produced by lithium, sodium, and potassium.

| Group 1 element | Symbol of element | Symbol of ion | Flame test colour |
|---|---|---|---|
| lithium | Li | $Li^+$ | red |
| sodium | Na | $Na^+$ | orange |
| potassium | K | $K^+$ | lilac |

## Exam tip OCR

- You need to know the flame test colours for lithium, sodium, and potassium.

It is easy to carry out a flame test, but care is needed to get clear results. It is important to wear eye protection.

The test is carried out using a flame test wire. This is usually a loop of nichrome alloy wire attached to a metal handle. The loop is cleaned by dipping it into hydrochloric acid or nitric acid, then holding it in the blue flame of a Bunsen burner. This is repeated until the flame colour does not change.

The loop is then dipped into the acid to moisten it, and then into the solid sample. The loop with its sample is held in the edge of the blue flame, and the colour observed.

## Did you know...?

Caesium was discovered in 1860 by the colour of the flame its compounds produce. It was named after the Latin word for 'heavenly blue' in recognition of its flame test colour. Robert Bunsen and Gustav Kirchhoff used their newly designed burner to make the discovery. They discovered rubidium, named after the Latin word for 'deep red', in 1861.

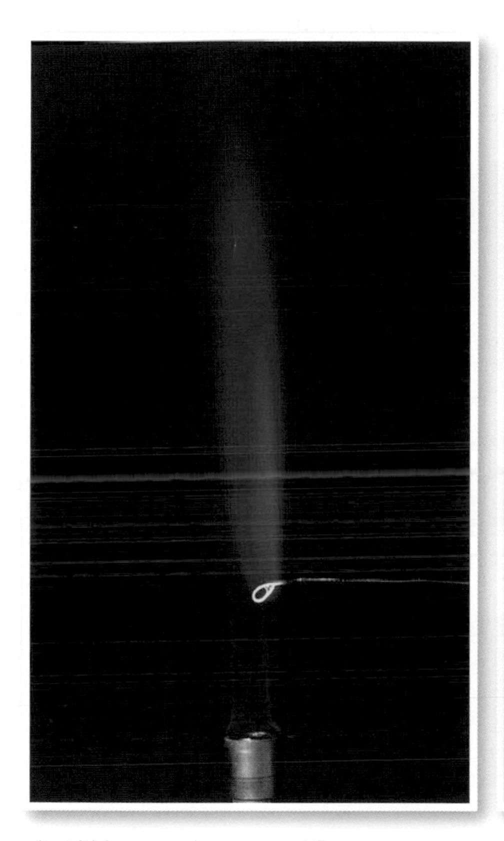

▲ Lithium produces a red flame

▲ Sodium produces an orange flame

▲ Potassium produces a lilac flame

## Questions

1 State the flame test colours for lithium, sodium, and potassium.

2 A white solid produces a lilac flame in a flame test. What metal does the compound contain?

E

3 Describe how to carry out a flame test.

4 A white solid is known to be a chloride. It produces a red colour in a flame test. Name the white solid and give its chemical formula.

C

# 9: The Group 7 elements

## Learning objectives

After studying this topic, you should be able to:

- recall the appearance and uses of chlorine, bromine, and iodine
- describe the reactions of Group 7 elements with Group 1 elements
- predict the properties of fluorine and astatine

## Key words

**Group 7, halogen, metal halide**

| 6 | 7 | 0 |
|---|---|---|
| | | 4 He helium 2 |
| 16 O oxygen 8 | 19 F fluorine 9 | 20 Ne neon 10 |
| 32 S sulfur 16 | 35.5 Cl chlorine 17 | 40 Ar argon 18 |
| 79 Se selenium 34 | 80 Br bromine 35 | 84 Kr krypton 36 |
| 128 Te tellurium 52 | 127 I iodine 53 | 131 Xe xenon 54 |
| [209] Po polonium 84 | [210] At astatine 85 | [222] Rn radon 86 |

Group 7 is between Groups 6 and 0 on the right of the periodic table

The halogens have very different appearances at room temperature

## Group 7

The **Group 7** elements are placed on the right of the periodic table. They are called the **halogens**, which means 'salt formers', because they react with metals to make salts.

Fluorine, chlorine, bromine, and iodine are all Group 7 elements.

Chlorine is used to sterilise tap water and swimming pool water. It kills harmful bacteria, making the water safe to drink or to swim in.

Chlorine is also used to make pesticides and plastics such as PVC, polyvinyl chloride. Iodine is used as an antiseptic to sterilise wounds.

Chlorine is used to sterilise swimming pool water

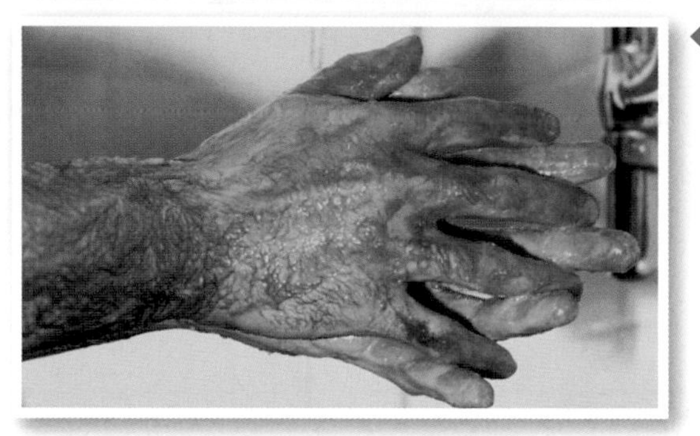

This surgeon is scrubbing up using an iodine-based antiseptic

## Coloured elements

The Group 7 elements are coloured and they occur in different states at room temperature.

- Chlorine is a green gas.
- Bromine is an orange liquid.
- Iodine is a grey solid.

When iodine is warmed, it changes easily into a purple vapour.

## Predicting properties

The melting points and boiling points of the Group 7 elements change in a regular way as you go down the group. It is possible to predict the physical properties of fluorine at the top of the group, and astatine at the bottom of the group, using the properties of the other Group 7 elements.

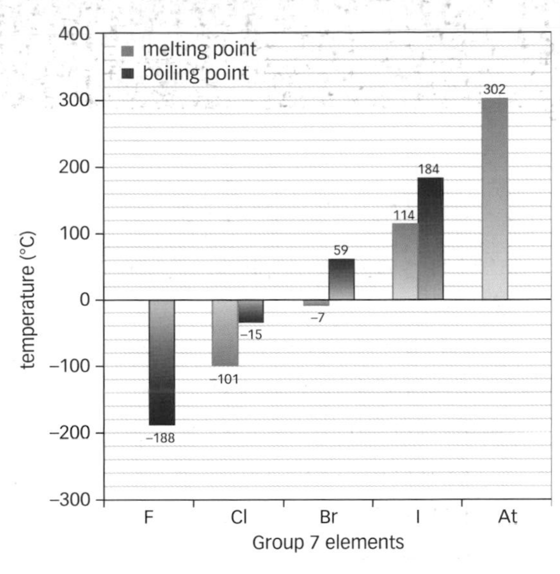


▲ The melting points and boiling points of the Group 7 elements

**A** Predict the approximate melting point of fluorine.

## Reactions with metals

The Group 7 elements react vigorously with the Group 1 elements, particularly if the metal is heated first. Clouds of a **metal halide** are produced. Fluorine produces a metal fluoride, chlorine produces a metal chloride, bromine produces a metal bromide, and iodine produces a metal iodide. The metal halide produced depends on the metal used, too. For example:

sodium + bromine → sodium bromide

The Group 7 elements have similar properties because the outer shells of their atoms all contain seven electrons.

◀ Potassium reacts vigorously with chlorine to make potassium chloride

## Balanced equations

Here is a general balanced symbol equation for the reaction of a Group 1 element, M, with a Group 7 element, $X_2$:

$$2M + X_2 \rightarrow 2MX$$

For example, lithium reacts with iodine to make lithium iodide, LiI:

$$2Li + I_2 \rightarrow 2LiI$$

## Questions

1 State a use for chlorine and a use for iodine.

2 Write the word equation for the reaction between potassium and iodine to make potassium iodide.

3 Write the word equation for the reaction between lithium and chlorine.

C

4 Use the graph to help you answer these questions:

(a) predict the boiling point of astatine

(b) at room temperature, what states will astatine and fluorine be in?

5 Write the balanced symbol equation for the reaction of potassium with chlorine.

# 10: Displacement reactions

## Learning objectives

After studying this topic, you should be able to:

- recall the reactivity of the Group 7 elements
- describe displacement reactions involving Group 7 elements
- explain why the Group 7 elements have similar properties
- predict the reactions of fluorine, astatine, and their compounds

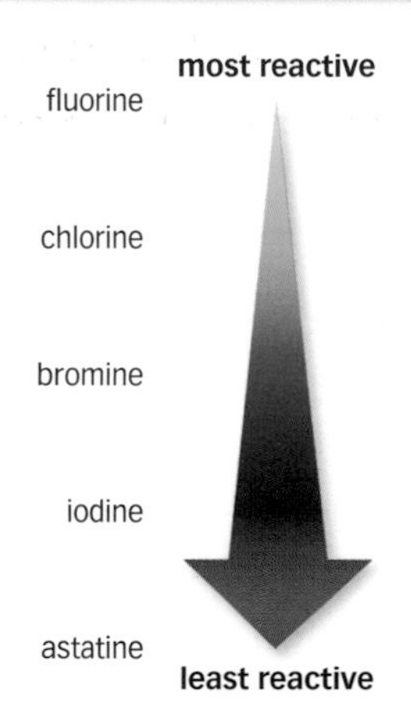


The reactivity of the Group 7 elements decreases down the group

Brown iodine is produced when chlorine is bubbled through potassium iodide solution

## Decreasing reactivity

The alkali metals in Group 1 become more reactive as you go down the group. On the other hand, the Group 7 elements (which each have seven electrons in their outer shell) become less reactive as you go down the group. This means that:

- Fluorine is more reactive than chlorine.
- Chlorine is more reactive than bromine.
- Bromine is more reactive than iodine.

## Displacement reactions

Group 7 elements can react with metal halides in solution. For example, chlorine reacts with sodium bromide solution:

chlorine + sodium bromide → bromine + sodium chloride

The chlorine 'pushes out' or **displaces** bromine from the sodium bromide solution. Notice that sodium seems to have 'swapped places'.

Reactions like this one are called **displacement reactions**. A more reactive Group 7 element will displace a less reactive Group 7 element from its metal halide. This means that:

- Chlorine will displace bromine from metal bromides, and iodine from metal iodides.
- Bromine will displace iodine from metal iodides.

The reaction will not work the other way around. For example, no reaction is seen if bromine is added to sodium chloride solution.

It does not matter which Group 1 element is involved. For example, bromine is still produced whether chlorine is bubbled through lithium bromide solution or through potassium bromide solution.

### Predicting reactions

Fluorine is the most reactive element in Group 7. It can displace all the other Group 7 elements from their compounds. On the other hand, astatine is the least reactive element in Group 7. It cannot displace any of the other Group 7 elements from their compounds. All the other Group 7 elements will be able to displace astatine. For example:

bromine + sodium astatide → astatine + sodium bromide

## Balanced equations

Here is a general balanced symbol equation for displacement reactions involving Group 7 elements. The more reactive Group 7 element is represented as $X_2$ and the metal halide as MY:

$$X_2 + 2MY \rightarrow Y_2 + 2MX$$

For example, chlorine reacts with potassium iodide solution, making iodine and potassium chloride:

$$Cl_2 + 2KI \rightarrow I_2 + 2KCl$$

## Gaining an electron

The Group 7 elements have similar chemical properties because, when they react, each atom gains an electron to become a negative ion. This has the stable electronic structure of a noble gas. The gain of electrons is **reduction**. Here is a general ionic equation for such a reduction, where $X_2$ stands for the symbol of the Group 7 element:

$$X_2 + 2e^- \rightarrow 2X^-$$

The easier it is for an atom of a Group 7 element to gain one electron, the more reactive the element is. For example, fluorine is more reactive than iodine because its atoms gain electrons more easily than iodine atoms do.

**A** Which is more reactive, iodine or fluorine?

**B** How can you tell that bromine is produced when chlorine is bubbled through sodium bromide solution?

**C** Bromine reacts with lithium iodide solution to form iodine and lithium bromide. Write the word equation for the reaction.

**D** Predict, with reasons, whether astatine will react with sodium fluoride solution.

### Questions

1 How does the reactivity of the Group 7 elements change as you go down the group?

2 Write the word equation for the reaction between chlorine and lithium iodide, forming iodine and lithium chloride.

E

3 What is a displacement reaction?

4 Write the word equation for the displacement reaction between bromine and sodium iodide.

C

5 In terms of electrons, what is reduction?

6 Write the balanced symbol equation for the reaction between chlorine, $Cl_2$, and lithium iodide, LiI.

A*

### Key words

**displace, displacement reaction, reduction**

# 11: The transition elements

## Learning objectives

After studying this topic, you should be able to:

- find the transition elements in the periodic table
- recall that compounds of transition elements are often coloured
- recall that transition elements and their compounds are often used as catalysts
- describe the thermal decomposition of carbonates of transition elements

A Give the names and symbols of two transition elements in the same period as copper and iron.

## Key words

**transition element, catalyst, thermal decomposition reaction**

Nickel is used as a catalyst in the manufacture of margarine

## Transition elements

Copper and iron are examples of **transition elements**. These elements are placed between Groups 2 and 3 in the periodic table. They are all metals and their properties are typical of metals. For example, in general they are shiny when cut, strong, and malleable (they can be bent or hammered into shape).

| 1 | 2 | | | | | | | | | | | 3 | 4 | 5 | 6 | 7 | 0 |
|---|---|---|---|---|---|---|---|---|---|---|---|---|---|---|---|---|---|
| | | | | | | | H | | | | | | | | | | He |
| Li | Be | | | | | | | | | | | B | C | N | O | F | Ne |
| Na | Mg | | | | | | | | | | | Al | Si | P | S | Cl | Ar |
| K | Ca | Sc | Ti | V | Cr | Mn | Fe | Co | Ni | Cu | Zn | Ga | Ge | As | Se | Br | Kr |
| Rb | Sr | Y | Zr | Nb | Mo | Tc | Ru | Rh | Pd | Ag | Cd | In | Sn | Sb | Te | I | Xe |
| Cs | Ba | La | Hf | Ta | W | Re | Os | Ir | Pt | Au | Hg | Tl | Pb | Bi | Po | At | Rn |
| Fr | Ra | Ac | | | | | | | | | | | | | | | |

The transition elements in the periodic table, shown here in yellow

## Coloured compounds and catalysts

The transition elements often form coloured compounds. For example:

- Copper compounds such as copper sulfate solution are often blue.
- Iron(II) compounds such as iron(II) sulfate solution are often light green.
- Iron(III) compounds such as iron(III) chloride solution are often orange-brown.

Compounds of transition elements are often coloured. Sodium is not a transition element, so sodium sulfate solution is colourless.

The transition elements and their compounds are often used as **catalysts**. For example, iron is the catalyst used to increase the rate of reaction between nitrogen and hydrogen in the Haber process, which makes ammonia.

## Thermal decomposition reactions

In many reactions, two reactants combine to form one or more products. In a **thermal decomposition reaction**, one substance breaks down to form two or more other substances when heated. Certain carbonates of transition elements decompose when heated. Here is a general word equation for the reaction:

metal carbonate → metal oxide + carbon dioxide

For example, copper carbonate easily breaks down when heated:

copper carbonate → copper oxide + carbon dioxide

$$CuCO_3 \rightarrow CuO + CO_2$$

Other carbonates decompose in a similar way, including iron(II) carbonate, $FeCO_3$, manganese(II) carbonate, $MnCO_3$, and zinc carbonate, $ZnCO_3$. A colour change is usually seen during the reaction. The carbon dioxide produced can be detected using a simple laboratory test. It turns limewater milky.

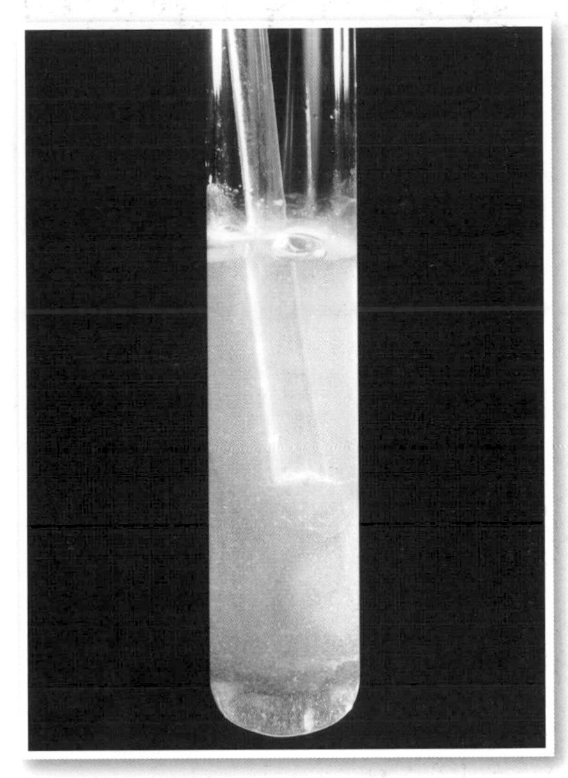

▲ Limewater turns milky when carbon dioxide is bubbled through it

### Exam tip

✔ You should be able to recall the typical colours of copper compounds and of iron(II) and iron(III) compounds.

### Questions

1. Use the periodic table to find the symbols for these transition elements: cobalt, gold, tungsten, mercury.
2. State two typical properties of the transition elements.

E

3. Suggest how you might tell iron(II) chloride and iron(III) chloride apart.
4. Describe two uses of transition elements as catalysts.
5. Write the word equation for the thermal decomposition of zinc carbonate, and explain how you would detect carbon dioxide.

C

6. Write balanced symbol equations for the thermal decomposition of manganese(II) carbonate, $MnCO_3$, and iron(II) carbonate, $FeCO_3$.

A*

## Learning objectives

After studying this topic, you should be able to:

- explain what a precipitation reaction is
- describe how sodium hydroxide solution is used to identify transition metal ions

## Key words

**precipitation reaction, precipitate**

## Exam tip OCR

- You need to be able to suggest ways to limit risks, and you should recognise the benefits of activities that have a known risk.

## Identifying metal ions

Forensic scientists analyse samples collected at crime scenes. They use many different laboratory tests involving complex equipment and chemical reactions. The scientist may want to see whether or not a particular substance is present in a sample. **Precipitation reactions** are one way to identify transition metal ions in solution.

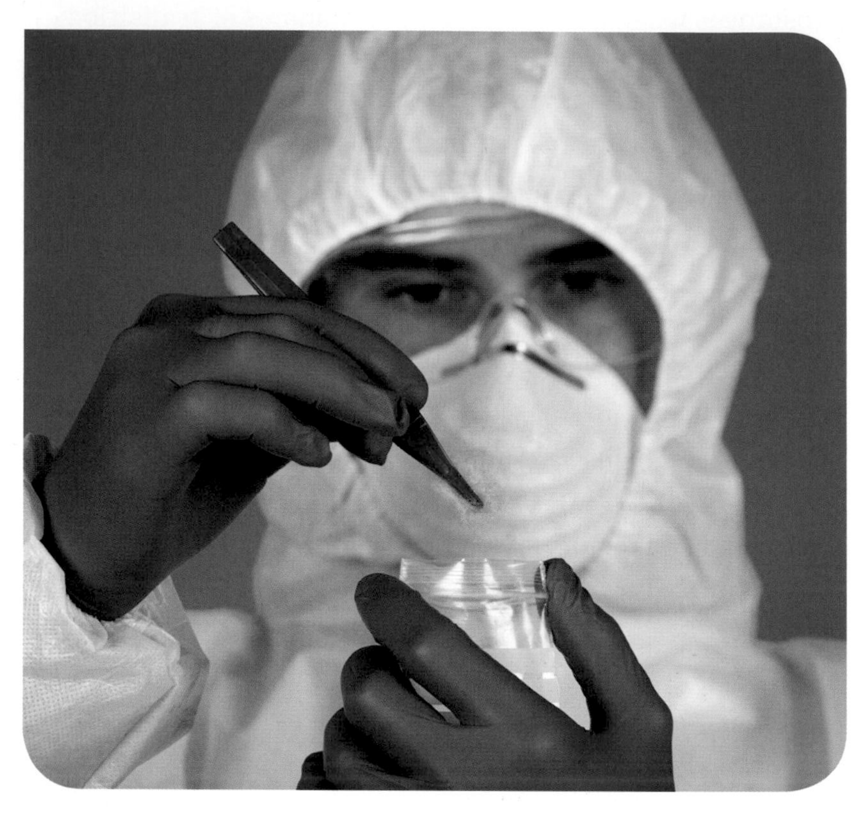

▲ This forensic scientist is collecting a sample of fibre so that it can be tested. The mask, gloves, and protective clothing stop the fibre sample being contaminated by hair and fibres from the scientist. They also protect him from any harmful substances that might be in the sample.

A precipitation reaction happens when a mixture of solutions reacts to make an insoluble solid. The insoluble solid is called the **precipitate**. It makes the mixture cloudy. Different metal ions give different coloured precipitates. This is how transition metal ions can be identified.

**A** Give one use of precipitation reactions in chemistry.
**B** Is a precipitate soluble or insoluble in water?
**C** What is a precipitation reaction?

## Hydroxide precipitates

When a compound of a transition element is dissolved in water, its metal ions separate out. These metal ions react with hydroxide ions to form coloured metal hydroxide precipitates. The table shows the typical colours seen.

| Name of metal ion | Formula of metal ion | Colour of metal hydroxide precipitate |
|---|---|---|
| copper(II) | $Cu^{2+}$ | blue |
| iron(II) | $Fe^{2+}$ | grey-green |
| iron(III) | $Fe^{3+}$ | orange-brown |

▲ Copper(II) ions form a jelly-like blue precipitate with sodium hydroxide solution

▲ Iron(II) ions form a grey-green precipitate with sodium hydroxide solution, and iron(III) ions form an orange-brown precipitate

### Balanced equations

Here is the balanced symbol equation for the reaction between copper(II) ions and hydroxide ions, $OH^-$:

$$Cu^{2+} + 2OH^- \rightarrow Cu(OH)_2$$

Notice that two $OH^-$ ions are needed to balance the charge on the copper(II) ion.

**D** What is the difference between the formula of the iron(II) ion and the iron(III) ion?

### Exam tip OCR

✔ You must be able to recall the colours of the three metal hydroxide precipitates.

### Questions

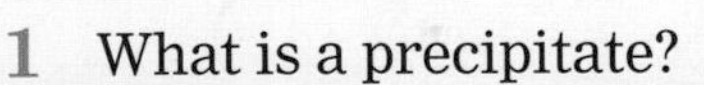

1 What is a precipitate?

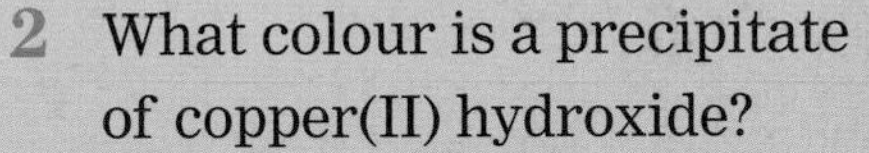

2 What colour is a precipitate of copper(II) hydroxide?

3 A forensic scientist has two unlabelled samples. One is iron(II) nitrate solution and the other is iron(III) nitrate solution. Explain how they could tell the two apart using a few drops of sodium hydroxide solution.

4 Write the balanced symbol equations for the reaction of iron(II) ions and iron(III) ions with hydroxide ions.

E ↓ C ↓ A*

## Learning objectives

After studying this topic, you should be able to:

- explain some uses of iron and copper
- describe the properties of metals
- link the properties of a metal to its uses
- explain the high melting and boiling points of metals
- describe metallic bonding

**A** Why are electrical wires made from copper?

Mercury is an unusual metal. Its melting point is –39 °C, so it is liquid at room temperature.

## Metals

Most elements are metals rather than non-metals. For example, copper and iron are metals. Copper is a good conductor of electricity. It is used to make brass and electrical wiring. Iron is used to make steel. This is used to make cars and bridges because it is strong.

The Millennium Bridge in Gateshead is made from 800 tonnes of steel. It is nicknamed the Blinking Eye because of the way it tilts to let ships through.

In general, metals are

- hard
- shiny or lustrous
- good conductors of heat and electricity.

Many metals have a high **tensile strength**, so they resist being stretched. They also have high melting points and boiling points.

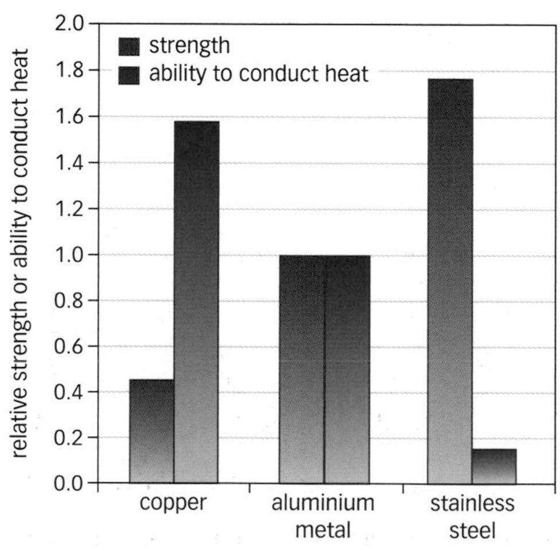


The relative strength of three metals, and their ability to conduct heat

B What does it mean if a metal has a high tensile strength?

## Metallic bonding

Ionic compounds are held together by ionic bonds and covalent compounds are held together by covalent bonds. Metals are held together by **metallic bonds**. These bonds are strong, so metals have high melting points and boiling points.

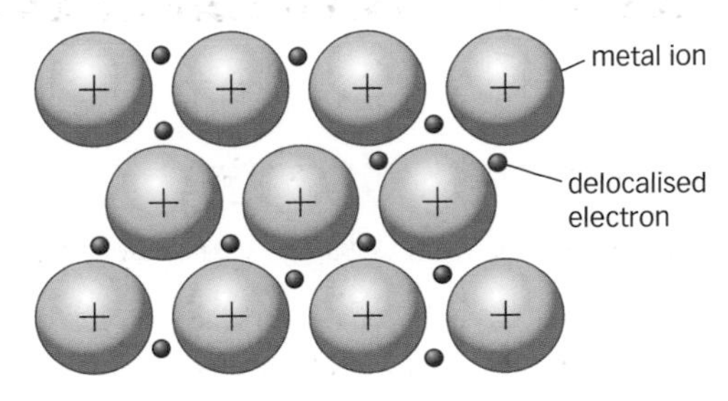


▲ Metallic bonding

**Key words**

**tensile strength, metallic bond**

### Ions and electrons

Metals contain positive metal ions packed closely together. These form when electrons leave the outer shell of each atom. The electrons become free to move within the structure of the metal. They form a 'sea' of delocalised electrons. Metallic bonding is the strong force of attraction between the sea of delocalised electrons and the closely packed positive metal ions. These forces are strong. It takes a lot of energy to overcome them, which is why metals have high melting points and boiling points.

C In general, are metallic bonds strong or weak?

D What are electrons that are free to move within the structure of a metal called?

**Exam tip** OCR

- ✔ In the exam, you need to be able to explain why a metal is suited to a particular use. You will usually be given data to analyse.
- ✔ Metals that are good conductors of heat are also good conductors of electricity, but make the property fit the use. For example, copper is used in wiring because it is a good conductor of electricity, not because it is a good conductor of heat.

### Questions

1 Why is steel used to make cars and bridges?

2 Name the bonds found in metals.

3 Use data from the graph to explain which metal (a) conducts heat the best, and (b) is the strongest.

E

4 Use data from the graph to explain why:

(a) Saucepans may be made from stainless steel with a copper base.

(b) Computer processors are cooled by blowing air over thin metal fins. Explain why copper may be used instead of aluminium, even though it is more expensive.

C

5 Describe metallic bonding, and explain why metals have high melting points and boiling points.

A*

## Learning objectives

After studying this topic, you should be able to:

- recall that some metals are superconductors at low temperatures
- describe the potential benefits of superconductors
- explain the drawbacks of superconductors

## Key words

**crystal, resistance, superconductor**

## Exam tip OCR

- You need to be able to identify some arguments for and against a scientific or technological development. Consider the impact of the development on different groups of people or on the environment. Recent developments in superconductors may create jobs in new industries but may also cause the loss of jobs in traditional industries.

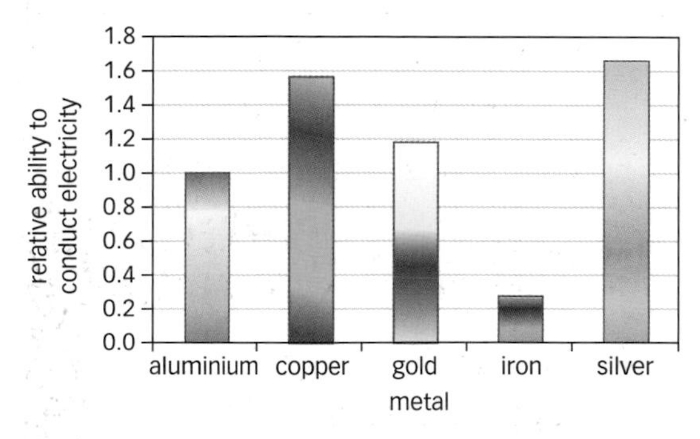


▲ Some metals conduct electricity more easily than others

## Metal crystals

The photograph may look like a bacterium seen through a microscope. However, it actually shows silver **crystals** growing on a piece of copper. Metals have a structure that contains crystals. This is because the particles in solid metals are packed closely together in a regular arrangement. This arrangement is repeated many times to produce crystals.

▲ Bacterium or crystals?

The metal crystals contain delocalised electrons. These are able to move easily through the structure. Metals conduct electricity when these electrons move from place to place. However, their path through the metal is not clear. Other electrons and the positive metal ions get in the way. This causes electrical **resistance**. Resistance makes metals heat up when they conduct electricity, wasting energy.

**A** Why do metal atoms form crystals?
**B** Why do metals have electrical resistance?
**C** Which metal is the best at conducting electricity?

## Superconductivity

The resistance of metals decreases as the temperature decreases. Some metals become **superconductors** at low temperatures. These are materials with little resistance to the flow of electricity, or even no resistance at all.

Superconductors are already being used and there are many exciting ideas for how they might be used in the future. For example, superconductors are used to make very powerful electromagnets. These are used in MRI scanners, allowing doctors to get detailed images of the inside of the body.

Superconductors could be used to make super-fast electronic circuits, leading to extremely powerful computers. Superconductors could also be used to replace the traditional metal cables that carry electricity from power stations to our homes, offices, and factories. There would be little or no loss of energy from superconducting cables, as they would have little or no resistance.

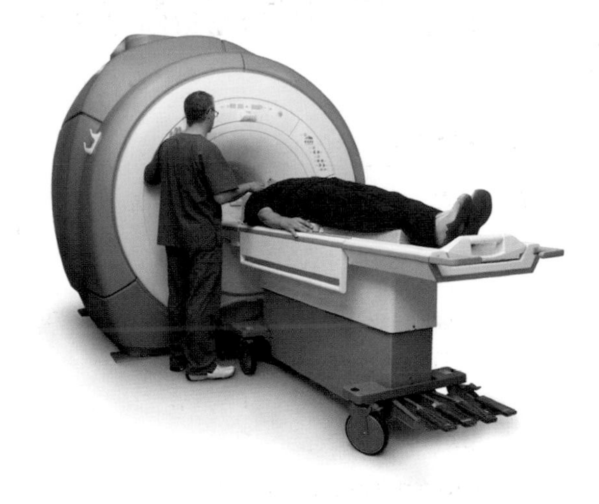

▲ MRI scanners use superconducting electromagnetics

**D** Why do doctors use MRI scanners on their patients?

### Drawbacks of superconductors

At the moment, superconductors only work at very low temperatures. For example, aluminium only becomes a superconductor with no resistance at –272 °C. This is only about 1 °C above the lowest possible temperature anywhere in the universe. The need for very low temperatures means that the applications of superconductors are limited. Scientists are researching many different materials to see if they can become superconducting at 20 °C, which is around room temperature.

### Did you know...?

Early in 2010, scientists made a superconducting substance with the chemical formula $(Tl_4Cd)Ba_2Ca_2Cu_7O_{13}$. The substance became a superconductor at –19 °C, about the temperature of a home freezer. It broke the record for the highest temperature superconductor.

▲ This experimental Japanese 'maglev' train floats 10 cm above its track using superconducting magnets. It has reached speeds above 580 km per hour.

### Questions

1 How are the particles arranged in a metal?

2 What is a superconductor?

3 Describe three potential benefits of superconductors.

4 Explain the drawbacks of superconductors.

# 15: Purifying water

## Learning objectives

After studying this topic, you should be able to:

- ✔ recall the different types of water resources in the UK
- ✔ recall some industrial uses of water, and explain why water should be conserved
- ✔ recall some pollutants and other substances that may be present in water before it is purified
- ✔ describe how water is purified
- ✔ explain the processes involved in purifying water

## Key words

**aquifer, pollutant, chlorination, sedimentation, filtration, distillation**

▲ Recycling materials is one way to conserve water

## Water as an industrial resource

It is easy to take water for granted. Turn the tap on and out it comes, purified and safe to drink. Originally, the water will have come from a reservoir or other resources, including lakes and rivers. It may also have come from an underground **aquifer**. Water from these resources contains many different dissolved substances. Many of these must be removed to make the water safe to drink or to use in industry.

▲ Reservoirs store large volumes of fresh water

Water is an important resource for many industrial chemical processes. It may be used as

- a solvent to dissolve other substances
- a coolant to stop overheating happening
- a cheap raw material.

Fresh water is a limited resource. About 97% of the water on the planet is salt water, so it is important to conserve fresh water. Industry uses huge amounts of it. For example, nearly 40 000 litres of water are needed to make a tonne of plastic.

## Substances in water

Water contains many different types of substance. These include insoluble materials such as leaves, and dissolved salts and minerals. It may also contain microbes and various **pollutants**. Common pollutants include

- nitrates from fertiliser that runs off fields
- lead compounds from old lead pipes
- pesticides from spraying crops in fields close to water resources.

Water must be purified before it can be used. Particles in the water are removed by means of sedimentation and filtration through beds of clean sand and gravel. Chlorine gas is then added to kill microbes in the water. This process is called **chlorination**.

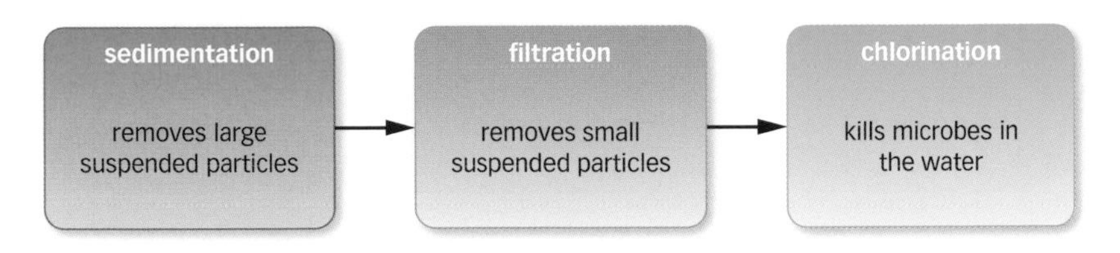


▲ The processes involved in water purification

**C** Why is water chlorinated?

### Water purification

Sedimentation allows large particles such as sand and soil to settle to the bottom over time. Filtration then removes the finer particles, including clay. Chlorination kills microbes in the water, helping to prevent disease. However, some soluble substances are not removed during purification. These may be poisonous, so strict limits are placed on how much of them can be in the water.

Very pure water can be produced from sea water by **distillation**. However, a lot of energy is needed to do this. It is usually too expensive to make large amounts of fresh water in this way.

**A** What is a reservoir?

**B** It has been estimated that a person living on their own in the UK uses 165 litres of water per day on average. However, a family of three uses 450 litres per day on average. Suggest why people living in a family group each use less water than a person living on their own.

### Exam tip OCR

✓ You must be able to interpret data about water resources in the UK, but you are not expected to recall any data about these.

### Questions

1. Name four types of water resources found in the UK.
2. State three uses of water in industrial chemical processes.

3. Explain where pollutants like nitrate, lead, and pesticides come from.
4. Name the three main processes involved in water purification.

5. Explain the three main processes involved in water purification.
6. Explain the drawback of distilling sea water in large amounts.

# 16: Testing water

## Learning objectives

After studying this topic, you should be able to:

- recall the tests for sulfate ions and halide ions in water
- write word equations for precipitation reactions
- write balanced symbol equations for precipitation reactions

A white precipitate of barium sulfate forms when barium chloride is added to water containing dissolved sulfate ions

## Testing water

Water contains several different types of substance before it is purified. These include insoluble materials, microbes, pollutants, and dissolved salts and minerals. The presence of some dissolved ions can be detected using simple tests.

Water arriving at a waterworks is tested regularly

## Testing for sulfate ions

Barium chloride solution is used to test for dissolved sulfate ions. A small sample of the water is put into a test tube, then a few drops of barium chloride solution are added. If sulfate ions are present, a white precipitate of barium sulfate forms. For example, this precipitation reaction happens if the water contains dissolved sodium sulfate:

sodium sulfate + barium chloride → sodium chloride + barium sulfate (white precipitate)

### Balanced symbol equations

The barium sulfate precipitate in the sulfate test is formed by the reaction between the barium ions from the barium chloride solution, and the sulfate ions from the sample. The other product depends upon the other ions present. For example, here is the balanced symbol equation for testing sodium sulfate solution with barium chloride solution:

$$Na_2SO_4 + BaCl_2 \rightarrow 2NaCl + BaSO_4$$

## Testing for halide ions

Silver nitrate solution is used to test for dissolved halide ions. These include chloride ions, bromide ions, and iodide ions. A small sample of the water is put into a test tube, then a few drops of silver nitrate solution are added. Different coloured precipitates are formed:

- chloride ions produce a white precipitate
- bromide ions produce a cream-coloured precipitate
- iodide ions produce a pale yellow precipitate.

For example, this precipitation reaction happens if the water contains dissolved sodium chloride:

sodium chloride + silver nitrate → sodium nitrate + silver chloride (white precipitate)

The precipitates formed when silver nitrate solution is added to solutions of halide ions

### Balanced symbol equations

The silver halide precipitate in the halide test is formed by the reaction between the silver ions from the silver nitrate solution and the halide ions from the sample. The other product depends upon the other ions present. For example, here is the balanced symbol equation for testing sodium chloride solution with silver nitrate solution:

$$NaCl + AgNO_3 \rightarrow NaNO_3 + AgCl$$

**A** Suggest why the water arriving at a waterworks should be tested.

**B** What do barium sulfate and silver chloride precipitates have in common?

### Questions

1. Name the solution used to test for sulfate ions, and give the colour of the precipitate formed if they are present.
2. Name the solution used to test for halide ions, and give the colours of the precipitates formed if they are present.

E

3. A sample of water is tested. A white precipitate forms when barium chloride solution is added, and also when silver nitrate solution is added. Identify the ions detected in these tests.

C

4. Write the word equation for the reaction between potassium sulfate solution and barium chloride solution.
5. Write the symbol equation for the reaction between potassium iodide solution, KI, and silver nitrate solution, $AgNO_3$.

A*

# Module summary

## Revision checklist

- The nucleus of an atom contains protons (positive) and neutrons (uncharged).
- Electrons have a negative charge and are very light.
- Mass number is the total number of protons and neutrons in the nucleus of an atom. Atomic number is the number of protons in the nucleus of an atom. Isotopes have the same atomic number but different mass numbers.
- Ionic bonds form when electrons are transferred from one atom to another.
- Positive and negative ions attract each other, forming giant ionic lattices.
- Ionic compounds have high melting points and conduct electricity when molten or in solution.
- Development of ideas about the structure of atoms and the periodic table shows how scientific theories change when new evidence is gathered.
- Covalent bonds are formed when atoms share pairs of electrons, forming molecules.
- Compounds with a simple molecular structure have low melting and boiling points and do not conduct electricity.
- The properties of substances can be predicted and explained using ideas about structure and bonding.
- Ions in ionic compounds and atoms in molecules have stable electronic structures (full outer shells).
- Alkali metals (Group 1) react vigorously with water to form hydrogen; they become more reactive going down the group.
- Halogens (Group 7) are coloured elements with low melting and boiling points. Halogens at the top of the group are reactive and displace less reactive halogens from metal halides.
- Transition metals have coloured compounds and are useful as catalysts in industry. Transition metal ions are identified by precipitation reactions with hydroxide ions.
- In metallic bonding, closely-packed ions are surrounded by a sea of delocalised electrons.
- Metals have many uses because they are hard, shiny and conduct electricity. Some metals become superconductors at low temperatures.
- Water from lakes, rivers, aquifers, and reservoirs is purified to remove solids, dissolved salts, microbes, and pollutants.
- Precipitation reactions test for dissolved ions in water.

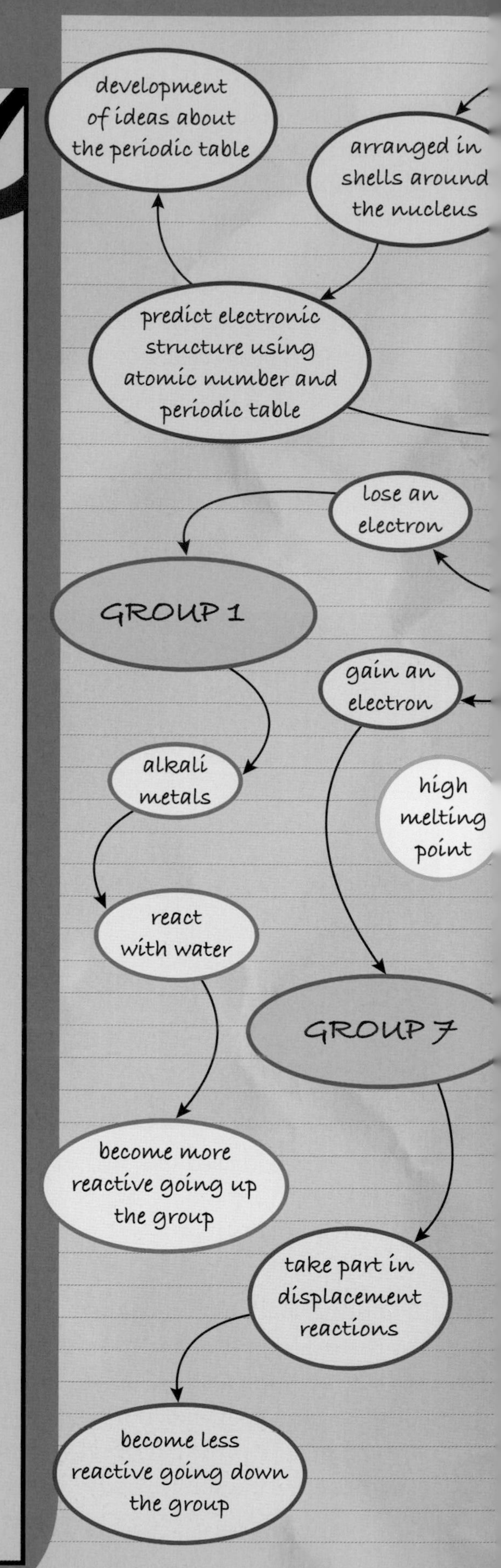




NOW USE THE C4 GRADE CHECKER ON PAGE 246

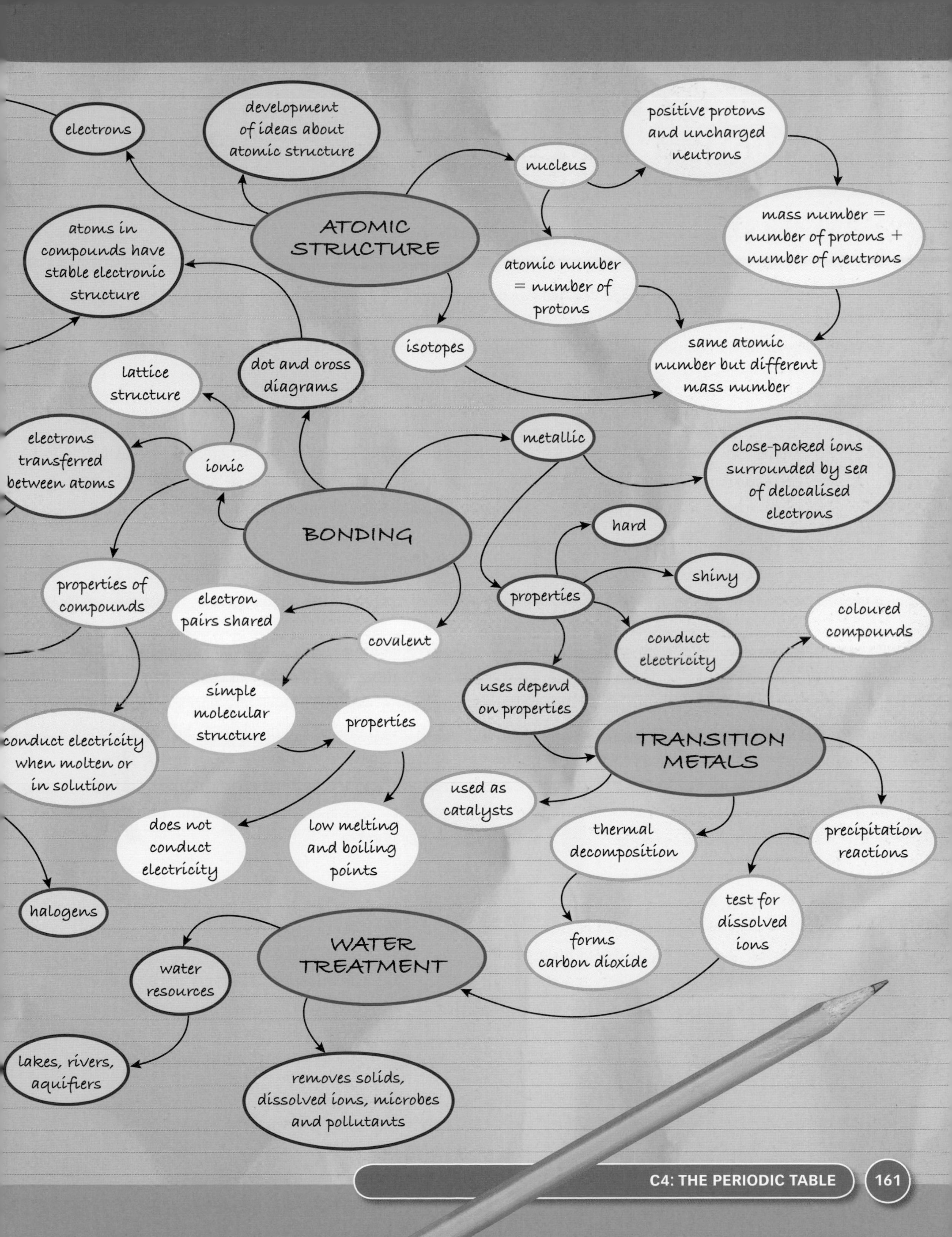
electrons
development of ideas about atomic structure
positive protons and uncharged neutrons
nucleus
atoms in compounds have stable electronic structure
ATOMIC STRUCTURE
mass number = number of protons + number of neutrons
atomic number = number of protons
isotopes
same atomic number but different mass number
lattice structure
dot and cross diagrams
electrons transferred between atoms
ionic
metallic
close-packed ions surrounded by sea of delocalised electrons
BONDING
hard
shiny
properties of compounds
electron pairs shared
properties
covalent
coloured compounds
conduct electricity
uses depend on properties
simple molecular structure
properties
conduct electricity when molten or in solution
TRANSITION METALS
used as catalysts
does not conduct electricity
low melting and boiling points
thermal decomposition
precipitation reactions
halogens
test for dissolved ions
WATER TREATMENT
forms carbon dioxide
water resources
lakes, rivers, aquifers
removes solids, dissolved ions, microbes and pollutants

OCR gateway Upgrade

## Answering Extended Writing questions

**QUESTION**

Metals have several important properties that can be explained by their structure and bonding. Describe the structure and bonding in a metal, and use this to explain the important properties of metals.

**The quality of written communication will be assessed in your answer to this question.**

Metals have metallic bonding. This is strong so metals are strong. They can be used in wires and bridges.

↓ E

*Examiner:* This answer includes the name of the type of bonding in metals. The structure of crystals is not described. The candidate knows that metals are strong, but could also mention high melting point and conducting electricity. The question doesn't ask about uses of metals – the final sentence is irrelevant and scores no marks.

Metals are strong, have high melting points and are conductors. Atoms are arranged close together and regularly. Metallic bonds between the atoms are strong which means metals are strong. There are electrons in the structure so metals conduct.

↓ C

*Examiner:* This candidate has made several good points. The properties of metals are listed, but the answer should indicate that metals are conductors of electricity and heat). The fact that strong metallic bonds also explain the high melting points of metals could also be mentioned. The answer mentions electrons, but should explain that they move when an electric current flows. Spelling, punctuation, and grammar are good.

In metals, the atoms are arranged regularly and closely packed. They are surrounded by a sea of delocalised electrons. The force between the atoms and the electrons is called metallic bonding. This is strong so the metal has high tensyl strength and high melting point. The delocalised electrons can move around so it conducts electricity.

*Examiner:* A good, well-structured answer that deals with all the points in the question. The only error is that metallic bonding is between electrons and the closely-packed positive ions (not atoms). There is one minor spelling mistake.

# Exam-style questions

1 Here is a diagram showing the structure of an atom.

A B

A01 a Name **A** and **B**.

A02 b This atom has an atomic number of 2. Use the periodic table to find the symbol of the element.

A01 c Complete this sentence by choosing the correct word from the list below:

Substances that contain only one type of atom are called _______.

elements compounds ions isotopes

E

2 a Which two of the following statements about ionic bonding are true?

A01

i Electron pairs are shared.

ii Molecules are formed.

iii Positive and negative ions attract each other.

iv Occurs between metals and non-metals.

A01 b Lead bromide is an ionic compound. Complete this sentence to predict some properties of lead bromide. Choose words from the list below.

Lead bromide conducts electricity when it is _______ or _______.

solid liquid gas in solution

3 This question is about Group 7 elements (the halogens).

A02 A03 a When chlorine is added to a solution of potassium bromide, an orange solution is formed.

i Name the orange solution.

ii What type of reaction is this?

C

iii Copy and complete this word equation for the reaction:

chlorine + potassium bromide → ____________

C

iv Explain what you can deduce about the reactivities of chlorine and bromine from the results of the experiment.

A01 b Halogen elements react by gaining electrons to form a halide ion.

i Complete this equation to show how chlorine forms chloride ions: $Cl_2$ + __$e^-$ → ___$Cl^-$

ii What name is given to reactions in which electrons are gained?

iii Halide ions (such as $Cl^-$) are more stable than halogen atoms. Explain why.

A*

## Extended Writing

4 Metals are very useful. What properties A01 do metals have, and what uses do they have because of these?

E

5 Analytical chemists often have to A01 analyse samples of unknown compounds to find out what elements are present. Describe how they use flame tests to do this.

C

6 Look at the diagram below. It shows the A01 bonding in carbon disulfide.

A02

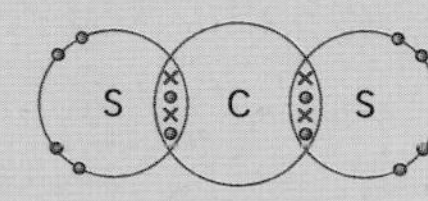


Describe the bonding in this molecule and use it to predict and explain two properties of carbon disulfide.

A*

A01 Recall the science

A02 Apply your knowledge

A03 Evaluate and analyse the evidence

# Quantitative analysis

| OCR specification match | | |
|---|---|---|
| C5a | 5.1 | Moles and molar mass |
| C5a | 5.2 | Reacting masses |
| C5b | 5.3 | Percentage composition |
| C5b | 5.4 | Empirical formulae |
| C5c | 5.5 | Concentration |
| C5c | 5.6 | Dilution and GDA |
| C5d | 5.7 | pH curves |
| C5d | 5.8 | Titrations |
| C5e | 5.9 | Measuring gases |
| C5e | 5.10 | Gas volumes |
| C5f | 5.11 | Equilibria |
| C5f | 5.12 | Making sulfuric acid |
| C5g | 5.13 | Strong and weak acids |
| C5g | 5.14 | Reactions of acids |
| C5h | 5.15 | Making insoluble compounds |
| C5h | 5.16 | Testing ions |
| Module summary | | |
| OCR Upgrade | | |

## Why study this module?

Most of the substances around you are likely to have been manufactured. Scientists and engineers must know how much of each raw material they need, and how much waste is likely to be made. There are many different ways to measure and calculate quantities in chemistry. Such quantitative analysis is useful to consumers like you, too.

In this module, you will investigate how to determine the formula of a compound. You will learn about ways to measure the concentration of acids and other solutions. Sulfuric acid is such an important industrial chemical that economists have often linked a country's economic success to its production of the acid. You will discover how it is made and how the reaction conditions are chosen to produce it economically. You will also find out why sulfuric acid is a strong acid but ethanoic acid is a weak acid.

Precipitation reactions are used in laboratory tests for metal ions and other ions. You will find out more about them and how they can produce useful insoluble salts.

### You should remember

1 How to measure the volume of gas produced in a reaction.

2 How to calculate the masses of the different substances involved in a reaction.

3 Food often contains many different ingredients, including additives.

4 Ammonia is made from nitrogen and hydrogen in a reversible reaction.

5 Acids and alkalis have different properties, and they can take part in neutralisation reactions.

6 Water can be tested using chemical reactions.

This amazing sight is not on an alien planet. Instead, it is a 1.5 m geyser formed by accident at Fly Ranch in Nevada, USA. A geothermal energy company drilled a test well in 1964 and reached a source of hot water. The water was not hot enough for their purposes, so they sealed the well. The seal did not hold, allowing hot water to reach the surface. Calcium carbonate and other minerals dissolved in the water precipitate out as it cools, forming these spectacular solid deposits.

# 1: Moles and molar mass

## Learning objectives

After studying this topic, you should be able to:

- recall the unit for amount of substance
- calculate the molar mass of a substance from its formula
- define the relative atomic mass of an element
- convert between moles, mass, and molar mass

## Key words

**mole, relative formula mass, relative atomic mass, molar mass**

▲ The amount of each of these different substances is one mole

## Did you know...?

If you could have 1 mole of £1 coins and could stack them on top of each other, they would reach from one side of our galaxy to the other. And then back again.

## Amount of substance

What do 18 g of water molecules have in common with 44 g of carbon dioxide molecules? The answer is that they both contain the same number of atoms and molecules. They each contain one mole of molecules, and they each contain three moles of atoms.

The **mole** is the unit for amount of substance. One mole of anything contains the same number of particles. There would be the same number of £1 coins in a mole of £1 coins as there would be carbon atoms in a mole of carbon atoms. That number is huge, around $6 \times 10^{23}$, or 6 followed by 23 zeros. In everyday life it would make no sense to use the mole, but it does for tiny particles like atoms and molecules.

## Mole calculations

No one could sit down to count out 1 mole of something. Even if they could count at the rate of one million objects per second, it would take longer than the age of the universe to complete the task. Fortunately, the mass of 1 mole of a substance is linked to its **relative formula mass**, $M_r$. To do this, you add together the **relative atomic masses**, $A_r$, for all the atoms in the formula for the substance.

### Defining relative atomic mass

Chemists have chosen the carbon-12 atom, $^{12}C$, as their standard atom. Its relative atomic mass is 12 exactly. The relative atomic mass of an element is the average mass of an atom of that element, compared with the mass of 1/12th of an atom of carbon-12. Chemists make it 1/12th the mass so that the mathematics is easier.

The mass of 1 mole of a substance is its relative formula mass in grams. This is called its **molar mass**. Its units are g/mol. The word 'mole' is abbreviated to 'mol' in units.

For example, the relative formula mass of water is 18. This means that 1 mole of water has a mass of 18 g – its molar mass is 18 g/mol. The relative formula mass of carbon dioxide is 44, so its molar mass is 44 g/mol. One mole of water contains the same number of molecules as one mole of carbon dioxide.

## More mole calculations

This equation links moles, mass, and molar mass:

number of moles = mass ÷ molar mass

It can be used to work out the number of moles of an element from the mass of that element, or the number of moles of a compound from the mass of that compound. For example, suppose you had 36 g of water. Its molar mass is 18 g/mol, so you would have 36 ÷ 18 = 2 mol of water.

If you know the number of moles of a compound, you can also work out the masses of each of the different elements present. For example, suppose you had 1 mol of carbon dioxide molecules. You would have 1 mol of carbon atoms and 2 mol of oxygen molecules (because the formula is $CO_2$). This would give you 1 × 12 = 12 g of carbon atoms, and 2 × 16 = 32 g of oxygen atoms. Notice that if you add 12 g and 32 g together, you get 44 g. This is the mass of 1 mol of carbon dioxide molecules.

### Worked example 1

What is the molar mass of magnesium hydroxide, $Mg(OH)_2$?

$A_r$ of Mg = 24, $A_r$ of O = 16, and $A_r$ of H = 1.

$M_r$ of magnesium hydroxide, $Mg(OH)_2$ = 24 + 2 × (16 + 1)
= 24 + 34 = 58.

So its molar mass = 58 g/mol.

### Worked example 2

What is the molar mass of water, $H_2O$?

$A_r$ of H = 1 and $A_r$ of O = 16.

$M_r$ of water, $H_2O$ = (2 × 1) + 16
= 2 + 16 = 18.

So its molar mass = 18 g/mol.

**A** How many moles of magnesium atoms are there in 12 g of magnesium?

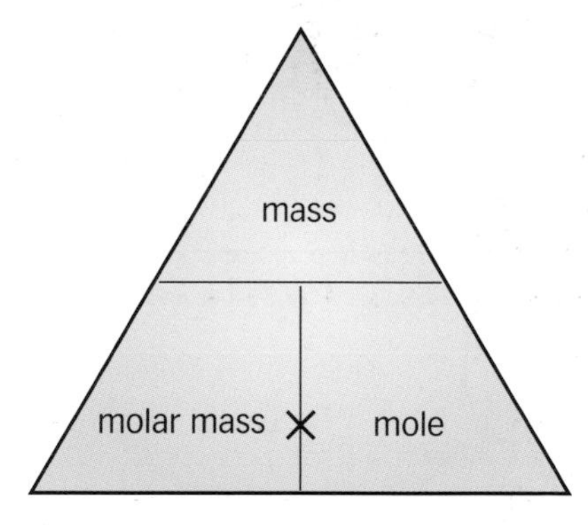


▲ This 'magic triangle' may help you with mole calculations

### Questions

Use these $A_r$ values to help you to answer the questions: H = 1, C = 12, O = 16, Na = 23, Ca = 40

1 What is the unit for amount of substance?

2 What is molar mass, and what are its units?

3 The formula for sodium hydroxide is NaOH.
(a) Calculate its relative formula mass.
(b) Calculate its molar mass.

E

4 Calculate the molar mass of
(a) calcium hydroxide, $Ca(OH)_2$
(b) calcium hydrogencarbonate, $Ca(HCO_3)_2$.

C

5 Define the relative atomic mass of an element.

6 Calculate the following:
(a) The number of moles of carbon atoms in 6 g of carbon.
(b) The number of moles of compound in 20 g of sodium hydroxide, NaOH.

A*

7 What is the mass of atoms of each element in 36 g of water, $H_2O$?

## Learning objectives

After studying this topic, you should be able to:

- use an understanding of conservation of mass to carry out calculations
- carry out calculations using balanced equations and the concept of the mole

## Key words

**conservation of mass, thermal decomposition**

Industrial lime kilns heat limestone (calcium carbonate) to make calcium oxide, which is used to make cement

**A** What mass of carbon dioxide is made from 50 g of calcium carbonate?

**B** What mass of calcium carbonate is needed to make 11 g of carbon dioxide?

## Conservation of mass

Mass is conserved during a chemical reaction. The total mass of reactants before the reaction is the same as the total mass of products formed during the reaction. This is because no atoms are created or destroyed during the reaction. They are only rearranged to form new substances. This is why in a balanced symbol equation the number and type of atom on each side of the equation must be the same. The idea of **conservation of mass** can be used to work out the mass of a substance released to the surroundings in a chemical reaction, or to work out the mass of a substance gained from the surroundings.

## Losing mass

In a **thermal decomposition** reaction, a single substance breaks down or decomposes to form two or more other substances. For example, calcium carbonate breaks down when heated, forming calcium oxide and carbon dioxide:

calcium carbonate → calcium oxide + carbon dioxide

$$CaCO_3 \rightarrow CaO + CO_2$$

Carbon dioxide is a gas, so it escapes into the air during the reaction. If 100 g of calcium carbonate is decomposed, 56 g of calcium oxide is left behind. The difference in mass is the mass of carbon dioxide released, in this case (100 – 56) = 44 g:

$$CaCO_3 \rightarrow CaO + CO_2$$
$$100\text{ g} \rightarrow 56\text{ g} + 44\text{ g}$$

The masses are directly proportional to each other. So, for example, 200 g of calcium carbonate would make 112 g of calcium oxide and 88 g of carbon dioxide.

These calculations can be very useful. Suppose you wanted to make 33 g of carbon dioxide. What mass of calcium carbonate would you need? Here is how you would work it out:

$$\text{mass of calcium carbonate needed} = \frac{100}{44} \times 33 = 75\text{ g}$$

## Gaining mass

In some reactions, a substance is gained during the reaction. For example, oxygen is gained from the air when magnesium is heated strongly:

magnesium + oxygen → magnesium oxide

$$2Mg + O_2 \rightarrow 2MgO$$

If 48 g of magnesium is heated, 80 g of magnesium oxide forms. The difference in mass is the mass of oxygen gained, in this case (80 – 48) = 32 g:

$$2Mg + O_2 \rightarrow 2MgO$$
$$48\text{ g} + 32\text{ g} \rightarrow 80\text{ g}$$

**C** What mass of oxygen is gained by 12 g of magnesium when it forms magnesium oxide?

Again, the masses are directly proportional to each other. So, for example, 24 g of magnesium would absorb 16 g of oxygen, forming 40 g of magnesium oxide.

Suppose you wanted to absorb 4 g of oxygen. What mass of magnesium would you need?

$$\text{mass of magnesium needed} = \frac{48}{32} \times 4 = 6\text{ g}$$

**D** What mass of magnesium is needed to absorb 20 g of oxygen? Show your working out.

## Mole calculations

It is possible to carry out reacting mass calculations using moles, the balanced equation, and the relative formula masses (or relative atomic masses, if appropriate).

### Worked example

Hydrogen reacts with oxygen to form water vapour:

$2H_2 + O_2 \rightarrow 2H_2O$

What mass of water can be made from 6 g of hydrogen gas?

$A_r$ of H = 1 and $A_r$ of O = 16

$M_r$ of $H_2$ = (2 × 1) = 2

amount of $H_2$ = 6 ÷ 2 = 3 mol

Looking at the equation, 2 mol of $H_2$ would make 2 mol of $H_2O$, so 3 mol of $H_2$ would make 3 mol of $H_2O$.

$M_r$ of $H_2O$ = (2 × 1) + 16 = 18, so its molar mass is 18 g/mol.

mass = number of moles × molar mass

= 3 × 18 = 54 g

## Exam tip

✔ Symbol equations are a model for what happens in chemical reactions. They let us appreciate that no atoms are created or destroyed during reactions.

## Questions

1. What is conservation of mass?
2. 3.1 g of copper carbonate decomposes to form carbon dioxide and 2.0 g of copper oxide. What mass of carbon dioxide is formed?
3. 2.4 g of carbon is burned in air, forming 8.8 g of carbon dioxide. What mass of oxygen is gained?

E

4. When 20.0 g of zinc is heated, 24.9 g of zinc oxide forms.
   (a) What mass of oxygen is gained?
   (b) What mass of oxygen would be gained by 50.0 g of zinc?

C

5. Nitrogen reacts with hydrogen to form ammonia:
   $$N_2 + 3H_2 \rightarrow 2NH_3$$
   $A_r$ of H = 1, $A_r$ of N = 14
   (a) Work out the molar masses of nitrogen, hydrogen, and ammonia.
   (b) How many moles are in 2.8 g of nitrogen?
   (c) How many moles of ammonia would form?
   (d) Work out the mass of ammonia formed.

A*

## Learning objectives

After studying this topic, you should be able to:

- work out the mass of an element in a known mass of a compound
- calculate the percentage by mass of an element in a compound using experimental data
- calculate the percentage by mass of an element in a compound from its formula

The front of the Wales Millennium Centre in Cardiff is made from steel coated with copper oxide

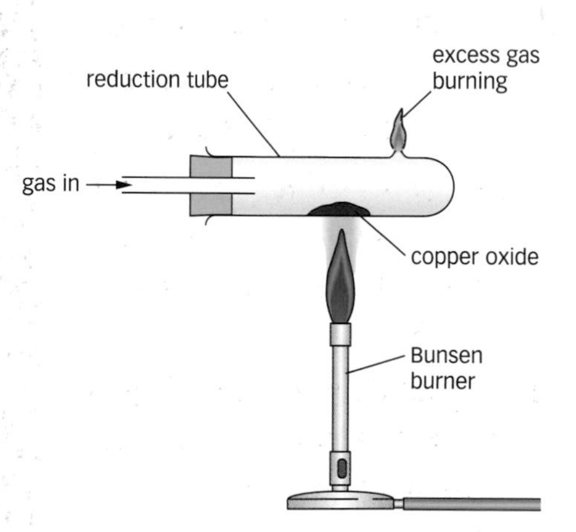


Copper oxide can be reduced to copper using hydrogen or methane

## Half and half?

The formula for copper oxide is CuO. You may think that, because the formula contains one Cu and one O, 50% of the mass of copper oxide would be due to copper and 50% to oxygen. If so, you would be wrong. This is because the atoms of different elements usually have different masses, and this needs to be taken into account. The way this is done involves the relative atomic mass of the elements in the substance, and its relative formula mass. It can also involve carrying out experiments.

## Masses from experimental data

Copper oxide can be reduced to copper using hydrogen:

copper oxide + hydrogen → copper + water

$$CuO + H_2 \rightarrow Cu + H_2O$$

The copper oxide must be hot for this to work, and the excess hydrogen has to be burnt off to make the experiment safe. The mass of the empty reduction tube is measured. After adding some copper oxide, the mass of the reduction tube with its contents is measured before the reaction and again after the reaction.

Here are some example results.

| | | mass (g) |
|---|---|---|
| A | empty reduction tube | 31.9 |
| B | reduction tube + copper oxide | 34.9 |
| C | reduction tube + copper | 34.3 |

The mass of copper oxide can be calculated using these results:

- mass of copper oxide (B – A) = 34.9 – 31.9 = 3.0 g

The masses of copper and oxygen in the original copper oxide can also be calculated:

- mass of copper (C – A) = 34.3 – 31.9 = 2.4 g
- mass of oxygen (B – C) = 34.9 – 34.3 = 0.6 g

Notice that the mass of copper, plus the mass of oxygen, equals the mass of copper oxide used.

## Percentage masses from experimental data

The masses from the copper oxide experiment can be used to calculate the percentage mass of each element in copper oxide. For example, for copper:

$$\text{percentage of copper} = \frac{\text{mass of copper}}{\text{mass of copper oxide}} \times 100$$

$$= \frac{2.4}{3.0} \times 100 = 80\%$$

**A** Use the method shown to calculate the percentage of oxygen in copper oxide.

**B** Explain why the percentages of copper and oxygen in copper oxide must equal 100%.

## Percentage masses from formulae and atomic masses

The formula of a compound, and the relative atomic masses of the elements it contains, can be used to calculate percentage masses. These are the steps needed:

1. Calculate the relative formula mass of the compound using the relative atomic masses.
2. Multiply the relative atomic mass of the element in the question by the number of its atoms in the compound.
3. Work out the percentage using the numbers from Steps 1 and 2.

### Worked example

Ammonium nitrate, $NH_4NO_3$, is used as an artificial fertiliser. Calculate the percentage by mass of nitrogen in ammonium nitrate. $A_r$ of N = 14, $A_r$ of H = 1, and $A_r$ of O = 16.

1. $M_r$ of $NH_4NO_3 = 14 + (4 \times 1) + 14 + (3 \times 16) = 80$
2. mass of N = 2 × 14 = 28 (notice that N appears twice in the formula)
3. percentage of N = $\frac{28}{80} \times 100 = 35\%$

### Exam tip OCR

✔ Make sure you understand how experimental data are used to find the masses of the different elements in a compound, and how these masses are used to find the percentage of each element in the compound.

### Questions

1 2.0 g of magnesium oxide, MgO, contains 1.2 g of magnesium. What mass of oxygen does it contain?

2 7.5 g of glucose, $C_6H_{12}O_6$, contains 0.5 g of hydrogen and 4.0 g of oxygen. What mass of carbon does it contain?

E

3 Use the answers to Questions 1 and 2 to calculate the following:

(a) The percentage by mass of oxygen in magnesium oxide.

(b) The percentage by mass of hydrogen in glucose.

(c) The percentage by mass of carbon in glucose.

C

Use these $A_r$ values to help you to answer Questions 4 and 5: H = 1, O = 16, Na = 23, S = 32, Cu = 63.5

4 Calculate the percentage by mass of sodium in sodium hydroxide, NaOH.

5 Calculate the percentage by mass of oxygen in copper sulfate, $CuSO_4$.

A*

# 4: Empirical formulae

## Learning objectives

After studying this topic, you should be able to:

- ✔ understand what an empirical formula shows
- ✔ work out the empirical formula of a compound given its chemical formula
- ✔ calculate the empirical formula of a compound

## Key words

**molecular formula, displayed formula, empirical formula**

▲ Butane is used as bottled gas for camping stoves

## Exam tip OCR

- ✔ The different types of formulae are models of molecules. Each one helps us understand a different aspect of a molecule, such as the number of atoms or elements it contains, or the way its atoms are joined together.

## Different formulae

You usually see a substance described by its **molecular formula**. This shows the number and type of each atom in a molecule. For example, the molecular formula of butane is $C_4H_{10}$. It shows that each butane molecule contains four carbon atoms and ten hydrogen atoms. Sometimes it is useful to describe a substance by its displayed formula. A **displayed formula** shows the atoms in a molecule and the bonds between them. The **empirical formula** of a compound is yet another type of formula. It is often the first formula to be worked out for a newly discovered compound.

```
     H   H   H   H
     |   |   |   |
 H — C — C — C — C — H
     |   |   |   |
     H   H   H   H
```

▲ The displayed formula of butane shows its atoms and the bonds between them

**A** What does the molecular formula for ethane, $C_2H_6$, tell you about an ethane molecule?

## Empirical formula

An empirical formula shows the simplest whole number ratio of each type of atom in a compound. For example, the molecular formula of butane is $C_4H_{10}$. You can divide both numbers by two to get a simpler formula: $C_2H_5$. This is the empirical formula for butane. The 2 and 5 in the formula are whole numbers, and you cannot make the formula any simpler while keeping whole numbers. For example, if you divide the formula again by 2 you get $CH_{2.5}$, and 2.5 is not a whole number.

**B** What is the empirical formula for ethane, $C_2H_6$?

The empirical formula can be worked out by looking for a number that is common to all the numbers in the chemical formula. For example, $C_3H_6$ can be divided by 3 to get the empirical formula $CH_2$.

## Empirical formulae from experimental data

The empirical formula of a compound can be calculated from experimental data. For example, magnesium is oxidised to magnesium oxide when it is heated in air. The reaction is usually carried out in a crucible. Here are some example results.

| | | mass (g) |
|---|---|---|
| A | empty crucible | 23.70 |
| B | crucible + magnesium | 24.42 |
| C | crucible + magnesium oxide | 24.90 |

From these results, the mass of magnesium is (24.42 – 23.70) = 0.72 g and the mass of oxygen is (24.90 – 24.42) = 0.48 g. The empirical formula can be worked out using these steps.

| | | |
|---|---|---|
| **1.** Write each element's symbol | Mg | O |
| **2.** Write each mass in g | 0.72 | 0.48 |
| **3.** Write each $A_r$ | 24 | 16 |
| **4.** Find the number of moles | $0.72 \div 24 = 0.03$ | $0.48 \div 16 = 0.03$ |
| **5.** Divide by the smallest number | $0.03 \div 0.03 = 1$ | $0.03 \div 0.03 = 1$ |
| **6.** Check for whole numbers, write the formula | MgO | |

You can also find the empirical formula if you know the percentage composition by mass of the compound (see the previous spread). Use the same method as shown above, but write the percentages as grams in Step 2. For example, assume you have 100 g of the compound, so 75% would be written as 75 g and 25% as 25 g.

▲ Magnesium is oxidised to magnesium oxide when it is heated in a crucible

### Exam tip OCR

✔ Take care at the last step not to round up or down too much. For example, $A_{0.75}X$ is not AX. Instead, multiply by four to get $A_3X_4$.

### Questions

1 What is an empirical formula?

2 Which of these formulae is an empirical formula, and why? $C_5H_{12}$, $C_6H_{12}O_6$, $Pb_2O_4$.

E

3 Write the empirical formulae for the other two substances in Question 2.

4 Work out the empirical formula for ethanoic acid, $CH_3COOH$.

C

Use these $A_r$ values to help you answer Questions 5 and 6:

H = 1, C = 12, O = 16, S = 32

5 An oxide of sulfur contains 50% sulfur and 50% oxygen by mass. Work out its empirical formula.

A*

6 A 2.3 g sample of compound X contains 1.2 g carbon, 0.3 g hydrogen, and 0.8 g oxygen. Work out its empirical formula.

## Learning objectives

After studying this topic, you should be able to:

- recall that the concentration of solutions is measured in $g/dm^3$ or $mol/dm^3$
- convert between $cm^3$ and $dm^3$
- carry out calculations involving concentrations in $mol/dm^3$

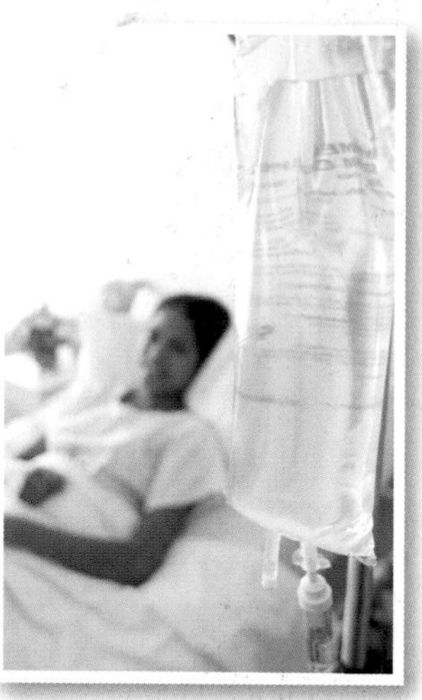

It is important that the concentration of an intravenous saline drip is correct

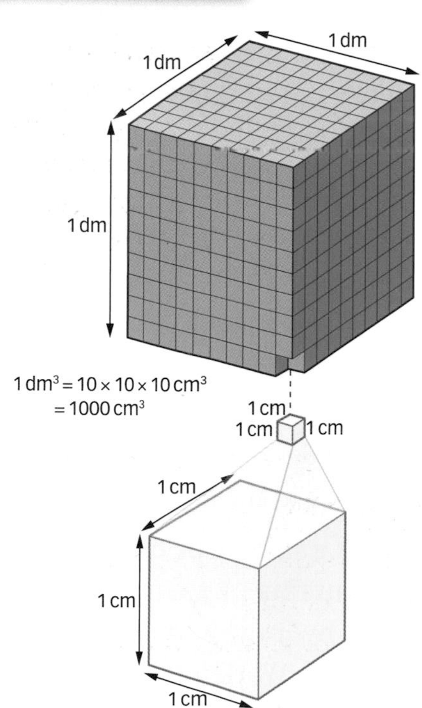


A cubic decimetre (blue) contains one thousand cubic centimetres (yellow)

## Concentration

Patients recovering in hospital may be given an intravenous drip. This is a sterile solution delivered slowly into a vein. There are different drips, but the most common is a normal saline drip. This contains sodium chloride at the same **concentration** as blood. It is important to match the concentration because there could be problems if the drip were too dilute or too concentrated. The concentration of sodium chloride in a normal saline drip is 9 $g/dm^3$ or 0.154 $mol/dm^3$. How are these concentrations worked out?

## Solutes and solutions

A **solution** consists of a **solvent**, which is usually water or another liquid, and a **solute**. The solute is the substance that dissolves in the solvent. A dilute solution contains relatively few solute particles. The more solute particles that are dissolved in the solvent, the more concentrated the solution becomes. A very concentrated solution will contain relatively large numbers of solute particles.

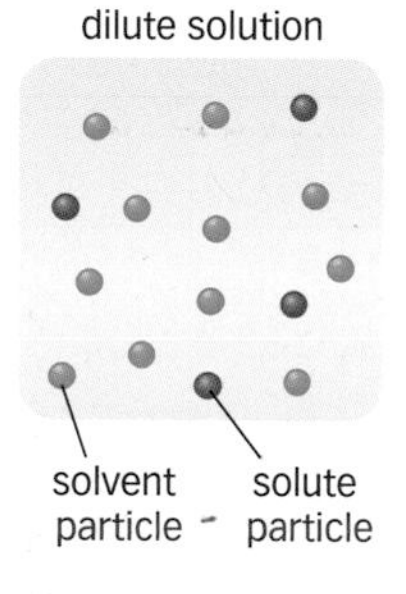


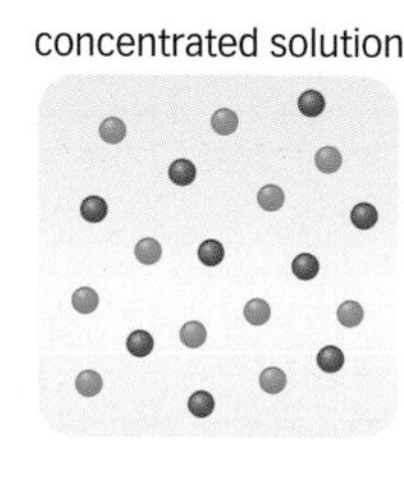


The more concentrated a solution is, the more crowded it is with solute particles

## Volumes

To work out the concentration of a solution, you need to know the volume of the solution, and the mass or amount of solute dissolved. This is why the concentration of sodium chloride in a normal saline drip is measured in $g/dm^3$ or $mol/dm^3$. It tells you that each cubic decimetre of solution contains 9 g of sodium chloride, which is 0.154 mol of sodium chloride.

The unit of volume used, the cubic decimetre or $dm^3$, is the same volume as 1000 $cm^3$. Volumes can be measured in either unit, $dm^3$ or $cm^3$, so it is important to be able to convert from one to the other.

**A** How many cubic centimetres are there in a cubic decimetre?

Here are the two conversions you need:

- to go from $cm^3$ to $dm^3$, divide by 1000
- to go from $dm^3$ to $cm^3$, multiply by 1000.

For example, 25 $cm^3$ is $25 \div 1000 = 0.025$ $dm^3$. Going the other way, 0.4 $dm^3$ is $0.4 \times 1000 = 400$ $cm^3$.

## Calculating concentrations

The concentration in mol/$dm^3$ is calculated by dividing the amount of solute by the volume of solution:

$$\text{concentration in mol/dm}^3 = \frac{\text{amount of solute in mol}}{\text{volume of solution in dm}^3}$$

### Worked example

What is the concentration of 500 $cm^3$ of a solution containing 4.5 g of sodium chloride?

amount of sodium chloride = mass ÷ molar mass = $4.5 \div 58.5$ = 0.077 mol

volume of solution = $500 \div 1000 = 0.5$ $dm^3$

concentration of solution = $0.077 \div 0.5 = 0.154$ mol/$dm^3$

### Did you know...?

The litre is the special name for cubic decimetre. The litre used to be defined as the volume of 1 kg of pure water at 4 °C. This meant that the litre was 1.000028 $dm^3$ until 1964, when it was redefined as 1 $dm^3$.

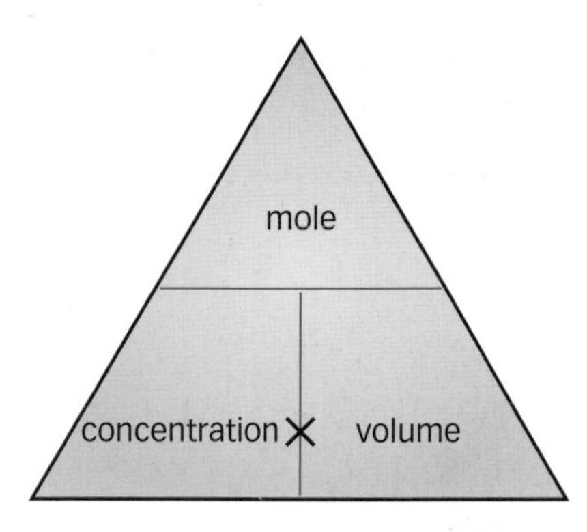


This 'magic triangle' may help you with concentration calculations

### Questions

1. Give two units for concentration of solution.
2. Are the solvent particles least crowded in a dilute solution or a concentrated solution?

↓ E

3. Convert the following volumes into $dm^3$.
   (a) 2000 $cm^3$ (b) 500 $cm^3$ (c) 120 $cm^3$
4. Convert the following volumes into $cm^3$.
   (a) 1.5 $dm^3$ (b) 0.1 $dm^3$ (c) 0.05 $dm^3$

↓ C

5. Calculate the concentration of 250 $cm^3$ solution containing 0.1 mol of solute.
6. Calculate the number of moles of solute in 2 $dm^3$ of a 0.5 mol/$dm^3$ solution.
7. Calculate the volume of a 2 mol/$dm^3$ solution that contains 0.5 mol of solute.

↓ A*

### Exam tip OCR

✔ Make sure you can rearrange the equation so you can find the amount of solute given the volume of solution and its concentration, and the volume of solution given its concentration and the amount of solute.

### Key words

**concentration, solution, solvent, solute**

## Learning objectives

After studying this topic, you should be able to:

- explain the need to dilute certain mixtures
- calculate the volumes needed for diluting solutions
- interpret information on food packaging about GDA
- carry out calculations based on GDA for sodium

## Key words

**guideline daily amount (GDA)**

A Explain why orange squash labels say 'dilute to taste'.

## Exam tip OCR

- Another way to work out the amount of water to add is to multiply the starting volume by the number of times the solution will be diluted, then subtract the starting volume.

▲ This scientist is carrying out accurate dilutions on various substances for analysis

## The need for dilution

You may dilute substances in your everyday life. For example, orange squash and fruit cordial are usually sold in a concentrated form. The labels tell you to 'dilute to taste', which means you should add enough water so that the taste is not too strong for you. Baby milk powder must be mixed with just the right volume of water to make it the right concentration, otherwise the baby could be harmed. Liquid medicines may be diluted to avoid giving an overdose.

Scientists need to dilute substances accurately so that they can carry out tests on them. If the concentration is too low or too high, their machines may give an inaccurate reading, or may not even detect the substance.

## Diluting solutions

A solution is diluted simply by adding water to it, then mixing. However, some calculations are needed if an accurate new concentration is needed. For example, suppose you had 20 $cm^3$ of a 1.0 $mol/dm^3$ solution and wanted to dilute it to a 0.1 $mol/dm^3$ solution. What would you do?

The volume of water to add can be calculated using this equation:

$$\text{volume of water to add} = \left(\frac{\text{starting concentration}}{\text{target concentration}} - 1\right) \times \text{starting volume}$$

In this example, you would need to add 180 $cm^3$ of water:

$$\text{volume of water to add} = \left(\frac{1.0}{0.1} - 1\right) \times 20$$
$$= (10 - 1) \times 20$$
$$= 9 \times 20 = 180\ cm^3$$

## Guideline daily amounts

Your body needs various nutrients in the right amounts to be healthy. For example, you may become overweight if you eat too much fatty or sugary food. On the other hand, you may become underweight if you do not eat enough. Nutritionists are scientists who study nutrients in food, how they are used by the body, and the relationships between diet, health, and disease. They have worked out a **guideline daily amount**, or **GDA**, for the amount of energy and various nutrients we need each day. The table on the right shows some GDAs.

| | Energy (calories) | Sugars (g) | Total fat (g) | Saturated fat (g) | Salt (g) |
|---|---|---|---|---|---|
| Children (5–10 years) | 1800 | 85 | 70 | 20 | 4 |
| Women | 2000 | 90 | 70 | 20 | 6 |
| Men | 2500 | 120 | 95 | 30 | 6 |

Food manufacturers often include information about GDAs on their food labels. These show the amount of energy, sugar, fat, saturated fat, and salt provided by the food. They may also show the percentage of the GDAs provided by the food. This is to help people choose food for a healthy diet. The GDA shown for an adult is usually the GDA for a woman. For example, a portion of food containing 5 g of saturated fat will provide 25% of an adult's GDA for saturated fat:

$$\text{percentage of GDA} = \frac{5}{20} \times 100 = 25\%$$

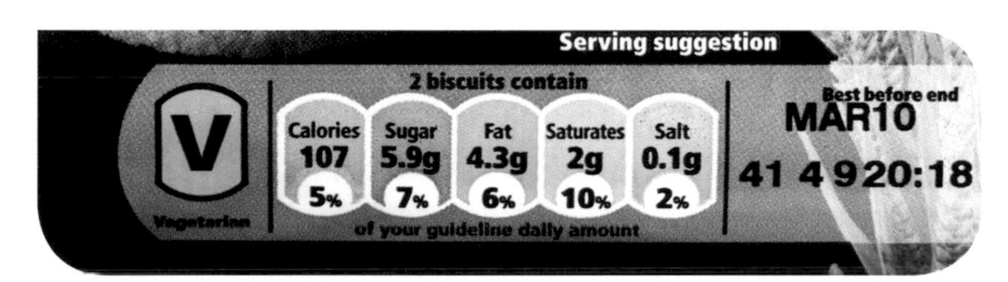


▲ This label on a packet of biscuits shows the percentage of various GDAs provided by two biscuits

## Salt and sodium in food

Common salt is sodium chloride, NaCl. The sodium ions it contains are needed for transmitting nerve signals and muscular contraction. All foods contain salt in varying amounts, and most processed foods contain added salt. However, too much salt can cause health problems including high blood pressure, leading to an increased risk of heart disease. The GDA for an adult is 6 g, which is about a teaspoon. Some food labels show the salt content of the food, some show its sodium content, and some show both.

The percentage by mass of sodium in sodium chloride is 39.3%. So 1 g of salt is the same as 1 × 39.3 ÷ 100 = 0.393 g of sodium. However, sodium ions can come from other sources in food, so this conversion may be inaccurate. For example, monosodium glutamate, commonly used to enhance the flavour of food, contains sodium ions.

### Questions

1 Explain why medicines may need to be diluted.

2 Look at the food label. Which substance was present in the smallest amount, and which in the largest amount?

E

3 A scientist wanted to dilute 10 $cm^3$ of a 0.5 $mol/dm^3$ solution to 0.1 $mol/dm^3$. How much water must be added?

4 The label on a packet of naan bread shows that each portion contains 1.8 g of salt. What percentage of an adult's GDA is this?

C

5 Food naturally contains sodium chloride, NaCl.

(a) Show that the percentage by mass of sodium in sodium chloride is about 39%. $A_r$ of Na = 23 and $A_r$ of Cl = 35.5

A*

(b) If the adult GDA for salt is 6 g, what is the adult GDA for sodium?

(c) Explain why this conversion may be inaccurate.

## Learning objectives

After studying this topic, you should be able to:

- describe and explain the changes that happen when an acid reacts with an alkali
- interpret simple pH curves
- sketch a pH titration curve

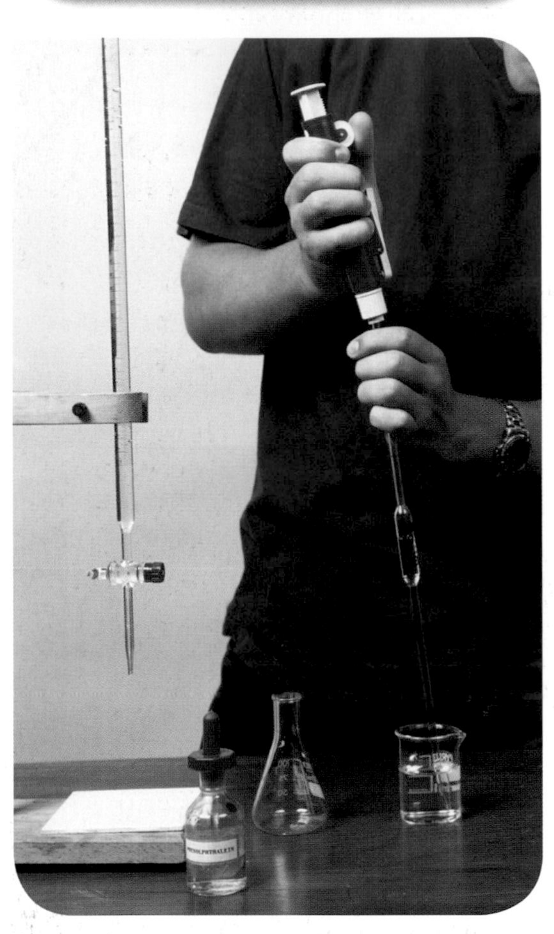

This is the apparatus needed to add acids to alkalis accurately. The conical flask will go underneath the burette.

**A** What is neutralisation?

**B** Why is it more usual to add an acid to an alkali, than to add an alkali to an acid?

## Neutralisation reactions

**Neutralisation reactions** happen when an acid reacts with an alkali:

acid + alkali → salt + water

**Acids** form solutions with a **pH** less than 7, and **alkalis** form solutions with a pH more than 7.

A **neutral** solution has a pH of 7. The pH decreases when an acid is added to an alkali, and it increases when an alkali is added to an acid.

Changes in pH can be followed using universal indicator and a pH colour chart. They can also be followed using a pH meter. This is a device that can measure and display the pH of a solution. Some designs consist of a pH electrode connected to a datalogger, while others are all-in-one devices.

## Investigating changes in pH

It is important that small volumes of acid or alkali can be added so that an accurate pH curve can be recorded. The **burette** is a long glass tube with a tap at the bottom. Graduations up the side show the volume of liquid delivered. Strong alkalis such as sodium hydroxide solution can damage glass. This means that, although a burette can be filled with an alkali, it is usual to fill one with an acid.

The alkali is measured using a **pipette**, filled safely using a **pipette filler**. The alkali is transferred to a conical flask. This is more convenient than a beaker. Its narrow neck and sloping sides means that the acid and alkali can be mixed without them sloshing over the sides.

With the pH electrode in the conical flask, the tap on the burette is opened to add acid slowly to the alkali. When investigating changes in pH, it is usual to add twice as much acid as the volume of alkali used. A graph of pH on the vertical axis against volume of acid added on the horizontal axis is called a **pH curve**.

## pH curves

The diagrams show typical pH curves involving a strong acid such as hydrochloric acid, and a strong alkali such as sodium hydroxide solution. In each case, the conical flask contained 25 $cm^3$ and the burette delivered a total of 50 $cm^3$.

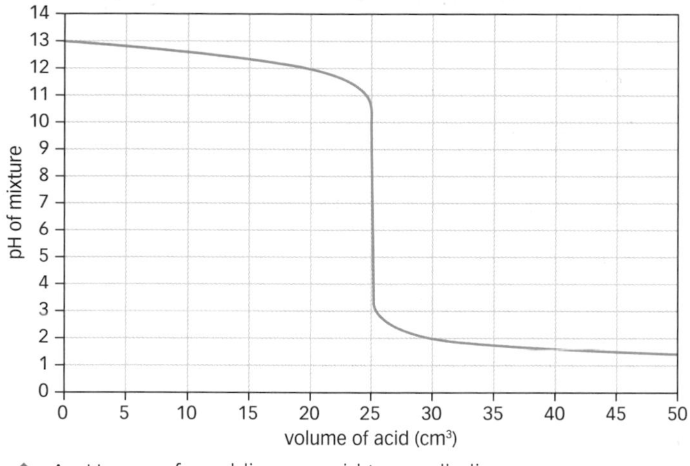

▲ A pH curve for adding an acid to an alkali

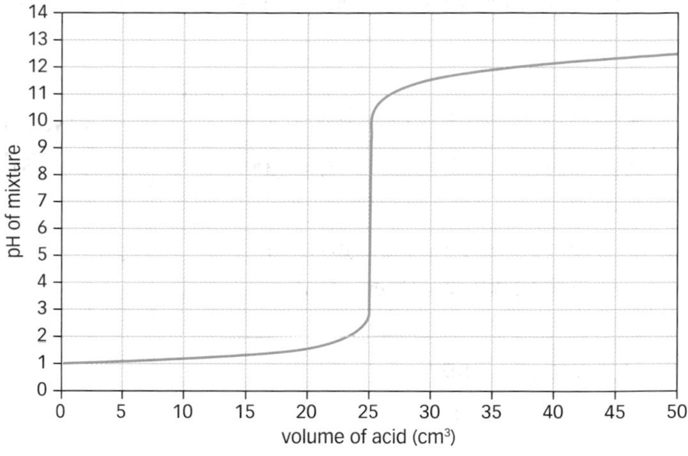

▲ A pH curve for adding an alkali to an acid

**C** In general, where does the sudden change of pH happen?

**D** What was the pH when 20 $cm^3$ of acid was added to 25 $cm^3$ of alkali?

## Key words

**neutralisation reaction, pH, burette, pipette, pipette filler, pH curve, end point**

Notice in both pH curves that the pH changes only gradually at first and at the end. The **end point** of the reaction is when the acid and alkali exactly neutralise each other. There is a sudden change in pH at the end point. These pH curves can be used to determine the volume of acid or alkali at the end point, and the pH when different amounts of acid or alkali are added.

## Exam tip

✔ The data obtained during a neutralisation reaction would be displayed in a table. However, it would be difficult to understand the changes in pH. A graph is a more appropriate format for displaying data here.

## Questions

1. What happens to the pH when an acid is added to an alkali?
2. Name the two pieces of glassware used to measure liquids when determining a pH curve.

E

3. Use the pH curve for adding acid to alkali to answer these questions.
   (a) What was the pH of the alkali at the start?
   (b) What was the pH of the mixture after 30 $cm^3$ of acid had been added?
   (c) What volume of acid was needed to exactly neutralise the alkali?

C

4. Use the pH curve for adding alkali to acid to answer these questions.
   (a) What was the pH of the mixture when 10 $cm^3$ of alkali had been added?
   (b) What volume of alkali was needed to reach a pH of 11.5?
   (c) What was the total volume of the mixture at the end point?

A*

# 8: Titrations

## Learning objectives

After studying this topic, you should be able to:

- describe the colour changes of common indicators
- describe how to carry out a titration
- explain the need for several consistent readings in a titration
- carry out concentration calculations using titration results

Litmus, a single indicator, is blue above pH 8.3 and red below pH 4.5

## Exam tip OCR

- Take care when describing phenolphthalein. It changes from pink to colourless, not clear.

Phenolphthalein, a single indicator, is pink above pH 10.0 and colourless below pH 8.2

## Indicators

The changes in pH in an experiment to find a pH curve can be followed using a pH meter, or using a suitable **indicator**. An indicator has different colours, depending on its pH. Phenolphthalein and litmus are **single indicators**. Each one contains a single substance, which gives a sudden change in colour at the end point:

- Phenolphthalein is pink in alkaline solutions and colourless in acidic solutions.
- Litmus is blue in alkaline solutions and red in acidic solutions.

Universal indicator is a **mixed indicator**. It contains several different indicators. Each one changes colour over a different range of pH values. This is why universal indicator gives a continuous colour change. It lets you find the approximate pH of a solution, but it is not suitable if you want to know the precise end point in a titration.

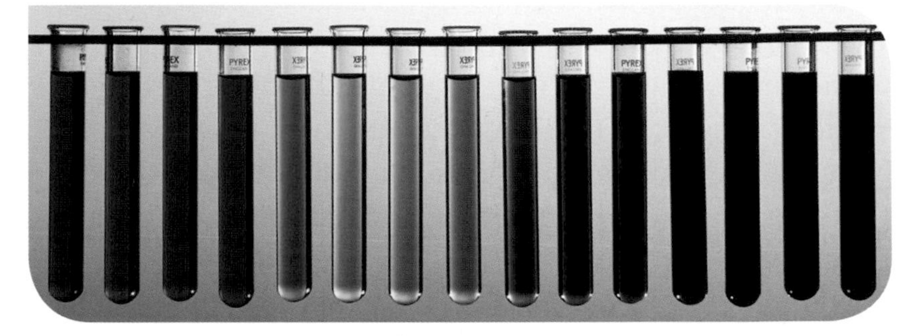

Universal indicator has a range of colours, from red in acids to purple in alkalis. Green means neutral.

**A** What is the difference between a single indicator and a mixed indicator?

## Titrations

**Titration** is a method to find the concentration of an acid or alkali using a neutralisation reaction. It is carried out in the same way as finding a pH curve. However, a single indicator such as phenolphthalein is used, instead of a pH meter or a mixed indicator such as universal indicator. Phenolphthalein gives a sudden colour change between pink and colourless at the end point.

A known volume of alkali, usually 25 $cm^3$, is added to a conical flask using a pipette and pipette filler. A few drops of phenolphthalein are added to produce a pink tinge. Acid is added slowly from a burette until the pink colour just disappears. The difference between the start and end readings on the burette gives the **titre**. This is the volume of acid needed to exactly neutralise the alkali.

The titre is used to work out the concentration of either the acid or the alkali, provided the concentration of the other one is known. It is important that the titre is accurate, otherwise the calculated concentration will be wrong. A titration is repeated until several consistent titres are obtained. Any **anomalous** titres (ones that are too high or too low compared to the others) are ignored, and the mean titre is calculated. The table shows how titration results may be recorded.

| | Run 1 | Run 2 | Run 3 | Run 4 |
|---|---|---|---|---|
| End reading ($cm^3$) | 21.2 | 41.0 | 21.2 | 41.2 |
| Start reading ($cm^3$) | 0.0 | 21.2 | 1.0 | 21.2 |
| Titre ($cm^3$) | 21.2 | 19.8 | 20.2 | |

## Titration calculations

Imagine that a titration is carried out involving 0.1 mol/$dm^3$ hydrochloric acid and 25.0 $cm^3$ of an unknown concentration of sodium hydroxide:

$$HCl + NaOH \rightarrow NaCl + H_2O$$

If the mean titre of acid is 24.0 $cm^3$, what is the concentration of the sodium hydroxide?

Convert all volumes to $dm^3$: the titre of HCl is 24.0 ÷ 1000 = 0.024 $dm^3$ and the volume of NaOH is 25.0 ÷ 1000 = 0.025 $dm^3$.

- amount of HCl = concentration × volume = 0.1 × 0.024 = 0.0024 mol

Looking at the balanced equation, you can see that 1 mol of HCl will react with 1 mol of NaOH. So 0.0024 mol of HCl will react with 0.0024 mol of NaOH.

- concentration of NaOH = amount ÷ volume = 0.0024 ÷ 0.025 = 0.096 mol/$dm^3$

### Key words

**indicator, single indicator, mixed indicator, titration, titre, anomalous**

**B** What is the titre in a titration?

### Exam tip

✔ Titration results are best shown in a table. The individual titres can be compared easily so that any anomalous values are identified.

### Questions

1 Describe the colours of universal indicator, litmus, and phenolphthalein in acids and alkalis.

2 Calculate the titre in Run 4 of the table on the left.

3 Explain why several consistent readings are needed in titrations.

4 Explain why a single indicator, rather than a mixed indicator, is used in titrations.

5 The results in the table on the left are from a titration in which 0.2 mol/$dm^3$ hydrochloric acid was added to 25.0 $cm^3$ of a sodium hydroxide solution. Calculate the concentration of the alkali.

# 9: Measuring gases

## Learning objectives

After studying this topic, you should be able to:

- describe how to measure the volume of gas produced in a reaction
- describe how to measure the mass of gas produced in a reaction

Results from crash tests help make cars safer

## Did you know...?

Sensors in an air bag make the decision to trigger the chemical reaction within 15 to 30 milliseconds of the start of a crash. It takes about 20 to 30 milliseconds for the bag to fill with hot nitrogen. Overall, the air bag is fully inflated within about 60 milliseconds of the crash starting.

**A** Why is upward displacement over water not so suitable for measuring the volume of carbon dioxide produced in a reaction?

## Making gases

A car air bag inflates in a fraction of second because it rapidly fills with hot nitrogen gas. Solid sodium azide decomposes when an electric current is passed through it, triggered by crash sensors in the air bag:

sodium azide → sodium + nitrogen

$$NaN_3 \rightarrow Na + 1\tfrac{1}{2}N_2$$

Many other chemical reactions release gases as one or more of their products. The course of these reactions can be followed by measuring the volume or mass of the gases produced.

## Measuring volumes

A **gas syringe** is a gas-tight glass syringe used for measuring the volumes of gases. It usually measures volumes up to 100 cm$^3$. Gas syringes can be used to measure the volume of any gas, including carbon dioxide, hydrogen, and oxygen. The syringe is connected to the reaction container by tubing. As the reaction occurs, gas fills the gas syringe and pushes the plunger out. Graduations on the side show the volume of gas contained in the gas syringe.

This gas syringe is being used to measure the volume of carbon dioxide produced when calcium carbonate reacts with hydrochloric acid

Gases can also be collected over water in an upturned measuring cylinder. This is filled with water, then turned upside down in a trough of water. Air pressure keeps the water inside the cylinder. A delivery tube is led from the reaction container and into the mouth of the measuring cylinder. As the reaction occurs, gas fills the measuring cylinder and pushes the water out. Graduations on the side show the volume of gas contained. The same method, called **upward displacement**, also works using an upturned burette. However, upward displacement is only accurate if the gas is not very soluble in water. It does not so work well for soluble gases such as carbon dioxide.

## Measuring masses

Gases do have mass, even though this is much less than for the same volume of a solid or liquid. For example, 100 $cm^3$ of carbon dioxide has a mass of 0.18 g at 25 °C and normal pressure. This means that the release of gas during a reaction can be followed using a balance. The balance has to have a suitable resolution, such as ±0.01 g, to be able to make precise readings. The reaction container, with its contents, is placed on the balance. As the reaction takes place, the gas escapes and the total mass of the container and its contents goes down. The difference in mass is due to the gas that has escaped.

This method is not suitable for all gases. The molar mass of carbon dioxide is 44 g/mol, but the molar mass of hydrogen is just 2 g/mol. This means that 100 $cm^3$ of hydrogen at 25 °C and normal pressure only has a mass of 0.008 g. This is too little to be measured by the sort of balances normally found in a school laboratory.

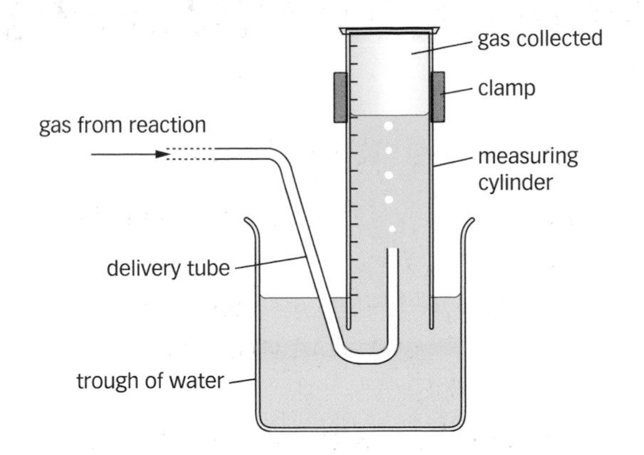


Measuring the volume of a gas in an upturned measuring cylinder or burette works well for gases such as hydrogen and oxygen

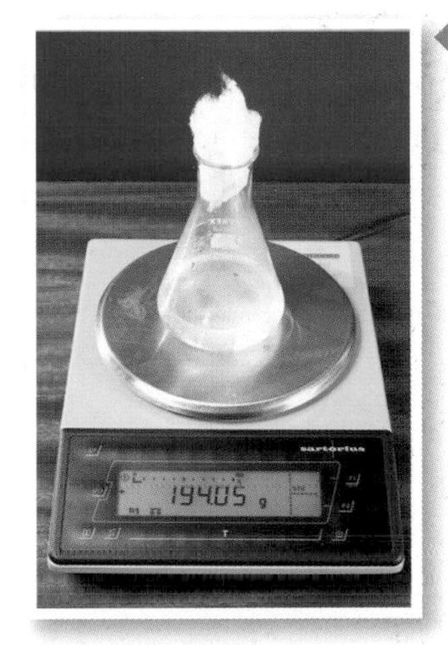


This balance is being used to measure the loss of carbon dioxide when calcium carbonate reacts with hydrochloric acid. The cotton wool stops any liquid escaping.

### Questions

1 Give two ways in which the volume of a gas produced in a reaction can be measured.

2 Describe how the mass of a gas produced in a reaction can be measured.

↓ E

3 Zinc carbonate decomposes when heated, producing zinc oxide and carbon dioxide. Describe two suitable ways to measure the gas produced.

4 Magnesium reacts with hydrochloric acid, producing magnesium chloride and hydrogen. Describe two suitable ways to measure the gas produced.

↓ C

5 Hydrogen peroxide rapidly decomposes in the presence of a suitable catalyst, producing water and oxygen. The molar mass of oxygen is 32 g/mol. Suggest why the production of oxygen could be followed using a gas syringe, upward displacement over water, or by measuring the mass lost during a reaction.

↓ A*

### Exam tip OCR

- ✔ Make sure you can recognise the apparatus used to collect the volume of gas produced in a reaction.
- ✔ Make sure you can describe how to measure the volume or mass of gas produced during a reaction.

### Key words

**gas syringe, upward displacement**

# 10: Gas volumes

## Learning objectives

After studying this topic, you should be able to:

- interpret graphs showing the volume of gas produced during a reaction
- explain why a reaction stops, in terms of a limiting reactant
- carry out calculations on the volume and amount of a gas

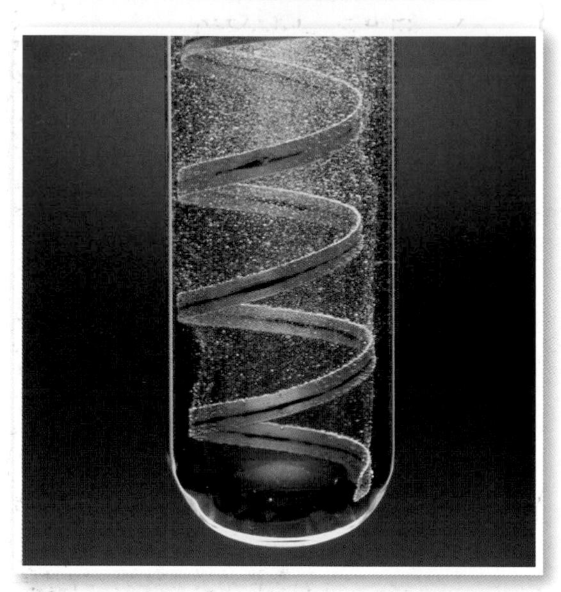

Magnesium ribbon reacts vigorously with hydrochloric acid

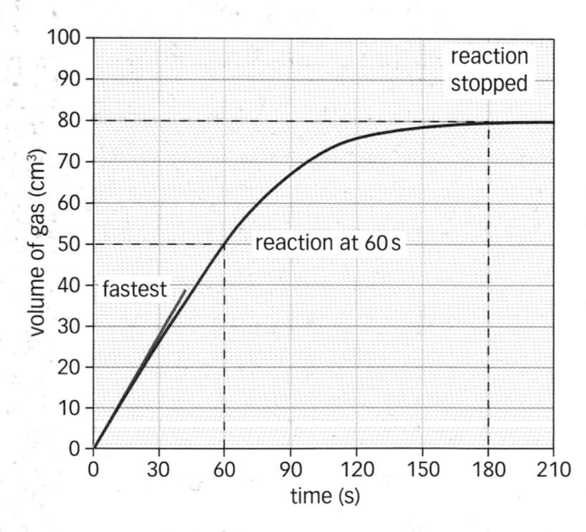


The volume of hydrogen produced as magnesium reacts with hydrochloric acid

## Following reactions

Magnesium reacts with hydrochloric acid, forming magnesium chloride and hydrogen:

| magnesium | + | hydrochloric acid | → | magnesium chloride | + | hydrogen |
|---|---|---|---|---|---|---|
| Mg | + | 2HCl | → | $MgCl_2$ | + | $H_2$ |

The reaction can be followed by measuring the volume of hydrogen produced at regular intervals. This can be done using a gas syringe or by upward displacement over water into an upturned measuring cylinder or burette. The graph below shows the results from one such experiment.

The graph has several important features. Notice that the line starts off steeply and eventually becomes horizontal. Its gradient decreases as the reaction continues. This shows that the rate of reaction decreases as the reaction carries on. It is fastest at the start, when the gradient is greatest. The reaction has stopped when the line becomes horizontal. The reaction has stopped by 180 s, as no more hydrogen is produced after that. The total volume of hydrogen produced in the reaction was 80 $cm^3$.

It is possible to work out how much gas was produced at a particular time. For example, by 60 s after the start of the reaction the volume of gas produced was 50 $cm^3$.

**A** How long did it take to produce 60 $cm^3$ of hydrogen?

## Limiting reactants

A reaction will stop when one of the reactants is all used up. You cannot tell from the graph whether the magnesium ribbon was all used up at the end, or the hydrochloric acid was all used up. If you could see that all the magnesium ribbon was used up at the end, the magnesium would be the **limiting reactant**. The hydrochloric acid would have been in **excess** – there would have been more than enough of it to react with all the magnesium. The graph on the next page shows what happens when different amounts of magnesium are used.

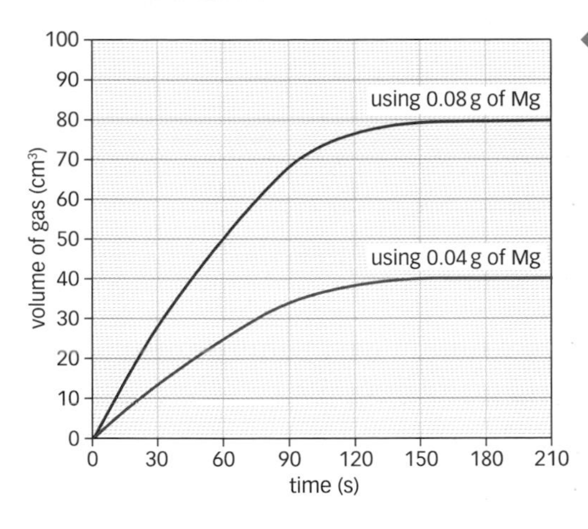


The volume of hydrogen produced when two different masses of magnesium are used

Notice that the more magnesium used, the steeper the gradient. This shows that the rate of reaction was greater. In addition, more hydrogen was made. The amount of product is directly proportional to the amount of reactant: 40 $cm^3$ of hydrogen was made when 0.04 g of magnesium was used, but double this was made when 0.08 g of magnesium was used. If the hydrochloric acid had been the limiting reactant, the same volume of hydrogen would have been produced in both reactions, and some magnesium would have been left in excess.

**B** What volume of hydrogen should be made if the experiment is repeated with 0.06 g of magnesium?

## Calculation involving gases

The amount of product formed depends upon the amount of the limiting reactant used in the reaction. The more reactant particles there are, the more of these particles react to make product particles. Remember that the mole shows the amount of a substance.

One mole of any gas occupies the same volume at room temperature and pressure, **rtp**. This **molar volume** is 24 $dm^3$. So 1 mol of hydrogen occupies 24 $dm^3$ at rtp, and so does 1 mol of carbon dioxide. The number of moles of gas can be calculated using this equation:

$$\text{amount of gas} = \text{volume of gas at rtp} \div \text{molar volume at rtp}$$

For example, 1.2 $dm^3$ of carbon dioxide at rtp contains $1.2 \div 24 = 0.05$ mol.

### Key words

**limiting reactant, excess, rtp, molar volume**

## Questions

1. How can you tell, from a graph of volume of gas against time, that a reaction has stopped?
2. Use the graph showing the volume of hydrogen produced by 0.04 g of magnesium to help you answer these questions.
   (a) What volume of gas was produced by 60 s?
   (b) How long did it take to produce 20 $cm^3$ of hydrogen?

E

3. Use the graph showing the volume of hydrogen produced by 0.04 g of magnesium to help you answer these questions.
   (a) What volume of gas was produced by 95 s?
   (b) How long did it take to produce 18 $cm^3$ of hydrogen?
4. What is a limiting reactant?

C

5. Calculate the amount, in moles, of each of the following gases at rtp:
   (a) 48 $dm^3$ of oxygen
   (b) 6 $dm^3$ of carbon dioxide
   (c) 120 $cm^3$ of hydrogen

A*

## Learning objectives

After studying this topic, you should be able to:

- ✔ describe that some reversible reactions reach equilibrium
- ✔ describe how a change in conditions may change the position of equilibrium
- ✔ explain why a reversible reaction may reach equilibrium
- ✔ explain how a change in conditions may change the position of equilibrium

▲ A see-saw can move in two directions and, correctly balanced, reaches equilibrium

### Exam tip OCR

- ✔ Not all the substances in an equilibrium mixture have to be at the same concentration.

**B** What can you say about the rate of the forward reaction and the rate of the backward reaction at equilibrium?

## Reversible reactions

In many reactions, the reaction can go in one direction only. In a **reversible reaction**, there is a forward reaction and a backward reaction. Instead of the usual arrow in the chemical equations, the symbol $\rightleftharpoons$ is used in reversible reactions. For example, the reaction between nitrogen and hydrogen to make ammonia is reversible:

$$\text{nitrogen} + \text{hydrogen} \rightleftharpoons \text{ammonia}$$

$$N_2 + 3H_2 \rightleftharpoons 2NH_3$$

There is a forward reaction between nitrogen and hydrogen to make ammonia, and a backward reaction in which ammonia decomposes to form nitrogen and hydrogen. In some reversible reactions these two reactions happen at the same time and at the same rate. These reversible reactions may reach **equilibrium**.

**A** What is the symbol for a reversible reaction?

## At equilibrium

When a reversible reaction reaches equilibrium, the rate of the forward reaction will be equal to the rate of the backward reaction. The concentrations of the reactants and products do not change. Note that, even though the concentrations of the reactants and products do not change at equilibrium, the forward and backward reactions are still happening. If the forward reaction is exothermic the backward reaction is endothermic, and vice versa.

## The position of equilibrium

The **position of equilibrium** is related to the ratio of the concentration of products to the concentration of reactants:

- It is on the left if the concentration of reactants is greater than the concentration of products.
- It is on the right if the concentration of products is greater than the concentration of reactants.

The position of equilibrium may change if there are changes in temperature, pressure, or the concentration of one of the reacting substances. This change in position of equilibrium will affect the composition of the mixture of substances.

### More about equilibrium

Imagine a reversible reaction starting off with just the substances on the left of the equation. These will react quickly because they are at a high concentration. The rate of the forward reaction will then decrease as these substances are used up.

However, as they are used up, the concentration of the substances on the right of the equation will increase. They will react increasingly quickly as their concentration increases. Eventually, the rate of the forward reaction will equal the rate of the backward reaction, and equilibrium is reached.

This works in a **closed system**, such as a stoppered flask or a beaker of liquid in which all the reactants stay in solution. If the system is open, the reacting substances may escape and equilibrium will not be reached.

### More about position of equilibrium

The equation shows the reaction to make sulfur trioxide:

$$2SO_2 + O_2 \rightleftharpoons 2SO_3$$

All the substances are gases. In a reversible reaction involving gases at equilibrium, if the pressure is increased the position of equilibrium moves to the side with the lower number of moles of gas molecules. In this example that will be to the right, as there are (2 + 1) = 3 mol of gas on the left and only 2 mol on the right.

If the concentration of one of the substances is increased, the position of equilibrium moves to the opposite side. In the example, the position of equilibrium will move to the right if the concentration of oxygen is increased.

If the concentration of one of the substances is decreased, the position of equilibrium moves to its side. In the example, the position of equilibrium will move to the right if the concentration of sulfur trioxide is decreased. This could happen simply by removing the sulfur trioxide as it is made.

If the temperature is increased, the position of equilibrium moves in the direction of the endothermic reaction. The production of sulfur trioxide is an exothermic process. Increasing the temperature moves the position of equilibrium to the left.

## Key words

**reversible reaction, equilibrium, position of equilibrium, closed system**

## Questions

1. How could you tell from its chemical equation that a reaction was reversible?
2. What happens to the concentrations of the reacting substances when a reversible reaction reaches equilibrium?
3. In general, what changes in conditions may change the position of equilibrium?

E

Questions 4 and 5 are about the reaction to make ammonia:

$$N_2 + 3H_2 \rightleftharpoons 2NH_3$$

4. If the concentration of ammonia is much greater than the concentration of nitrogen and hydrogen, is the position of equilibrium on the right or on the left?
5. Describe the composition of the equilibrium mixture when the position of equilibrium is on the left.

C

6. Hydrogen can be made by reacting methane with steam: $CH_4(g) + H_2O(g) \rightleftharpoons CO(g) + 3H_2(g)$ Describe and explain the effect on the position of equilibrium of increasing the concentration of methane, and increasing the pressure.

A*

# 12: Making sulfuric acid

## Learning objectives

After studying this topic, you should be able to:

- recall the raw materials used to make sulfuric acid
- recall the conditions used in the Contact Process
- describe the reactions involved in the Contact Process
- explain the conditions used in the manufacture of sulfuric acid

## Key words

**Contact Process**

## Exam tip OCR

- The percentage uses of sulfuric acid could be shown in a bar chart or in a pie chart. In this case however, the 'phosphate fertilisers' would be so much higher than the others, it would be difficult to appreciate the percentage use of the other uses.

Sulfur burns with a bright blue flame in oxygen, producing clouds of sulfur dioxide gas

## Sulfuric acid

Over a million tonnes of sulfuric acid are made in the UK each year, about 20 kg for each person in the country. Car batteries contain it, but most sulfuric acid is used to make other substances. The pie chart shows some of these uses.

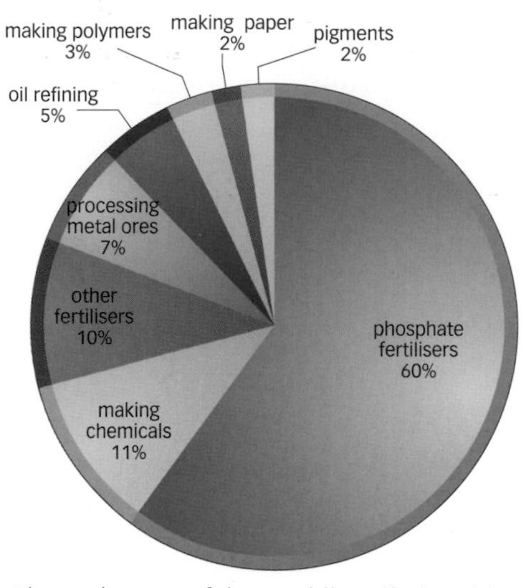


The main uses of the world's sulfuric acid

**A** What is the main use of sulfuric acid in the world?

Sulfur deposits are found naturally around volcanoes. Most sulfur today is extracted from crude oil and natural gas, which helps to reduce emissions of sulfur dioxide.

## The Contact Process

The **Contact Process** is used to make sulfuric acid. Three raw materials are needed:

- sulfur
- air (which provides oxygen)
- water.

There are three stages in the Contact Process. In stage 1, sulfur is burned in air to produce sulfur dioxide gas:

$$\text{sulfur} + \text{oxygen} \rightarrow \text{sulfur dioxide}$$
$$S + O_2 \rightarrow SO_2$$

In stage 2, sulfur dioxide and oxygen react together to produce sulfur trioxide gas:

$$\text{sulfur dioxide} + \text{oxygen} \rightleftharpoons \text{sulfur trioxide}$$

$$2SO_2 + O_2 \rightleftharpoons 2SO_3$$

Notice that the reaction is a reversible reaction. The conditions are chosen carefully to achieve a sufficient daily yield of sulfur trioxide at a reasonable cost. The temperature used is around 450 °C. The pressure used is atmospheric pressure, rather than a high pressure. Vanadium(V) oxide, $V_2O_5$, is used as a catalyst. It increases the rate of the reaction without being used up in the reaction.

In the final stage, the sulfur trioxide reacts with water to produce sulfuric acid:

$$\text{sulfur trioxide} + \text{water} \rightarrow \text{sulfuric acid}$$

$$SO_3 + H_2O \rightarrow H_2SO_4$$

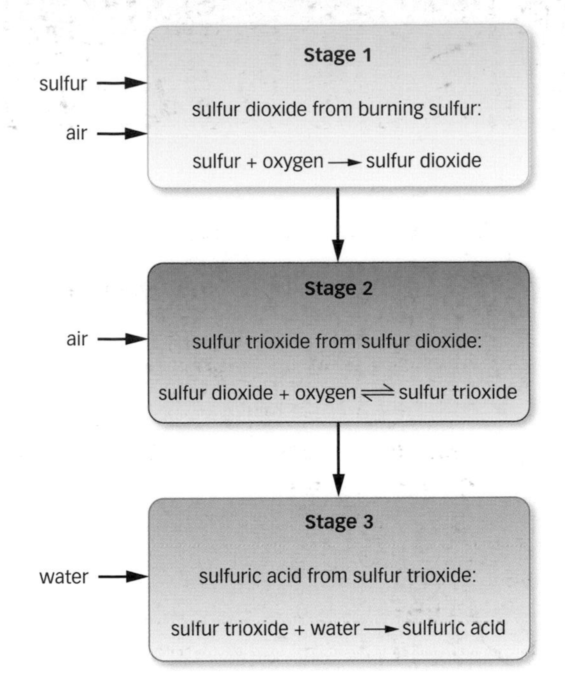


▲ The three stages of the Contact Process

**B** State the chemical formula of the catalyst used in the Contact Process.

## Choosing suitable conditions

As the second stage in the Contact Process involves a reversible reaction, the conditions needed must be chosen carefully.

The forward reaction, which produces sulfur trioxide, $SO_3$, is exothermic. This means that a greater yield can be obtained by reducing the temperature. On the other hand, the rate of reaction will be low if the temperature is too low. So an optimum temperature of around 450 °C is used. It is high enough to give a reasonable rate of reaction without decreasing the yield too much.

At atmospheric pressure, the position of equilibrium is far to the right. Almost all the sulfur dioxide is converted into sulfur trioxide. A high pressure would increase the yield of sulfur trioxide. However, the extra cost involved would not be worth it as there would only be a small increase in yield.

The $V_2O_5$ catalyst does not change the position of equilibrium, but it does increase the rate of reaction.

## Questions

1 Name the three raw materials used in the Contact Process.

2 Sulfur trioxide is made in the second stage of the Contact Process.
(a) Write a word equation for the reaction in this stage.
(b) Name the acid made by the Contact Process.

E

3 State the temperature and pressure used in the Contact Process.

4 What is the function of $V_2O_5$ in the Contact Process?

C

5 Write balanced symbol equations for the three stages of the Contact Process.

6 Explain the choice of conditions used in the Contact Process.

A*

## Learning objectives

After studying this topic, you should be able to:

- name some strong and weak acids
- describe acids in terms of pH, hydrogen ion concentration, and electrical conductivity
- explain the differences between strong and weak acids

## Key words

**weak acid, strong acid, hydrogen ions, ionise, electrode, electrolysis, electrical conductivity**

Vinegar contains ethanoic acid, a weak acid

**A** At the same concentration, acid A has a pH of 2 and acid B has a pH of 5. Which one is the weak acid, and why?

## Eating acid

Vinegar contains ethanoic acid. This gives vinegar the pleasant, sharp taste that makes it a popular addition to fish and chips. While you can safely consume vinegar, it would be very unwise to swallow hydrochloric acid. This is because ethanoic acid is a **weak acid**, but hydrochloric acid is a **strong acid**. Nitric acid and sulfuric acid are strong acids, too.

## Differences in pH

Strong acids and weak acids have different pH values, even if they are at the same concentration. A strong acid will have a lower pH than a weak acid of the same concentration. Its solution will be more strongly acidic. The differences in pH can be measured using universal indicator and a pH colour chart, or using a pH meter. The reason for these differences is due to **hydrogen ions** produced by the acids.

Acids **ionise** in water to produce hydrogen ions, $H^+$. Strong acids such as hydrochloric acid completely ionise in solution. Weak acids such as ethanoic acid only partially ionise in water. Many of their molecules do not release hydrogen ions. Unlike the reaction for strong acids, the reaction is reversible, and an equilibrium mixture is produced.

### Acid strength and concentration

The concentration of an acid is a measure of the number of moles of acid in 1 $dm^3$. The higher the concentration, the more moles of acid are dissolved in the same volume of water. The strength of an acid is a measure of how ionised the acid is in water. The stronger the acid, the more ionised it is. Hydrochloric acid is completely ionised in water, whereas ethanoic acid is only partially ionised in water:

$$HCl \rightarrow H^+ + Cl^- \quad \text{(hydrochloric acid)}$$

$$CH_3COOH \rightleftharpoons CH_3COO^- + H^+ \quad \text{(ethanoic acid)}$$

So, for a given concentration of acid, the concentration of hydrogen ions is greater in hydrochloric acid than it is in ethanoic acid. This is why the pH of a strong acid is much lower than the pH of a weak acid at the same concentration.

## Electrolysis of acids

Solutions containing ions conduct electricity because the ions are free to move. They carry charge from one electrode to the other. Acids in solution contain hydrogen ions and other ions, so they conduct electricity. Hydrogen gas is made at the negative **electrode** during the **electrolysis** of either hydrochloric acid or ethanoic acid. This is because the positively charged hydrogen ions are attracted to the negatively charged electrode, where they gain electrons and become hydrogen gas:

$$2H^+ + 2e^- \rightarrow H_2$$

**B** What gas is produced at the negative electrode during the electrolysis of acids?

However, although the electrolysis of both strong and weak acids produces hydrogen gas, there is a difference in the **electrical conductivity** of the acids. It is more 'difficult' for current to pass through a weak acid than it is through a strong acid at the same concentration. For a given potential difference or voltage, the current will be lower through the weak acid. The electrical conductivity of ethanoic acid is lower than it is for hydrochloric acid at the same concentration, because ethanoic acid solution contains fewer hydrogen ions.

The ammeter on the left is measuring the current through ethanoic acid, and the one on the right is measuring the current through hydrochloric acid at the same concentration

### More on conductivity

Ethanoic acid is less conductive than hydrochloric acid at the same concentration because it is a weak acid, whereas hydrochloric acid is a strong acid. There is a lower concentration of hydrogen ions to carry the charge through ethanoic acid than there is in hydrochloric acid.

### Exam tip OCR

- ✔ The lower the pH, the more strongly acidic a solution is.
- ✔ Make sure you understand the difference between strength and concentration in acids.
- ✔ The pH scale was developed using hydrogen ion concentrations. These vary so much that a linear scale would be had to understand. The pH scale uses 'logarithms', which reduces the gap between very large and very small numbers.

### Questions

1 Which will have the lower pH, 0.1 mol/dm³ ethanoic acid or 0.1 mol/dm³ hydrochloric acid?

2 Which will have the lower electrical conductivity, 0.1 mol/dm³ ethanoic acid or 0.1 mol/dm³ hydrochloric acid?

E

3 In terms of ions and ionisation, explain why ethanoic acid is a weak acid but hydrochloric acid is a strong acid.

4 Explain why hydrogen is produced at the negative electrode during the electrolysis of acids.

C

5 Explain the difference between acid strength and acid concentration. Include balanced equations referring to ethanoic acid and hydrochloric acid.

A*

## Learning objectives

After studying this topic, you should be able to:

- describe and explain the similarities and differences in the reactions of ethanoic acid and hydrochloric acid

The heating element in this electric kettle is coated with limescale

### Exam tip OCR

- Remember that a lighted splint. ignites hydrogen with a 'pop'. Carbon dioxide turns limewater milky.

**A** Which acid, at the same concentration, will produce the faster reaction with calcium carbonate, ethanoic acid or hydrochloric acid?

## Acid attack

Tap water contains dissolved mineral ions. Over time, particularly in hard water areas, **limescale** can build up in washing machines, kettles, and irons. This a deposit of calcium carbonate, which coats the surface of the heating elements in these appliances. Limescale looks unpleasant and it makes the appliances less efficient, wasting electricity.

Luckily, weak acids come to the rescue. Descalers contain a weak acid such as citric acid. This reacts with the limescale without damaging the metal of the heating element. A strong acid such as hydrochloric acid would remove the limescale. Unfortunately, it would also damage the metal. Why do weak and strong acids react so differently?

## Rate of reaction

Both ethanoic acid and hydrochloric acid react with magnesium, producing hydrogen. They also react with calcium carbonate, this time producing carbon dioxide. However, the reactions with ethanoic acid are slower than the reactions with hydrochloric acid, even if the two acids are at the same concentration. This is because ethanoic acid is a weak acid, but hydrochloric acid is a strong acid.

At the same concentration, there are fewer hydrogen ions in ethanoic acid. As a result, there are fewer collisions between hydrogen ions and the reactant particles.

The reaction between magnesium and hydrochloric acid produces large bubbles very quickly, keeping the magnesium powder at the surface. The reaction with ethanoic acid is slower and produces more, but smaller, bubbles. These spread through the mixture, giving it a milky appearance.

## Volume of gas

You might think that the reactions with ethanoic acid would produce smaller volumes of gas than they would with hydrochloric acid. However, if the volume and concentration of the acids are the same, and the same mass of magnesium or calcium carbonate is used, the same volume of gas is produced. It is just produced faster in reactions with hydrochloric acid, a strong acid. The volume of hydrogen or carbon dioxide produced depends on the amount of reactants present in the reaction mixture. It does not depend on the strength of the acid.

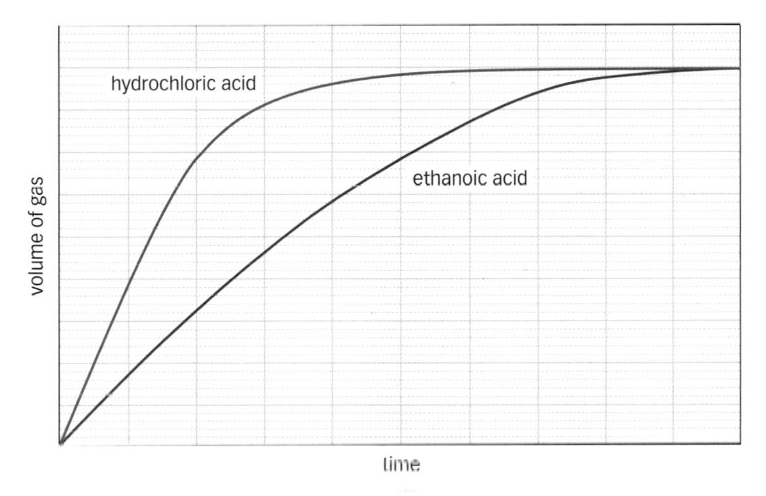


▲ The reactions between magnesium and ethanoic acid or hydrochloric acid produce the same final volume of hydrogen, provided the concentrations, masses, and volumes of reactants are the same

### More about rates and volumes

Hydrochloric acid reacts more quickly with magnesium and calcium carbonate than ethanoic acid does because it is a strong acid. At the same acid concentration, hydrochloric acid has a greater concentration of hydrogen ions, so the frequency of collisions between hydrogen ions and reactant particles is greater.

Ethanoic acid is only partially ionised in solution, producing an equilibrium mixture. As hydrogen ions become used up in reactions with reactant particles, the position of equilibrium moves further to the right, releasing more hydrogen ions into solution. This continues until all the acid has reacted, so the volume of gas produced is the same as for a strong acid.

**Key words**

**limescale**

**B** How can you tell from the graph that the rate of reaction between hydrochloric acid and magnesium is different from the rate of reaction between ethanoic acid and magnesium?

### Questions

1. Describe a non-food use of weak acids.
2. Ethanoic acid and hydrochloric acid both react with calcium carbonate to produce carbon dioxide. Compare the rates of reaction and volumes of gas produced in each reaction.

E

3. Explain why a strong acid such as hydrochloric acid might be inappropriate for use as a kettle descaler.
4. What does the volume of carbon dioxide produced in the reaction between calcium carbonate and an acid depend on?

C

5. Explain why a weak acid reacts with calcium carbonate more slowly than a strong acid does, but still produces the same volume of gas under the same conditions.

A*

# 15: Making insoluble compounds

## Learning objectives

After studying this topic, you should be able to:

- ✔ describe what happens in precipitation reactions
- ✔ describe the stages involved in the preparation of an insoluble salt by precipitation
- ✔ recognise and use state symbols in ionic equations
- ✔ explain the speed of precipitation reactions

## Key words

**precipitation reaction, precipitate, state symbol**

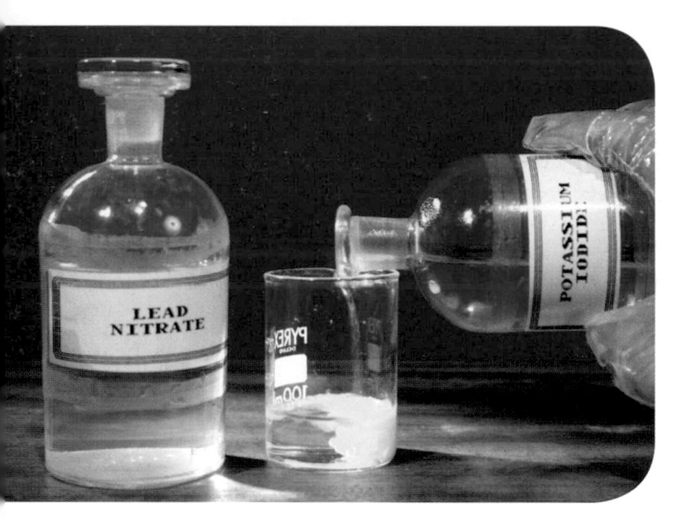


▲ Insoluble yellow lead iodide is formed in a precipitation reaction between lead nitrate solution and potassium iodide solution

**A** Why is potassium nitrate formed in the reaction between lead nitrate solution and potassium iodide solution?

## Precipitation reactions

A **precipitation reaction** happens when two different solutions react together to form an insoluble substance. For example, one happens when lead nitrate solution and sodium iodide solution are mixed together. They react to form potassium nitrate and insoluble lead iodide. The lead iodide forms a bright yellow **precipitate** of tiny particles. Precipitation reactions are a useful way to make insoluble compounds.

## Swapping places

Ionic substances, such as lead nitrate and potassium iodide, contain ions. These ions are joined together by strong ionic bonds in solids. They are in fixed positions and are unable to move from place to place. On the other hand, the ions can move from place to place when they are in solution.

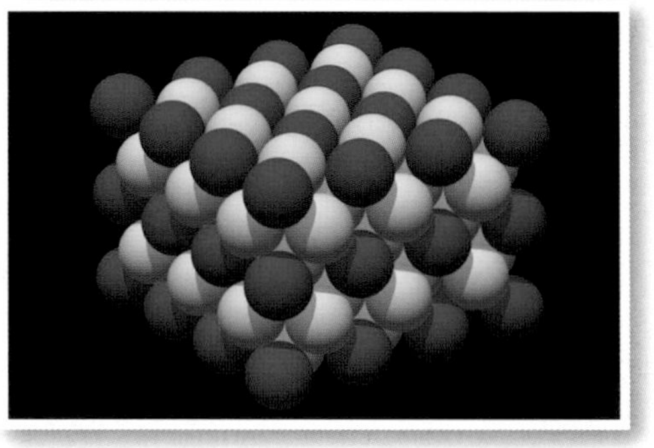

◀ The ions in solid ionic substances are held in fixed positions

Most precipitation reactions involve ions from two different solutions reacting with each other. The ions must collide with other ions so that they can react to form a precipitate. A precipitate will form if the new combination of ions produces an insoluble compound, such as lead iodide. It is as if the ions have swapped places:

lead nitrate + potassium iodide → potassium nitrate + lead iodide

### Fast reactions

The collision frequency between ions in solution is very large. There is a high chance that different ions will collide with each other and cause a reaction, so precipitation reactions are extremely fast. Precipitates form as soon as two suitable solutions are mixed together.

## State symbols in equations

**State symbols** are used in balanced symbol equations. They show the state of each substance: (s) means solid, (l) means liquid, (g) means gas, and (aq) means aqueous solution or dissolved in water. Here is the equation for the formation of lead iodide:

$$Pb(NO_3)_2(aq) \quad + \quad 2KI(aq) \quad \rightarrow \quad 2KNO_3(aq) \quad + \quad PbI_2(s)$$

You can tell from the state symbols (s) and (aq) that lead iodide, $PbI_2$, forms a solid and the other three substances are in aqueous solution.

## Making an insoluble salt

Insoluble compounds can be prepared or made by precipitation. There are three main stages:

1. A suitable combination of solutions is mixed together to form a precipitate of the required insoluble compound.
2. The mixture is filtered to separate the precipitate from the other reactants and product.
3. The precipitate is washed with water while it is on the filter paper. It is then dried, for example, in a warm oven.

### Questions

1 What is a precipitation reaction?

2 Describe and explain the meaning of the four state symbols.

E

3 Describe what happens to the ions in a precipitation reaction.

4 Magnesium carbonate is an insoluble compound made by reacting magnesium chloride solution with sodium carbonate solution.

(a) Write the word equation for the reaction.

(b) Outline how you could prepare a dry sample of magnesium carbonate using a precipitation reaction.

C

5 Explain the speed of precipitation reactions.

A*

**Exam tip** OCR

✔ Take care not to confuse (l) and (aq). For example, NaCl(l) means molten or liquid sodium chloride and NaCl(aq) means sodium chloride dissolved in water.

①

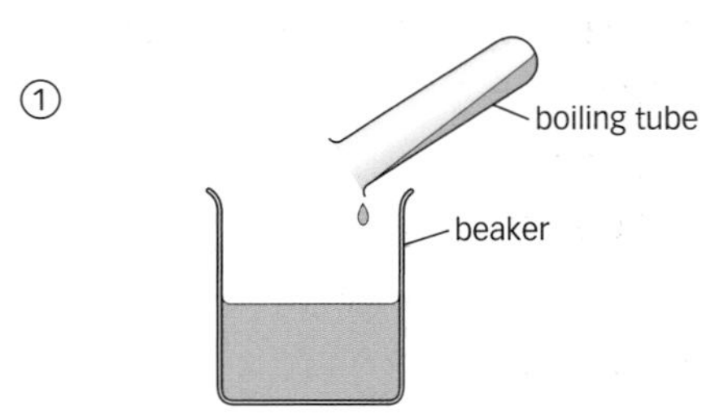


Mix two suitable solutions

②

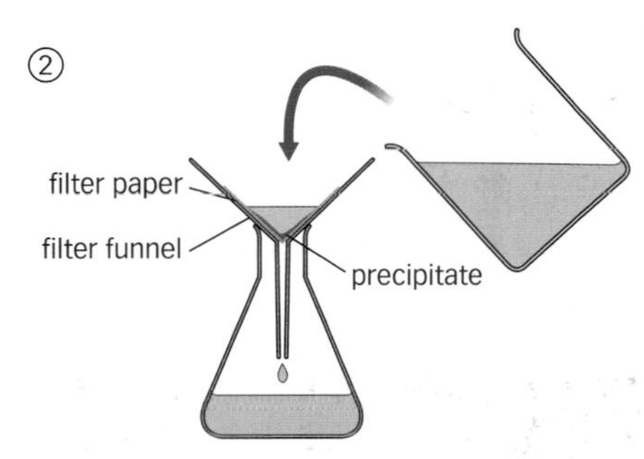


Filter to separate the precipitate

③

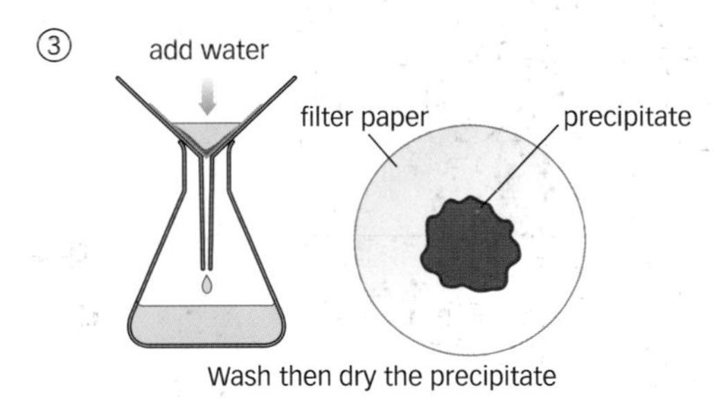


Wash then dry the precipitate

▲ Dry insoluble compounds can be prepared using precipitation reactions

**Exam tip** 

✔ Make sure you can label the apparatus needed to make an insoluble compound by precipitation.

# 16: Testing ions

## Learning objectives

After studying this topic, you should be able to:

- describe the laboratory tests for halide ions and sulfate ions
- construct word equations for precipitation reactions
- identify the reactants and products in ionic equations
- construct ionic equations for precipitation reactions
- explain the idea of spectator ions

## Key words

**ionic equation**, **spectator ions**

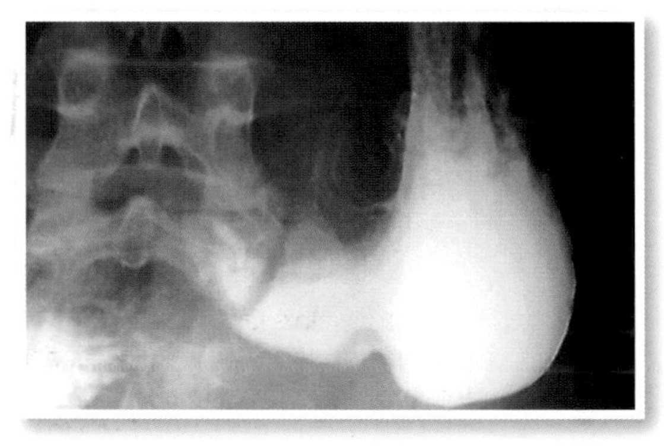

This X-ray image shows a patient's stomach on the right, lined with barium sulfate

**A** How can you tell from the equations above that barium sulfate forms a precipitate?

A white precipitate when barium chloride solution is added shows the presence of sulfate ions

## A medical precipitate

Barium sulfate is an insoluble compound that can be prepared by precipitation reactions. X-rays pass through it with difficulty, making it useful for medical investigations. A patient who needs an X-ray image of their digestive system is given a 'barium meal', a drink containing barium sulfate. The outline of the digestive system shows up in the X-ray image, giving doctors the information they need.

## Making barium sulfate

Barium sulfate can be prepared by mixing together solutions of barium chloride and sodium sulfate:

barium chloride + sodium sulfate → sodium sulfate + barium sulfate

$$BaCl_2(aq) + Na_2SO_4(aq) \rightarrow Na_2SO_4(aq) + BaSO_4(s)$$

The balanced symbol equation can be simplified so that it just shows the ions that react together to make the precipitate:

$$Ba^{2+}(aq) + SO_4^{2-}(aq) \rightarrow BaSO_4(s)$$

This sort of equation is called an **ionic equation**. The barium ions and sulfate ions in this example are the reactants, and the barium sulfate precipitate is the product. Note that barium ions and sulfate ions will make a precipitate together, even if they came from different soluble compounds than barium chloride and sodium sulfate. This is the basis of a simple test to see if a solution contains sulfate ions.

## Testing for sulfate ions

A sample of the test solution is placed in a test tube, then acidified with a few drops of hydrochloric acid. A few drops of barium chloride solution are added. If sulfate ions are present, a white precipitate of barium sulfate forms.

## Testing for halide ions

Halide ions are ions produced by the halogens, elements in Group 7 of the periodic table. They include chloride ions, $Cl^-$, bromide ions, $Br^-$, and iodide ions, $I^-$. The presence of these ions can be detected using lead nitrate solution:

- Chloride ions give a white precipitate.
- Bromide ions give a cream precipitate.
- Iodide ions give a yellow precipitate.

A sample of the test solution is placed in a test tube, then acidified with a few drops of nitric acid. A few drops of lead nitrate solution are added. If a precipitate forms, its colour lets you identify the halide ion present.

**B** Describe the colours of lead chloride, lead bromide, and lead iodide.

## More about ionic equations

You should be able to write the ionic equation for a precipitation reaction, if you are given the ions present and the identity of the products. For example, lead nitrate and sodium chloride react together to form sodium nitrate and a white precipitate of lead chloride. The mixture contains aqueous $Pb^{2+}$, $NO_3^-$, $Na^+$, and $Cl^-$ ions. It must be the aqueous $Pb^{2+}$ and $Cl^-$ ions that react together to make lead chloride, so the ionic equation is:

$$Pb^{2+}(aq) \quad + \quad 2Cl^-(aq) \quad \rightarrow \quad PbCl_2(s)$$

The other ions, $Na^+$ and $NO^{-3}$, are called **spectator ions**. They do not take part in the reaction, but instead form the other product, aqueous sodium nitrate.

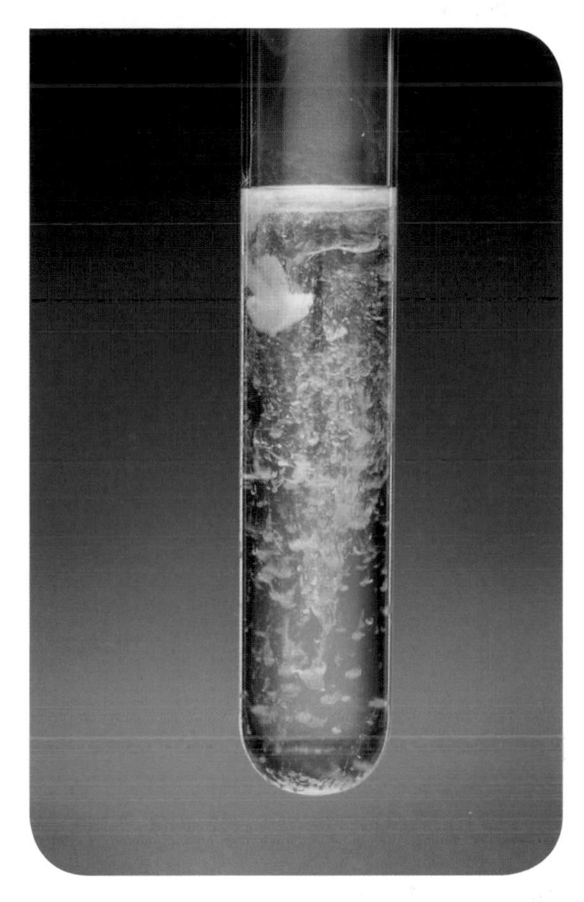

The formation of a yellow lead iodide precipitate when lead nitrate solution is added shows the presence of iodide ions

### Questions

1. Describe the colours of the precipitates formed when lead nitrate solution is added to solutions containing chloride, bromide, or iodide ions.
2. Describe the test for sulfate ions.

E

3. A solution produces a cream-coloured precipitate with lead nitrate solution, and a white precipitate with barium chloride solution. Which ions must it contain?

C

4. What is a spectator ion?
5. Lead nitrate and potassium bromide solution react together to form potassium nitrate and a precipitate of lead bromide.
   (a) Identify the ions present in the mixture.
   (b) Write the ionic equation for the formation of the precipitate, including state symbols.
   (c) Explain why potassium nitrate also forms.

A*

### Exam tip OCR

- ✔ You may be asked to interpret the results of tests using barium chloride solution and lead nitrate solution.
- ✔ Precipitation reactions are useful for identifying the ions present in unknown substances. This can be useful in forensic science.

# Module summary

## Revision checklist

- The amount of a substance is measured in moles. Molar mass = formula mass in grams.
- Number of moles = $\frac{\text{mass}}{\text{molar mass}}$
- Mass is conserved in chemical reactions.
- Reacting masses in reactions are calculated using formula masses and ratios in equations.
- Percentage composition of elements in compounds $= \frac{\text{mass of element}}{\text{mass of compound}} \times 100$
- The empirical formula of a compound is the simplest ratio of atoms (or moles) of elements in the compound.
- The concentration of a solution is how much solute is dissolved in 1 $dm^3$ of the solution. In solutions, number of moles = concentration (in mol/$dm^3$) × volume.
- In titrations, acids are slowly added to alkalis. The pH changes suddenly at the end point of the reaction. Indicators demonstrate this.
- Solutions are measured out using burettes and pipettes. The amount of gas produced as a product of a reaction is measured using a gas syringe.
- Titrations are used to calculate concentrations of solutions.
- Number of moles of a gas = $\frac{\text{volume of gas}}{\text{molar volume (24 dm}^3\text{)}}$
- Reversible reactions are shown by the ⇋ symbol.
- Reactions reach equilibrium if rate of forward reaction = rate of backward reaction. Changes in concentration, pressure, and temperature affect equilibrium position.
- The Contact Process (forming sulfuric acid from sulfur, air, and water) involves a reversible reaction. A temperature of 450 °C, atmospheric pressure, and a $V_2O_5$ catalyst are used.
- Strong acids completely ionise in water (forming $H^+$ ions). Weak acids partly ionise and have higher pH values.
- Strong acids have a higher concentration of $H^+$ ions than weak acids. They react faster with magnesium and carbonates and have greater conductivity.
- In precipitation reactions, two solutions react together to make an insoluble substance.
- Precipitation reactions are used to test for sulfate ions, halide ions, or to prepare insoluble salts.

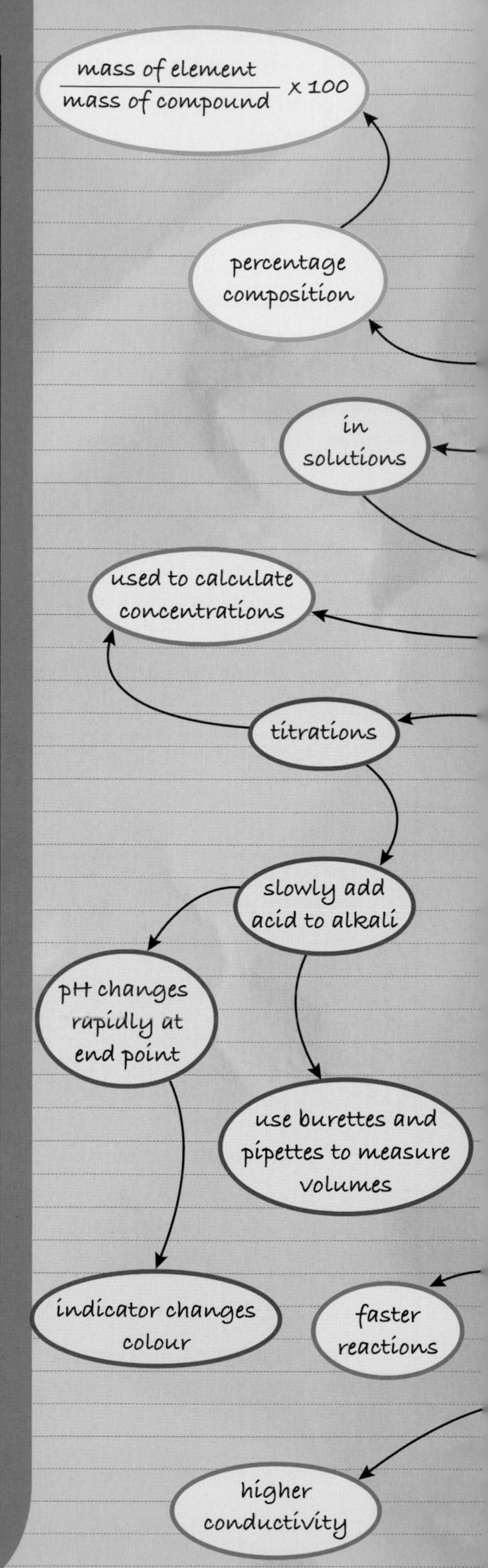




NOW USE THE C5 GRADE CHECKER ON PAGE 248

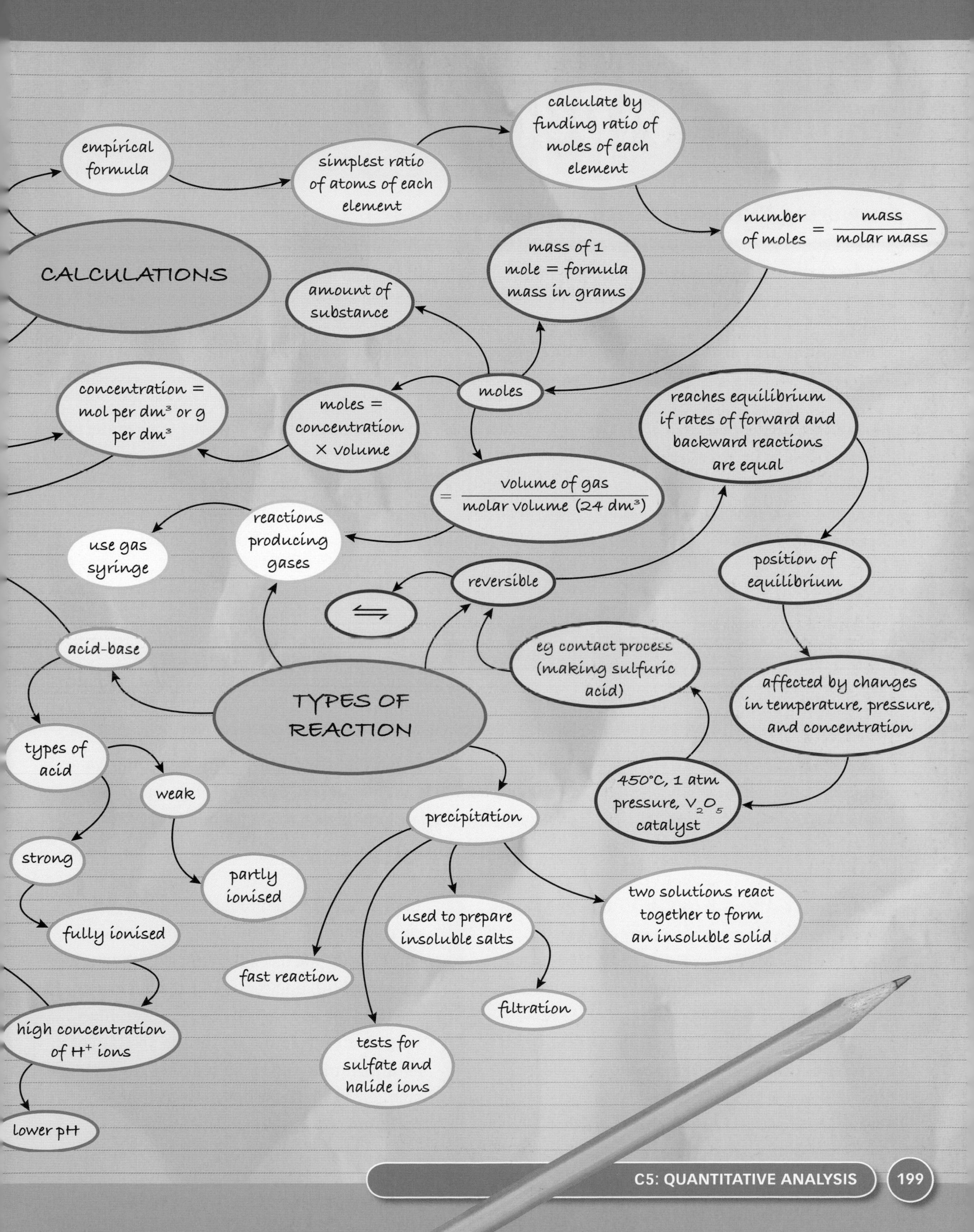
empirical formula
simplest ratio of atoms of each element
calculate by finding ratio of moles of each element
number of moles = mass / molar mass
CALCULATIONS
mass of 1 mole = formula mass in grams
amount of substance
moles
concentration = mol per dm³ or g per dm³
moles = concentration × volume
reaches equilibrium if rates of forward and backward reactions are equal
= volume of gas / molar volume (24 dm³)
reactions producing gases
use gas syringe
reversible
⇌
position of equilibrium
acid-base
eg contact process (making sulfuric acid)
affected by changes in temperature, pressure, and concentration
TYPES OF REACTION
types of acid
weak
450°C, 1 atm pressure, V₂O₅ catalyst
precipitation
strong
partly ionised
two solutions react together to form an insoluble solid
used to prepare insoluble salts
fully ionised
fast reaction
filtration
high concentration of H⁺ ions
tests for sulfate and halide ions
lower pH

## Answering Extended Writing questions

QUESTION

Sulfuric acid is manufactured by the Contact Process, which uses sulfur as the raw material from which to produce sulfuric acid. Describe how sulfur is converted into sulfuric acid and explain what conditions are used in the Contact Process.

**The quality of written communication will be assessed in your answer to this question.**

As well as sulfur you need air and oxygen. Getting a lot of sulfuric acid is difficult because the reaction is revursable so special conditions are used.

E

Examiner: The answer mentions one other raw material (air), but the second one is water. The candidate has correctly explained the problem about the reversible reaction (note spelling). If the conditions had been listed, this might have been an answer in the D–C band.

Sulfur dioxide reacts to make sulfuric acid in the Contact Process using high temperature and a vanadium catalyst. These help the reaction to be fast, oxygen is also needed for burning the sulfur.

C

Examiner: This is an incomplete answer. The candidate needs to make clear that the Contact Process only refers to the conversion of sulfur dioxide to sulfur trioxide. There is not much detail about the conditions – the candidate should give the actual temperature and mention that atmospheric pressure is used. The catalyst is actually vanadium oxide. The final step (adding water to sulfur trioxide) is not mentioned. Spelling is fine, but a full stop should be used to split the last sentence into two.

First of all sulfur is burnt in air to make sulfur dioxide. This will react with more oxygen to make sulfur trioxide in the Contact Process, but the reaction is in equilibrium. You would expect a high pressure and high temperature to move the position of equilibrium and make the yeald greater but that would be expensive so 450 °C and normal pressure is used, also a $V_2O_5$ catalyst. Finally the sulfur trioxide reacts with water to make sulfuric acid.

A*

Examiner: The candidate includes a lot of correct facts about the conditions and goes some way towards explaining why they are chosen. A full answer would also mention the need for increased rate. However, high temperature actually moves the position of equilibrium to the left and makes the yield smaller, so 450 °C is chosen to give the optimum combination of rate and yield.

# Exam-style questions

1 Calcium carbonate breaks down when it is heated to make calcium oxide + carbon dioxide.

calcium carbonate → calcium oxide + carbon dioxide

Asif heated 10 g of calcium carbonate. He was left with 5.6 g of calcium oxide.

A02 **a** Calculate the mass of carbon dioxide given off in this experiment.

A02 **b** Is mass conserved in this reaction? Explain your answer.

↓ E

2 Alkanes are hydrocarbons that are used as fuels. Different alkanes have different formulae.

A01 **a** **i** What is meant by the term empirical formula?

A02 **ii** The chemical formula of butane is usually written as $C_4H_{10}$. Write down its empirical formula.

A02 **b** 11.6 g of butane ($C_4H_{10}$) contains 9.6g of carbon. Calculate the percentage by mass of carbon in butane.

↓ C

3 William has a bottle of sodium hydroxide of unknown concentration. He takes 25 cm³ (0.025 dm³) and titrates it against hydrochloric acid with a concentration of 0.1 mol/dm³.

A02 **a** 18.2 cm³ of hydrochloric acid was the average volume needed to neutralise the sodium hydroxide. The hydrochloric acid and sodium hydroxide react in a 1:1 ratio.

**i** Calculate the average amount in moles of hydrochloric acid used in the titration.

**ii** What amount in moles of sodium hydroxide reacts with this amount of hydrochloric acid?

**iii** Use your result from **ii** and other information from the question to calculate the concentration of the sodium hydroxide in mol/dm³.

↓ A*

## Extended Writing

4 A02 Write instructions for a titration to find out what volume of sulfuric acid is needed to neutralise a solution of sodium hydroxide.

↓ E

5 A01 Hydrochloric acid (HCl) is a strong acid and ethanoic acid ($CH_3COOH$) is a weak acid. Explain the meaning of these terms and describe how the properties of hydrochloric acid are different to those of ethanoic acid.

↓ C

6 A02 The Contact Process is a method for manufacturing sulfuric acid. The key stage is a reaction between sulfur dioxide and oxygen:

$$2SO_2 + O_2 \rightleftharpoons 2SO_3$$

A temperature of 450 °C and atmospheric pressure are used, along with a $V_2O_5$ catalyst. Explain why these conditions are chosen.

↓ A*

A01 Recall the science

A02 Apply your knowledge

A03 Evaluate and analyse the evidence

# Chemistry out there

| OCR specification match | | |
|---|---|---|
| C6a | 6.1 | Electrolysis |
| C6a | 6.2 | Electrolysis of solutions |
| C6b | 6.3 | Fuel cells |
| C6b | 6.4 | Uses of fuel cells |
| C6c | 6.5 | Redox reactions |
| C6c | 6.6 | Redox and rusting |
| C6d | 6.7 | Fermentation |
| C6d | 6.8 | Alcohols |
| C6e | 6.9 | The ozone layer |
| C6e | 6.10 | Looking after the ozone layer |
| C6f | 6.11 | Water hardness |
| C6f | 6.12 | Removing water hardness |
| C6g | 6.13 | Natural fats and oils |
| C6g | 6.14 | Emulsions and soap |
| C6h | 6.15 | Detergents |
| C6h | 6.16 | Cleaning and washing |
| Module summary | | |
| OCR Upgrade | | |

## Why study this module?

Chemistry is all around you, at home as well as in industry. Some industrial chemicals have brought us great benefits, while others have caused unforeseen problems. In this module, you will learn about electrolysis and its uses. Fuel cells, used in spacecraft and submarines, hold the promise of cleaner electricity for use in vehicles. You will learn about fuel cells, alternative fuels such as bioethanol, and the chemistry behind them.

CFCs were widely used in the second half of the last century, but were found to damage the ozone layer that protects life on Earth from harmful ultraviolet light. You will discover why they were used, how they cause damage, and the steps taken to reverse this. Closer to home, you will find out about hard and soft water, natural fats and oils, and the chemistry behind keeping yourself clean.

### You should remember

1. Electrolysis can be used to purify copper and to make useful chemicals from sodium chloride.
2. Heat energy is released by burning fuels.
3. Rusting is a slow reaction between iron, oxygen and water.
4. Oxidation involves reacting with oxygen, gaining or losing electrons.
5. Alcohols react with acids to make an ester and water.
6. Bromine water is used in a laboratory test for unsaturation in hydrocarbons.
7. Emulsions are mixtures in which one liquid is finely dispersed in another liquid.
8. Emulsifiers are molecules with a water hydrophilic part and a hydrophobic part.

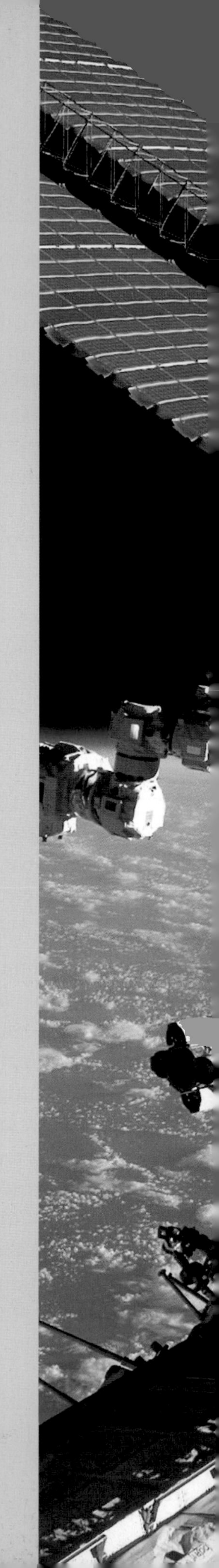

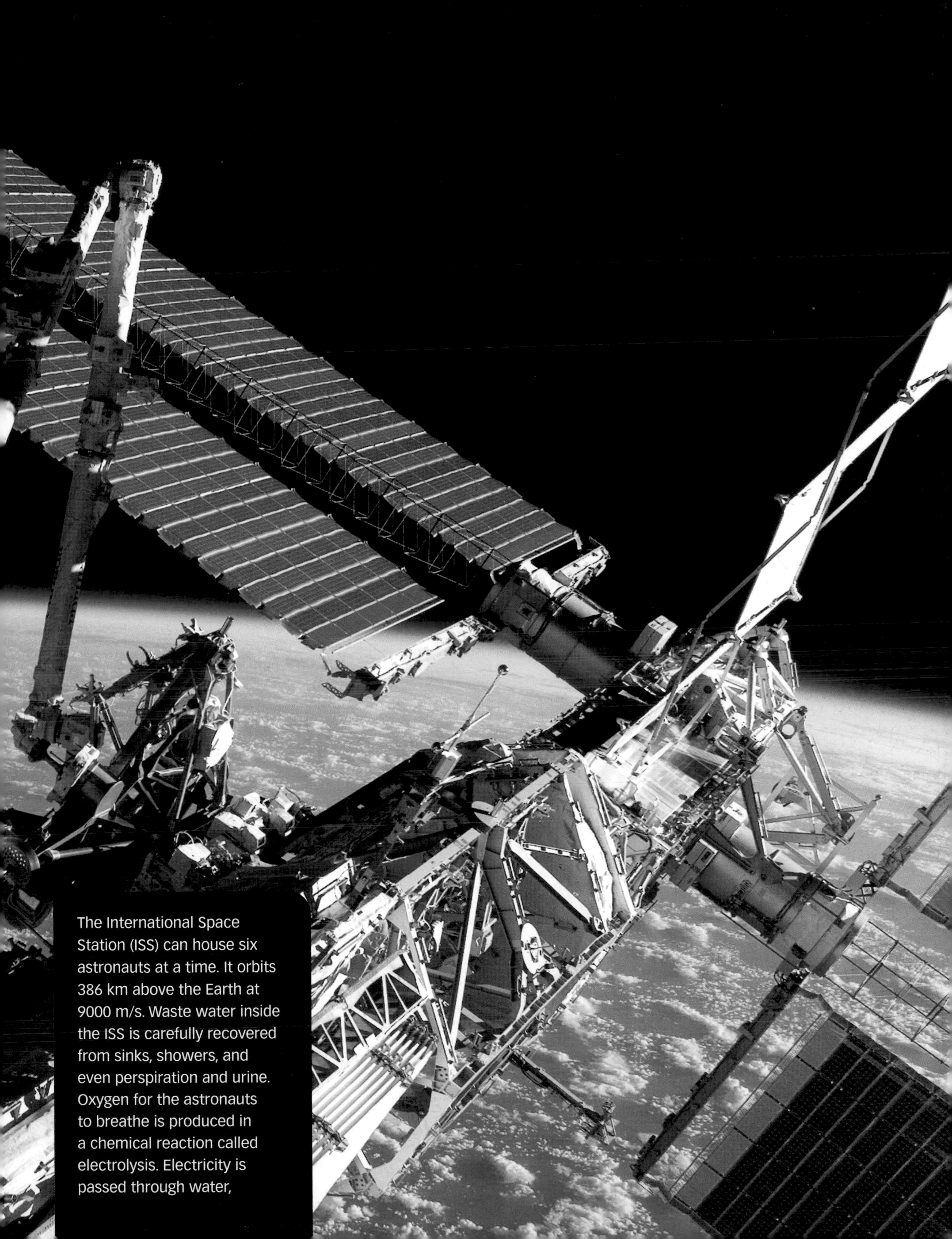

The International Space Station (ISS) can house six astronauts at a time. It orbits 386 km above the Earth at 9000 m/s. Waste water inside the ISS is carefully recovered from sinks, showers, and even perspiration and urine. Oxygen for the astronauts to breathe is produced in a chemical reaction called electrolysis. Electricity is passed through water,

## Learning objectives

After studying this topic, you should be able to:

- describe and explain electrolysis
- predict the products of electrolysis of molten ionic compounds
- write half equations for electrode processes

## Key words

**electrolyte, cathode, cation, anode, anion, discharged, half equation**

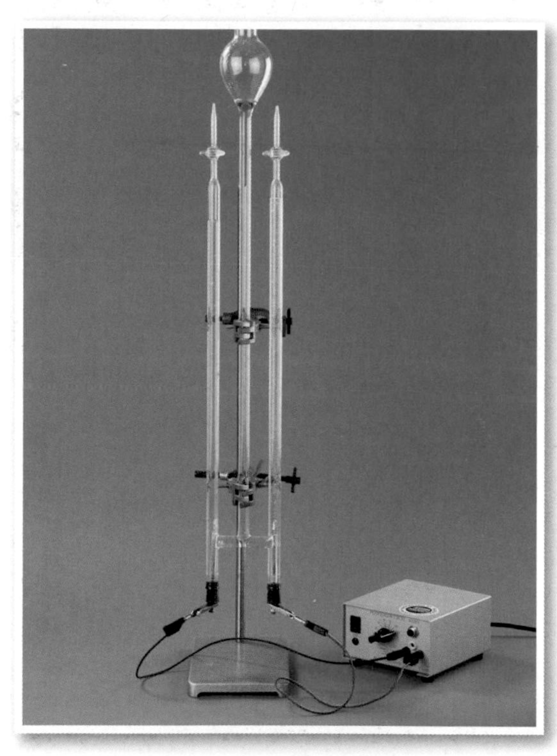

The Hofmann voltameter uses platinum electrodes and a d.c. power pack to electrolyse acidified water

**A** What is electrolysis?

## Splitting water

Alessandro Volta invented the battery in 1800. Scientists were soon keen to build their own 'voltaic pile' and find out what happens when an electric current is passed through a substance. William Nicholson built England's first voltaic pile in 1800 and tested it with water. He discovered bubbles of hydrogen coming from one wire and oxygen from the other wire. The race was on to see what else electrolysis could reveal.

## Electrolysis of water

Electrolysis is the decomposition of a liquid by passing an electric current through it. The liquid that is decomposed or broken down is called the **electrolyte**. It must be able to conduct electricity, and this can only happen if its ions are free to move:

- The negative electrode is called the **cathode**. Positively charged ions called **cations** move towards it.
- The positive electrode is called the **anode**. Negatively charged ions called **anions** move towards it.

The potential difference across the two electrodes causes the ions to move and so carry charge through the liquid. When the ions reach the electrodes, they may be **discharged**, releasing elements as products:

- Cations (positively charged ions) are discharged at the cathode, the negative electrode.
- Anions (negatively charged ions) are discharged at the anode, the positive electrode.

Water contains small concentrations of hydrogen ions, $H^+(aq)$, and hydroxide ions, $OH^-(aq)$, discharged at the electrodes during electrolysis. The hydrogen ions are discharged at the cathode, producing hydrogen, $H_2(g)$. The hydroxide ions are discharged at the anode, producing oxygen, $O_2(g)$.

### Half equations

A **half equation** describes what happens at an electrode during electrolysis. For example, during the electrolysis of water:

- At the cathode: $2H^+(aq) + 2e^- \rightarrow H_2(g)$
- At the anode: $4OH^-(aq) - 4e^- \rightarrow O_2(g) + 2H_2O(l)$

Notice that the atoms and the charges are balanced.

## Molten electrolytes

1807 was an exciting year for Humphry Davy. He passed an electric current through molten potassium hydroxide and was rewarded by being the first person to see potassium metal. Later the same year, Davy discovered sodium in a similar way using molten sodium hydroxide.

When ionic substances are solid, their ions are in fixed positions and cannot move from place to place. The substances cannot be electrolysed. However, they can be electrolysed if they are heated until they melt, because the ions in a molten liquid can move. During electrolysis, metal ions or hydrogen ions move to the negative electrode, and non-metal ions (apart from hydrogen ions) move to the positive electrode. For example, when molten aluminium oxide is electrolysed, aluminium forms at the negative electrode and oxygen at the positive electrode.

Humphrey Davy's discoveries led to the development of ways to manufacture metals such as aluminium on a huge scale. This has brought economic benefits, but also an environmental damage through mining and the production of waste materials.

### Questions

1 What are anions, cations, and electrolytes?

2 When molten lead bromide is electrolysed, lead forms at the cathode and bromine at the anode. Predict the products formed at each electrode during the electrolysis of molten lead iodide and molten sodium chloride.

E

3 Explain why there is a flow of charge during electrolysis.

4 Explain why ionic substances can be electrolysed when molten as liquids but not when solid.

C

5 Write half equations for the electrode reactions that happen when the following are electrolysed:

(a) $PbBr_2$(l) (contains $Pb^{2+}$(l) ions and $Br^-$(l) ions).

(b) $PbI_2$(l) (contains $Pb^{2+}$(l) ions and $I^-$(l) ions).

(c) NaCl(l) (contains $Na^+$(l) ions and $Cl^-$(l) ions).

A*

### Exam tip OCR

✔ You should be able to describe the chemical tests for hydrogen and oxygen. Hydrogen burns with a 'pop' when lit with a lighted splint. Oxygen relights a glowing splint.

### Did you know...?

Humphry Davy danced around his laboratory in delight when he discovered potassium.

▲ This engraving shows the electric battery built at the Royal Institution in London for Davy's experiments. It was the most powerful battery at the time.

### More half equations

Half equations can be written for the electrode reactions during electrolysis of molten compounds. For example, for the electrolysis of molten aluminium oxide:

- At the cathode:
  $Al^{3+}(l) + 3e^- \rightarrow Al(l)$
- At the anode:
  $2O^{2-}(l) - 4e^- \rightarrow O_2(g)$

# 2: Electrolysis of solutions

## Learning objectives

After studying this topic, you should be able to:

- describe the apparatus needed to electrolyse solutions in school
- describe the electrolysis of copper(II) sulfate solution
- describe the factors affecting the amount of substance formed during electrolysis
- write half equations for electrode processes in solutions
- carry out simple calculations relating amount of substance to current and time during electrolysis

▲ Pencil lead is graphite, so even pencils can be used as electrodes. During the electrolysis of aqueous potassium iodide, yellow-brown iodine is released at the anode and hydrogen is released at the cathode.

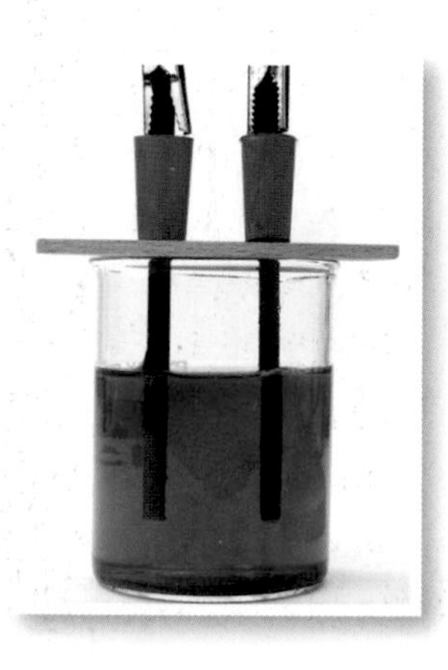

◀ Copper(II) sulfate solution can be electrolysed using a d.c. power supply and carbon electrodes. Copper is deposited at the anode, on the left here, and oxygen is released at the cathode.

## Simple electrolysis

The electrolysis of molten liquids can be difficult to do because of the high temperatures involved. On the other hand, the electrolysis of aqueous solutions of ionic substances is usually quite easy to do. This is because aqueous solutions are just substances dissolved in water.

The electrolyte is poured into a beaker. Two graphite rods are used as electrodes. Graphite is a form of carbon that conducts electricity and it does not usually react with the electrolysis products. A d.c. power supply, or simply an ordinary battery, supplies the potential difference needed for a current to flow through the electrolyte.

## Electrolysis of aqueous copper(II) sulfate

Copper(II) sulfate, $CuSO_4$, is an ionic compound that dissolves in water to form a blue aqueous solution. During the electrolysis of aqueous copper(II) sulfate, copper is formed at the cathode, which becomes coated with copper. Oxygen is formed at the anode, and bubbles of this gas are released from the surface of the anode.

## Some other solutions

Copper is less reactive than hydrogen, so it is produced at the cathode during the electrolysis of aqueous copper(II) sulfate. However, if the metal in solution is more reactive than hydrogen, or there is no metal in solution, hydrogen gas is produced instead.

If the aqueous solution contains chloride ions, chlorine is produced at the anode. If not, oxygen gas is produced instead. The table shows five examples of how this works.

| Electrolyte | Product at cathode | Product at anode |
|---|---|---|
| $CuCl_2$(aq) | copper | chlorine |
| $CuSO_4$(aq) | copper | oxygen |
| $KNO_3$(aq) | hydrogen | oxygen |
| NaOH(aq) | hydrogen | oxygen |
| $H_2SO_4$(aq) | hydrogen | oxygen |

**A** What products would be formed during the electrolysis of hydrochloric acid, HCl?

## More about electrolysis

Water contains $H^+(aq)$ ions and $OH^-(aq)$ ions. It may be easier for these to be discharged during electrolysis than for the ions from the solute to be discharged. This is the case with ions from reactive metals, like $Na^+(aq)$ ions and $K^+(aq)$ ions. Similarly, it is easier for $OH^-(aq)$ ions to be discharged than for $NO_3^-(aq)$ ions and $SO_4^{2-}(aq)$ ions to be discharged.

Copper forms at the cathode during the electrolysis of aqueous copper(II) sulfate, but hydrogen forms there during the electrolysis of aqueous potassium nitrate:

- $Cu^{2+}(aq) + 2e^- \rightarrow Cu(s)$ . . . and . . . $2H^+(aq) + 2e^- \rightarrow H_2(g)$

Chlorine forms at the cathode during the electrolysis of aqueous copper(II) chloride, but oxygen forms there during the electrolysis of aqueous copper sulfate:

- $2Cl^-(aq) - 2e^- \rightarrow Cl_2(g)$ . . .
  and . . . $4OH^-(aq) - 4e^- \rightarrow O_2(g) + 2H_2O(l)$

## How much?

The amount of any product made during electrolysis is affected by the electric current used, and the time taken for the process. The amount made is **directly proportional** to the current, and to the time. For example, during the electrolysis of aqueous copper(II) sulfate, twice as much copper will be made if the current or time is doubled. Four times as much copper will be made if both the current and time are doubled.

## How much again?

The amount of charge transferred during electrolysis is equal to the current multiplied by the time. For example, 1 g of hydrogen is produced from acid when 2 A flows for 800 minutes. In the same time, 0.5 g would be made using a current of 1 A. It would take 80 minutes to make 1 g if a current of 20 A is used.

### Key words

**directly proportional**

### Questions

1. Describe the apparatus needed to carry out electrolysis in the school laboratory.

2. Describe what happens during the electrolysis of aqueous copper(II) sulfate.
3. Describe the products formed during the electrolysis of aqueous sodium hydroxide, NaOH(aq), and dilute sulfuric acid, $H_2SO_4(aq)$.

4. Describe the factors affecting the amount of product formed during electrolysis.
5. Dilute sulfuric acid contains $H^+(aq)$, $SO_4^{2-}(aq)$, and $OH^-$ ions. Write half equations for the electrode reactions that happen when it is electrolysed.
6. In an electrolysis experiment using a current of 1.0 A, 1.0 g of copper was deposited in 60 minutes.

   (a) How long will it take to deposit 3.0 g of copper if a current of 2 A is used?
   (b) What current will be needed to deposit 10.0 g of copper in 4 hours?

## Learning objectives

After studying this topic, you should be able to:

- outline how a hydrogen-oxygen fuel cell works
- write the word and symbol equations for the overall reaction in a hydrogen-oxygen fuel cell
- draw and interpret an energy level diagram for the reaction between hydrogen and oxygen
- explain the changes that take place at each electrode in a hydrogen-oxygen fuel cell

## Key words

**fuel cell, exothermic reaction, energy level diagram**

**A** What is an exothermic reaction?

## Grove's gaseous voltaic battery

In 1800, William Nicholson passed an electric current through water using platinum electrodes and produced hydrogen and oxygen. This gave scientists an idea. They thought that the process might be reversed. It might be possible to pass hydrogen and oxygen over platinum electrodes and produce an electric current, with water as the waste product. In 1839, the Welsh scientist William Grove invented a device that did this. Grove's 'gaseous voltaic battery' produced an electric current as long as hydrogen and oxygen were bubbled into it. His invention was the first hydrogen-oxygen **fuel cell**.

## Hydrogen and oxygen

Hydrogen is a highly flammable gas. It reacts explosively with oxygen if it is ignited with a spark or flame, producing water vapour. The reaction is an **exothermic reaction** because it releases energy to the surroundings. It is difficult to use hydrogen as a fuel in this way. Some vehicles use hydrogen instead of petrol in their engines, but this is less efficient than using a fuel cell.

### Energy level diagrams

The energy involved in a chemical reaction can be represented using an **energy level diagram**. This is a graph in which the energy contained in the reactants and products is plotted on the vertical axis, and the course of the reaction on the horizontal axis. Each energy level is shown as a horizontal line. The higher up it is, the more energy the substance contains.

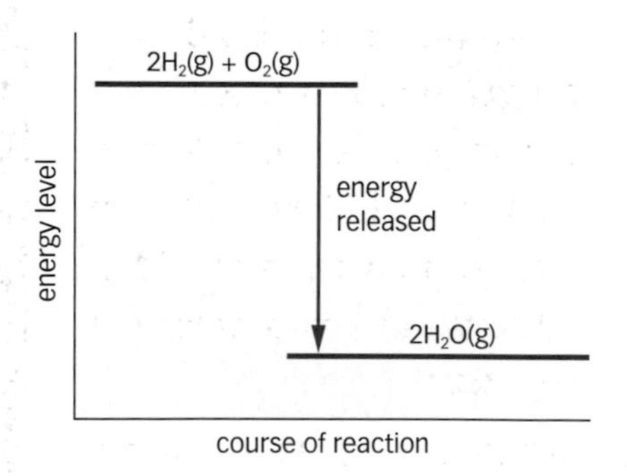


▲ The reactants, hydrogen and oxygen, contain more energy than the product, water. The downwards arrow shows that energy is released to the surroundings during the reaction.

## The hydrogen-oxygen fuel cell

A fuel cell produces electrical energy efficiently from an exothermic reaction. There are many different types using different fuels, such as methanol. The hydrogen-oxygen fuel cell uses hydrogen as its fuel. When oxygen or air is supplied, the hydrogen and oxygen react together. The energy released in the reaction is used to produce a potential difference or voltage. This causes a current to flow when the fuel cell is connected in a circuit.

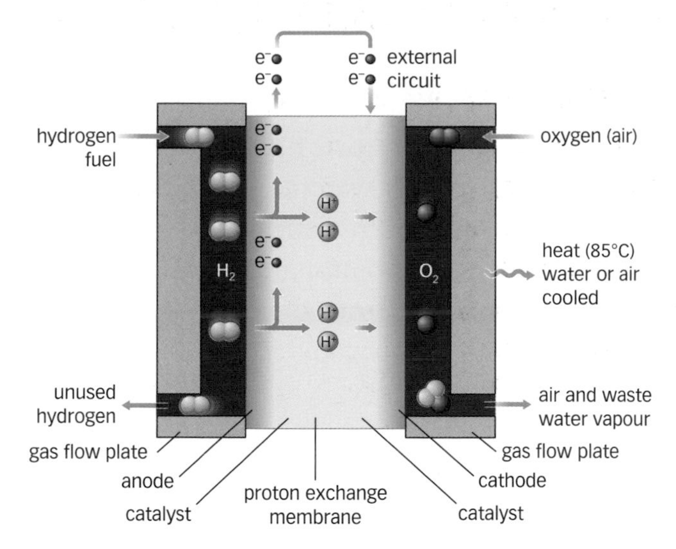


▲ A cross-section through a typical hydrogen-oxygen fuel cell. The potential difference can be increased by stacking several cells together.

The overall reaction that takes place in the hydrogen-oxygen fuel cell is the same as the one that happens when hydrogen explodes in air:

hydrogen + oxygen → water

$$2H_2(g) + O_2(g) \rightarrow 2H_2O(g)$$

However, the design of the fuel cell ensures that the way in which the hydrogen and oxygen react is not the same, but instead happens in a controlled way.

## Electrode reactions

Hydrogen molecules lose electrons at the anode and become hydrogen ions:

$$H_2(g) - 2e^- \rightarrow 2H^+(aq)$$

This is an oxidation reaction because the hydrogen molecules lose electrons. The electrons pass through the electrical circuit, and the hydrogen ions pass through a special artificial membrane. This lets hydrogen ions through, but not hydrogen or oxygen gas.

When the hydrogen ions reach the cathode, they combine with oxygen and electrons from the electrical circuit:

$$4H^+(aq) + O_2(g) + 4e^- \rightarrow 2H_2O(g)$$

This is a reduction reaction because electrons are gained. The overall reaction is an example of a redox reaction: reduction happens at the cathode while oxidation happens at the anode.

### Exam tip OCR

✔ You do not need to be able to explain the detailed working of a fuel cell.

**B** Name the waste product of a hydrogen-oxygen fuel cell.

### Questions

1. Name the fuel used in a hydrogen-oxygen fuel cell, and the type of energy produced.
2. Write the word equation for the reaction between hydrogen and oxygen.

↓ E

3. Write the balanced symbol equation for the reaction between hydrogen and oxygen.
4. What is created from the energy released by the reaction between the fuel and oxygen in a fuel cell?

↓ C

5. Draw an energy level diagram to represent the decomposition of calcium carbonate to form calcium oxide and carbon dioxide, which is an endothermic reaction.
6. Write half-equations for the reactions in a hydrogen-oxygen fuel cell, and explain whether they represent reduction or oxidation.

↓ A*

# 4: Uses of fuel cells

## Learning objectives

After studying this topic, you should be able to:

- describe some advantages and uses of fuel cells
- explain why the car industry is developing fuel cells
- explain why the use of fuel cells still produces pollution

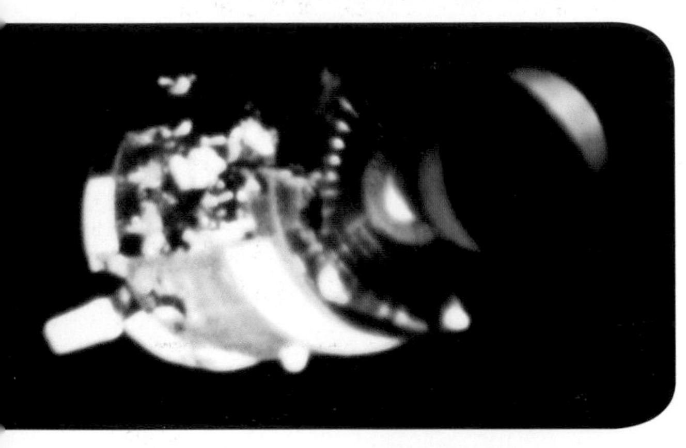

This is Apollo 13's damaged CSM. The explosion blew an entire panel off the side of the module and nearly cost the lives of the astronauts.

### Exam tip OCR

- The development and use of hydrogen fuel cells affects different people in different ways. For example, it provides new opportunities for some while threatening the jobs of people working in businesses that rely on conventional engines.

**A** Give three reasons why a fuel cell might be used on a spacecraft.

## Apollo 13

A near-disaster in space in April 1970 kept millions of people around the world anxiously watching their television sets for the latest news. Two days into a mission to the Moon, an oxygen tank on Apollo 13's Command Service Module (CSM) exploded because of a fault with the electrical wiring. Vital oxygen needed for the on-board hydrogen-oxygen fuel cells was lost.

The three astronauts had to escape into the Lunar Module, attached to the CSM. This was only intended to carry two people for two days, rather than three people for four days. Its design had been changed four years earlier, and it now relied on batteries for its electricity rather than fuel cells. This meant that electricity, and water for drinking and for cooling the spacecraft's systems, was limited. The astronauts had a cold and miserable journey home, but landed safely six days after blast off. Fuel cells are very important for providing electrical power in spacecraft.

## Fuel cells in space

Satellites and the International Space Station in orbit above the Earth usually get their electricity from solar panels. These convert sunlight into electricity, and on-board batteries store it so that enough electrical energy is always available. This is not practical for spacecraft like Apollo and the Space Shuttle. They rely on hydrogen-oxygen fuel cells for their electricity. Fuel cells have several advantages that make them suited to use on spacecraft. They

- are lightweight
- are compact
- have no moving parts
- provide drinking water for the astronauts.

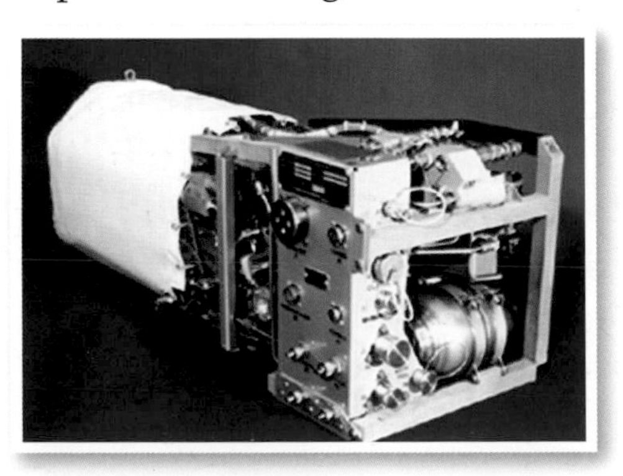

A fuel cell used in NASA's Space Shuttle. It measures just 35 cm × 38 cm × 114 cm, and has a mass of 118 kg. It is over 70% efficient and can supply 12 kW of electricity continuously.

## Fuel cells on Earth

There is a lot of interest in hydrogen-oxygen fuel cells from the car industry. Most vehicles use petrol or diesel. These hydrocarbon fuels are made from crude oil, which is a **non-renewable resource**: it takes a very long time to make and it is being used up faster than it can form. In addition, the combustion of hydrocarbon fuels produces carbon dioxide. This is a **greenhouse gas** that has been linked with **global warming** and climate change. Vehicles using hydrogen-oxygen fuel cells, rather than petrol or diesel engines, have several advantages:

- Water vapour is the only waste product.
- There are no emissions of carbon dioxide from the car.
- There is a large potential source of hydrogen – water can be decomposed by electrolysis.

**B** Give two advantages of using a fuel cell in a car.

### Some drawbacks

Conventional methods of generating electricity rely on using a fuel to turn water into steam. The steam spins turbines, which turn generators. These stages make the process less efficient than a fuel cell, which has fewer stages. A fuel cell transfers electrical energy directly as needed, whenever fuel and oxygen are supplied to it. The hydrogen-oxygen fuel cell is less polluting than, for example, a coal-fired power station.

**C** Give two advantages of using a fuel cell to generate electricity, compared to conventional methods.

At the moment, most hydrogen production involves fossil fuels. Hydrogen can be produced by reacting steam with coal or natural gas. The electrolysis of water produces hydrogen, too, but fossil fuels are used to generate most of the world's electricity. Fuel cells often contain poisonous catalysts. These have to be disposed of safely at the end of the useful life of the fuel cell.

▲ This bus uses a hydrogen-oxygen fuel cell to generate electricity. The electricity charges batteries and powers an electric motor.

**Key words**

**non-renewable resource, greenhouse gas, global warming**

**Questions**

1. Give two important uses of hydrogen-oxygen fuel cells on spacecraft.
2. Explain why the use of fossil fuels is linked with climate change.

↓ E

3. Describe the advantages of using a fuel cell to provide electrical energy on a spacecraft.
4. Explain why the car industry is developing fuel cells.

↓ C

5. To what extent is it true that hydrogen-oxygen fuel cells do not produce pollution? Explain your answer.

↓ A*

## Learning objectives

After studying this topic, you should be able to:

- recall the reactivity of magnesium, zinc, iron, and tin from most to least reactive
- describe oxidation and reduction in terms of gain or loss of oxygen
- predict and explain displacement reactions between metals and metal salt solutions
- explain oxidation and reduction in terms of loss or gain of electrons
- explain displacement reactions as examples of redox reactions

## Key words

**redox reaction, reduction, oxidation, oxidising agent, reducing agent, displacement reaction, reactivity series**

▲ The reaction between powdered aluminium and iron oxide is very vigorous. So much heat is released in the exothermic reaction, called the thermite reaction, that the iron which is produced melts.

## Reduction and oxidation

A mixture of aluminium and iron(III) oxide is often called a thermite mixture. When it is heated strongly enough, a very vigorous exothermic reaction happens:

aluminium + iron(III) oxide → aluminium oxide + iron

This is an example of a **redox reaction**. Reduction and oxidation happen at the same time.

- Iron(III) oxide loses oxygen and is reduced to iron.
- Aluminium gains oxygen and is oxidised to aluminium oxide.

**Reduction** is the removal of oxygen from a substance. **Oxidation** is the gain of oxygen by a substance, or the reaction of a substance with oxygen.

**A** In the following reaction, which substance is reduced and which is oxidised?

copper oxide + magnesium → copper + magnesium oxide

## Redox and electrons

Oxidation and reduction can happen even when oxygen is not involved. In these reactions, electrons are lost or gained:

- Oxidation is loss of electrons.
- Reduction is gain of electrons.

These processes happen in electrode reactions during electrolysis or the operation of a fuel cell, but they can also happen in test tube reactions.

An **oxidising agent** is a substance that can remove electrons from other substances, oxidising them. On the other hand, a **reducing agent** is a substance that can give electrons to other substances, reducing them. Look at this example of a redox reaction involving electrons:

| $Cl_2(g)$ | + | $2Fe^{2+}(aq)$ | → $2Cl^-(aq)$ + $2Fe^{3+}(aq)$ |
|---|---|---|---|
| chlorine | + | iron(II) | → chloride + iron(III) |
| reduced | | oxidised | |
| oxidising agent | | reducing agent | |

Chlorine gains electrons and is reduced to chloride ions, while iron(II) ions lose electrons and are oxidised to iron(III) ions.

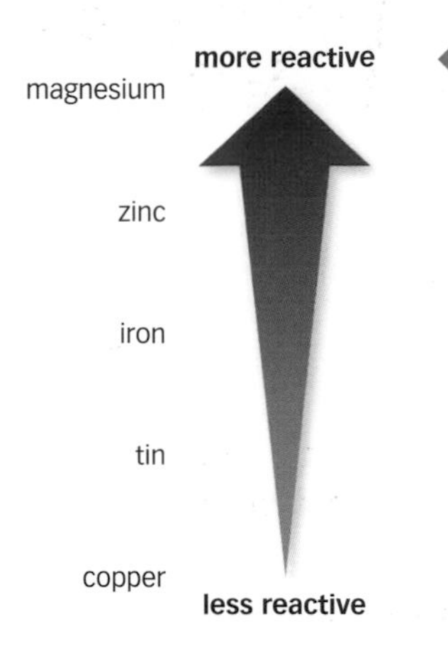

The reactivity series for five common metals

> **B** Aluminium is a powerful reducing agent. In the following reaction, which substance is reduced and which is oxidised?
>
> aluminium + iodine → aluminium iodide

## Displacement reactions

When a strip of zinc foil is dipped into copper(II) sulfate solution, it becomes coated in copper and the blue colour of the solution gradually fades. This is because of a type of redox reaction called a **displacement reaction**:

zinc + copper(II) sulfate → zinc sulfate + copper

The temperature rises because it is an exothermic reaction. One way of understanding displacement reactions involves the **reactivity series** of metals. A more reactive metal can displace a less reactive metal from its compounds. In this example, zinc is more reactive than copper, so it displaces or pushes out copper from copper(II) sulfate. The sulfate has swapped places from the less reactive metal to the more reactive one.

> **C** Will a displacement reaction happen between magnesium and tin nitrate solution? Explain your answer.

## Displacement as redox

Displacement reactions can also be explained in terms of redox reactions involving electrons. For example, consider again the reaction between zinc and copper(II) sulfate solution:

$$Zn(s) + CuSO_4(aq) \rightarrow ZnSO_4(aq) + Cu(s)$$

The sulfate ions act as spectator ions. You can write two half equations, one for what happens to zinc, and one for what happens to the copper(II) ions:

$Zn(s) - 2e^- \rightarrow Zn^{2+}(aq)$ zinc atoms are oxidised

$Cu^{2+}(aq) + 2e^- \rightarrow Cu(s)$ copper(II) ions are reduced

### Questions

1 In terms of oxygen, what are oxidation and reduction? (E)

2 For each of the following mixtures, explain whether a displacement reaction will happen. Write a word equation for each reaction.
(a) Zinc + tin chloride solution, $SnCl_2(aq)$.
(b) Zinc + magnesium chloride solution, $MgCl_2(aq)$. (C)

3 A pale green solution containing iron(II) ions $Fe^{2+}(aq)$ gradually turns orange-brown as iron(III) ions $Fe^{3+}(aq)$ form. Explain whether iron(II) ions are oxidised or reduced.

4 Magnesium displaces tin from tin chloride solution. (A*)
(a) Write a balanced symbol equation for the reaction.
(b) Explain which reactant is reduced and which reactant is oxidised.

## Learning objectives

After studying this topic, you should be able to:

- recall the requirements for iron and steel to rust
- understand that rusting is a redox reaction
- describe and explain methods of rust prevention
- explain why rusting is a redox reaction
- explain how sacrificial protection and tin plating work

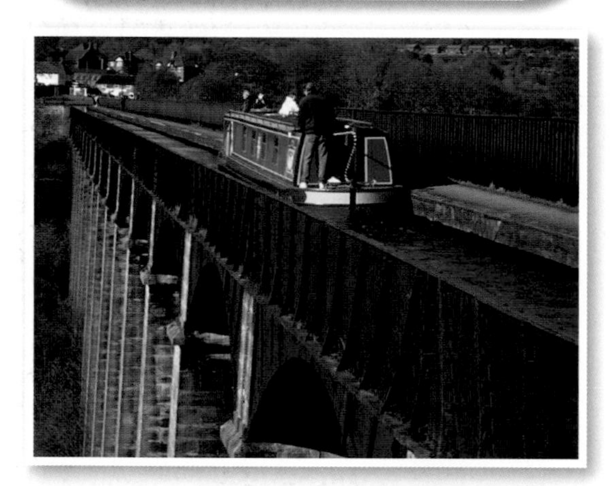

The use of cast iron on Pontcysyllte Aqueduct is an example of applying a scientific idea.

**B** Give the chemical name for rust.

### Did you know...?

It has been estimated that rusting costs national economies around the world up to 5% of the value of their goods and services. In the US alone, this means at least $270 billion a year.

## Rusting

The Pontcysyllte Aqueduct on the Llangollen Canal in Wales is an amazing structure. It has carried canal traffic in a 307 m cast iron trough over the River Dee 38 m below since 1805. Cast iron contains around 4% carbon. This makes the metal hard, and too brittle for most uses. However, the carbon protects the iron from rusting away. It reacts to form insoluble products that keep air and water away from the metal.

Iron and steel rust when they react with both water and oxygen. Air contains oxygen, so the metal will rust in damp air. Rusting is a redox reaction because the iron atoms are oxidised to iron(III) ions, and the oxygen atoms are reduced to oxide ions. Rusting damages iron and steel objects, so various ways to prevent it happening have been developed.

**A** What two substances are needed for iron and steel to rust?

## Hydrated iron(III) oxide

Rust is hydrated iron(III) oxide:

iron + oxygen + water → hydrated iron(III) oxide

During rusting, the iron atoms lose electrons to become iron(III) ions, $Fe^{3+}$. This means that they are oxidised. On the other hand, the oxygen atoms gain electrons to become oxide ions, $O^{2-}$. This means that they are reduced.

## Preventing rust

Many methods of rust prevention rely on stopping air and water reaching the surface of the iron. These include:

- coating the metal part with oil or grease
- painting the surface of the metal part
- plating the surface with zinc (galvanising)
- plating the surface with tin.

It is important to choose the correct method. For example, it is sensible to oil a bicycle chain rather than painting it, but it is not sensible to oil car bodywork instead of painting it.

Some iron alloys are resistant to rusting. Stainless steel contains chromium. This oxidises to chromium oxide when exposed to the air, forming a thin film on the surface of the steel. The layer stops air and water reaching the metal below.

▲ Stainless steel is used in kitchen sinks because it does not rust

**C** Describe two ways in which steel can be coated to protect it from rusting.

## Sacrificial protection

**Galvanising** involves coating the surface of the iron or steel object with a layer of zinc. The zinc coating stops air and water reaching the metal below, but it also does something else. Zinc is more reactive than iron, so it is more likely to be oxidised. It sacrifices itself to protect the iron below. Galvanising is used to protect car body panels before painting, and to protect metalwork sited outdoors.

▲ This farm gate is made from galvanised steel to prevent it rusting

### More about sacrificial protection

Sacrificial protection does not just work with zinc. It works with other metals that are more reactive than iron, such as magnesium. The more reactive metal loses electrons more readily than iron does, so it is more readily oxidised. This even works if the reactive metal is just attached to the iron or steel. Ships have zinc or magnesium blocks bolted onto their hulls under the waterline. These protect the hull but they gradually corrode away and have to be replaced.

**D** Put iron, tin, and zinc into an order of reactivity, starting with the most reactive.

Tin plating is used to protect the inside of steel food cans from rusting. Just like zinc plating, the layer of tin stops air and water reaching the surface of the iron. However, unlike zinc, tin is less reactive than iron. This means that tin loses electrons less readily than iron does, so it is less readily oxidised. Unfortunately, the result is that the iron rusts even faster if the tin layer is broken or scratched.

**Key words**

galvanising

### Questions

1. Explain why rusting involves oxidation.
2. Why do oil, grease, and paint prevent iron from rusting?

↓ E

3. Explain how galvanising works.

↓ C

4. Explain why rusting is a redox reaction.
5. Zinc plating and tin plating both protect iron from rusting. Explain how they do this, and describe any differences between the two methods.

↓ A*

## Learning objectives

After studying this topic, you should be able to:

- ✔ recall the molecular and displayed formulae of ethanol
- ✔ describe how ethanol can be made by fermentation
- ✔ explain the conditions used in fermentation

## Key words

**alcohol, ethanol, fermentation, yeast, fractional distillation, denatured**

```
    H  H
    |  |
H — C — C — O — H
    |  |
    H  H
```

▲ This is the displayed formula of ethanol

## Exam tip OCR

- ✔ Make sure you can recall the main uses of ethanol.

## Did you know...?

In 2010 divers discovered bottles of beer and champagne in a shipwreck off the Finland coast. The corks were still intact and no seawater had got into the bottles. At around two hundred years old, the bottles contain the world's oldest drinkable alcoholic drinks.

## Alcohol

Alcoholic drinks such as wine and beer contain an **alcohol** called **ethanol**. Although most people just call it alcohol, ethanol is just one of a family of similar substances. These all contain carbon, hydrogen, and oxygen bonded together. The alcohols are not hydrocarbons, because they contain oxygen atoms in addition to carbon and hydrogen atoms. They often burn well, and ethanol is used as fuel for cars. It is also used as a solvent in perfumes and aftershaves.

◀ Ethanol burns well and is used as a fuel for cars. Modern cars can run on a mixture of 95% petrol and 5% ethanol without any modifications to the engine.

**A** How many carbon, hydrogen, and oxygen atoms does a molecule of ethanol contain?

## Making alcohol

**Fermentation** is an ancient method of making ethanol. People have been using it for thousands of years to produce alcoholic drinks. It is still used for this purpose, but nowadays it is also used to make ethanol on an industrial scale.

Fermentation relies on microscopic single-celled fungi called **yeast**. Yeast contains enzymes that can convert glucose to carbon dioxide and ethanol, as long as oxygen is kept out:

$$\text{glucose} \rightarrow \text{carbon dioxide} + \text{ethanol}$$
$$C_6H_{12}O_6 \rightarrow 2CO_2 + 2C_2H_5OH$$

It is easy to make ethanol in the school laboratory. Glucose is dissolved in a conical flask of warm water to make glucose solution. Yeast is added, and a bung and delivery tube fitted. The end of the delivery tube is dipped into a test tube of water or limewater to act as an airlock. It is important to keep the temperature between 25 °C and 50 °C, so the apparatus should be left in a warm place.

As fermentation begins, carbon dioxide bubbles out of the airlock and eventually all the original air inside is replaced with carbon dioxide.

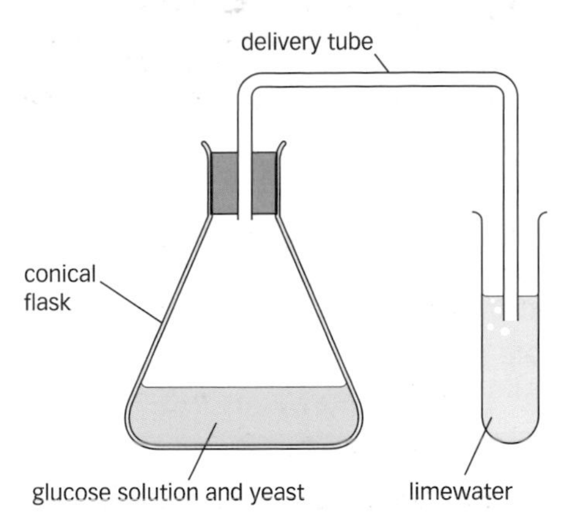


▲ Apparatus used to demonstrate fermentation. The limewater turns milky as carbon dioxide bubbles through it.

**B** Describe the conditions needed for fermentation.

### Exam tip OCR

✔ Make sure you can describe the conditions needed for fermentation.

Fermentation is a slow reaction, so several days at least are needed to make sufficient ethanol to investigate. The ethanol eventually kills the yeast, which sinks to the bottom. Ethanol is separated from the mixture by **fractional distillation**. This works because water boils at 100 °C, but ethanol boils at only 78 °C.

▲ Wine is made by fermentation on a huge scale at this vineyard in California

## Choosing the conditions

Yeast is made up of living organisms. They will become inactive if the temperature is too low, so it should not fall below about 25 °C. Enzymes are natural catalysts. They are proteins that need to maintain a particular shape, otherwise they will not work. If enzymes become too hot, their shape changes irreversibly. They become **denatured** and stop working, so the temperature should not rise above about 50 °C. If air gets inside during fermentation, the oxygen causes the ethanol to oxidise to ethanoic acid, the acid found in vinegar.

### Questions

1 Explain why the alcohols are not hydrocarbons.

2 Give three main uses of ethanol.

↓ E

3 Write the molecular formula of ethanol, and draw its displayed formula.

4 Describe how ethanol can be made by fermentation. Include the word equation for the reaction in your answer.

↓ C

5 Explain the range of temperatures chosen for fermentation, and why air must be kept out. Include the balanced symbol equation for fermentation in your answer.

## Learning objectives

After studying this topic, you should be able to:

- describe how ethanol can be made by the hydration of ethene
- evaluate the merits of making ethanol by fermentation or hydration
- recall the general formula for alcohols
- draw the displayed formulae for alcohols containing from one to five carbon atoms

## Making ethanol

Fermentation is not the only way to make ethanol. It can also be made by reacting ethene with steam. The two gases are passed over a hot phosphoric acid catalyst, causing a **hydration reaction** to happen:

$$\text{ethene} + \text{water} \rightarrow \text{ethanol}$$
$$C_2H_4 + H_2O \rightarrow C_2H_5OH$$

This method of producing ethanol is used for ethanol intended for industrial use, rather than ethanol intended for alcoholic drinks.

Ethanol made by fermentation is a renewable fuel. The sugars needed for the process come from plants, including sugar cane and sugar beet. As long as new crops are planted after other crops have been harvested for use in fermentation, we should not run out of ethanol. However, ethanol made by hydration of ethene is a non-renewable fuel. The ethene needed for the process comes from cracking crude oil fractions. Since crude oil itself is a non-renewable resource, ethanol made in this way is also non-renewable.

## Exam tip OCR

- The increasing use of farmland for growing crops for fuel rather than for food might lead to increased food prices, or even to food shortages.

H H
C=C → H—C—C—O—H
H H
+
H—O—H

▲ Ethanol can be made by hydration of ethene

## More about making ethanol

The two methods of making ethanol, fermentation and hydration, each have different advantages and disadvantages. The table summarises the main differences between them.

| Feature | Fermentation of sugars | Hydration of ethene |
|---|---|---|
| conditions | low temperatures and normal pressure | high temperatures and high pressures |
| raw materials | sugars from plants | ethene from crude oil |
| purity of product | low – needs filtering and distillation | high – no by-products are made |
| percentage yield | low – about 15% | high – around 100% |
| atom economy | 51% | 100% |
| type of process | batch | continuous |

**A** Which method produces ethanol that is non-renewable?

Fermentation is a more sustainable process than the hydration of ethene: it requires renewable raw materials and relatively little energy. On the other hand, its percentage yield, atom economy (the percentage of atoms in the reactants that become atoms in the desired product), and purity of the product are all lower than for the hydration process. In addition, fermentation is a batch process, which makes it more labour-intensive and means automation is more difficult.

## More alcohols

Ethanol is just one example from a family of compounds called the alcohols. All alcohols contain a hydroxyl group, –OH. The alcohols are named after the number of carbon atoms they contain, just as the alkanes and alkenes are. Their names usually end in 'anol'. The table shows the names and formulae of some alcohols.

| Name of alcohol | Number of carbon atoms | Molecular formula | Displayed formula |
|---|---|---|---|
| methanol | 1 | $CH_3OH$ | H<br>\|<br>H—C—O—H<br>\|<br>H |
| ethanol | 2 | $C_2H_5OH$ | H H<br>\| \|<br>H—C—C—O—H<br>\| \|<br>H H |
| propanol | 3 | $C_3H_7OH$ | H H H<br>\| \| \|<br>H—C—C—C—O—H<br>\| \| \|<br>H H H |
| butanol | 4 | $C_4H_9OH$ | H H H H<br>\| \| \| \|<br>H—C—C—C—C—O—H<br>\| \| \| \|<br>H H H H |
| pentanol | 5 | $C_5H_{11}OH$ | H H H H H<br>\| \| \| \| \|<br>H—C—C—C—C—C—O—H<br>\| \| \| \| \|<br>H H H H H |

Notice that that all the molecular formulae have this general formula:

$$C_nH_{2n+1}OH$$

### Key words

**hydration reaction**

**B** What does 'atom economy' mean?

**C** Write the molecular formula for eicosanol, an alcohol with 20 carbon atoms.

### Questions

1. Write the word equation for the hydration of ethene.
2. Describe how ethanol is produced by the hydration of ethene.

E

3. Ethanol can replace petrol. What method should be used to make ethanol so that the use of fossil fuels is reduced? Explain your answer.
4. Write the balanced symbol equation for the hydration of ethene.

C

5. Worldwide production of ethanol is around 80 billion litres per year. Around 95% of this is made by fermentation rather than by hydration of ethene. Suggest why fermentation is the preferred way to make ethanol.
6. Write the molecular formula for hexanol, an alcohol with six carbon atoms, and draw its displayed formula.

A*

C6

# 9: The ozone layer

### Learning objectives

After studying this topic, you should be able to:

- explain what ozone is and how it forms
- describe the uses of CFCs
- describe how CFCs can deplete the ozone layer
- describe, with the help of equations, how chlorine radicals react with ozone

These scientists are carrying out research to do with the ozone layer in the Antarctic. The large helium balloon will take scientific equipment 20 km up into the atmosphere. Airliners typically cruise 9 km to 12 km up.

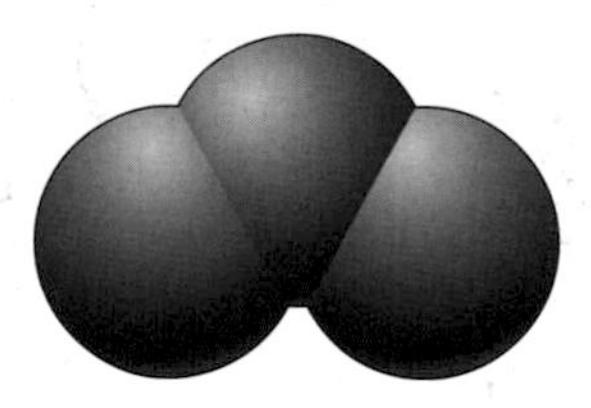

Ozone molecules contain three oxygen atoms

```
     Cl
     |
F —  C — Cl
     |
     Cl
```

The displayed formula of a typical CFC

**A** What is ozone?

## Ozone

**Ozone** is a form of oxygen with the formula $O_3$ rather than the more usual $O_2$. It is a pale blue gas with a choking smell similar to bleach. Ozone can cause breathing problems when it is in the air at ground level. However, ozone high up in the atmosphere is vital to the health and survival of living things.

## The ozone layer

Ozone is found throughout the atmosphere, but it is most concentrated in the stratosphere. This is a part of the atmosphere 10 km to 50 km above the Earth's surface, also called the **ozone layer**. However, there is not really a layer of ozone, just a region where it is most concentrated.

Sunlight contains ultraviolet light, as well as visible light. Ultraviolet light causes oxygen, $O_2$, to react to form ozone, $O_3$. This happens in stages, but overall three oxygen molecules produce two ozone molecules:

$$3O_2 \rightarrow 2O_3$$

The ozone layer absorbs most of the ultraviolet light in sunlight, stopping it from reaching the Earth's surface.

## CFCs

**CFC** is the shortened name for **chlorofluorocarbon**. CFCs contain carbon atoms with chlorine and fluorine atoms attached. There are many different types of CFC, and they were used in refrigerators and aerosol cans in the past. Unfortunately, CFCs damage the ozone layer.

## Damaging the ozone layer

CFC molecules spread out in the atmosphere after release. Ultraviolet light in the stratosphere breaks carbon-chlorine bonds in CFC molecules. For example:

$$CCl_3F \rightarrow CCl_2F + Cl$$

Highly reactive chlorine atoms called chlorine radicals are released. These react with ozone molecules and break them down to form oxygen, $O_2$. Chlorine atoms are formed in this reaction, and they can go on to react with even more ozone molecules. The reactions reduce the concentration of ozone in the ozone layer, letting more ultraviolet light reach the Earth's surface.

Increased levels of ultraviolet light can cause medical problems, including:

- skin cancer
- an increased risk of sunburn
- an increased risk of eye cataracts
- faster ageing of skin, causing more wrinkles and age spots.

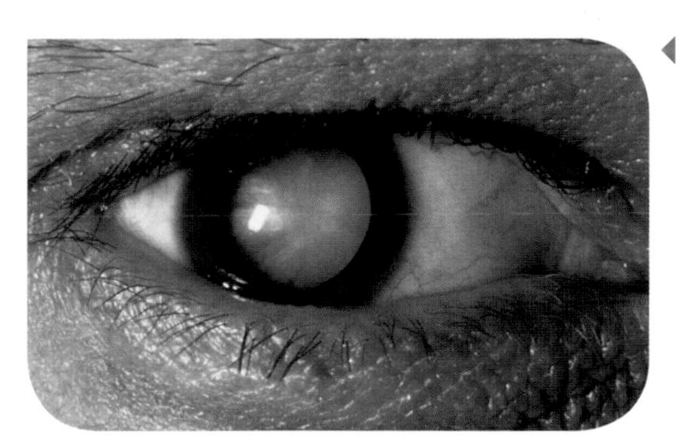

This person has an eye cataract. The proteins in the lens of the eye have become damaged, causing the lens to go cloudy.

**Key words**

**ozone, ozone layer, chlorofluorocarbon (CFC), chlorine radical, chain reaction**

**B** What is the ozone layer?

**C** Give two medical problems caused by increased levels of ultraviolet light.

## More about ozone and chlorine radicals

A covalent bond consists of a shared pair of electrons. Ultraviolet light can break a covalent bond in such a way that each atom from the bond gets one of the electrons. This produces highly reactive radicals. The extra electron is shown as a dot in formulae. For example, this equation shows how a CFC molecule can form a chlorine radical:

$$CCl_3F \xrightarrow{\text{UV light}} \bullet CCl_2F + \bullet Cl$$

$$F{-}CCl_2{-}Cl \longrightarrow F{-}CCl_2\bullet + \bullet Cl$$

Radicals can form when covalent bonds break

Once formed, these chlorine radicals can go on to damage many ozone molecules. There are two steps:

**Step 1** Chlorine radical reacts with ozone $\bullet Cl + O_3 \rightarrow \bullet ClO + O_2$

**Step 2** New radical reacts with more ozone $\bullet ClO + O_3 \rightarrow 2O_2 + \bullet Cl$

Notice that a chlorine radical goes in at step 1 and another one comes out at step 2. This sets up a **chain reaction** in which a lot of ozone is destroyed. Steps 1 and 2 can be combined to give an overall equation:

$$2O_3 \rightarrow 3O_2$$

**Exam tip** OCR

✔ Remember to show a free radical with a dot in chemical equations.

### Questions

1. Give two uses of CFCs.
2. What is a chlorine radical?

E

3. Describe the effect of ultraviolet light on CFCs. What happens as a result?
4. What happens if the concentration of ozone in the ozone layer decreases?

C

5. Explain how a radical forms.
6. Describe how a chain reaction is set up between chlorine radicals and ozone. Include symbol equations in your answer.

A*

# 10: Looking after the ozone layer

## Learning objectives

After studying this topic, you should be able to:

- describe some properties of CFCs
- explain why the use of CFCs has been banned in the UK
- describe some replacements for CFCs
- describe and explain how scientists' attitude to CFCs has changed

## Exam tip OCR

- The development and use of CFCs seemed like a wonderful scientific advance at the time. It was only much later that the environmental impact of CFCs was discovered, leading to an almost complete ban in their use.

The CFCs in old refrigerators must be removed before the refrigerator can be recycled. The insulating foam may also contain CFCs.

## More about CFCs

There are many different types of chlorofluorocarbon or CFC, but they all have similar properties. These include:

- low boiling points, so they are usually gases
- they are insoluble in water
- they are chemically inert, so they do not react easily with other substances.

These properties make CFCs very versatile. In the last century, CFCs were widely used as **refrigerants**. These are the substances in refrigerators and air conditioning systems that carry the heat away. CFCs were also widely used as aerosol **propellants**, the substances that push the can's contents out as a spray when the button is pressed. Unfortunately, the properties that make CFCs so useful also made them a big problem for the ozone layer.

## The Montreal Protocol

The inertness of CFCs means that they are removed from the stratosphere only very slowly. Experiments, calculations, and observations made in the 1970s showed that CFCs were destroying ozone and depleting the ozone layer. At a meeting in Montreal in Canada in 1987, 24 countries signed a treaty that placed strict limits on the production and use of CFCs. Since then, almost all countries have signed it. The treaty has been revised several times. Now all but the most vital uses of CFCs have been banned. As a result, the ozone layer is beginning to show signs of recovery.

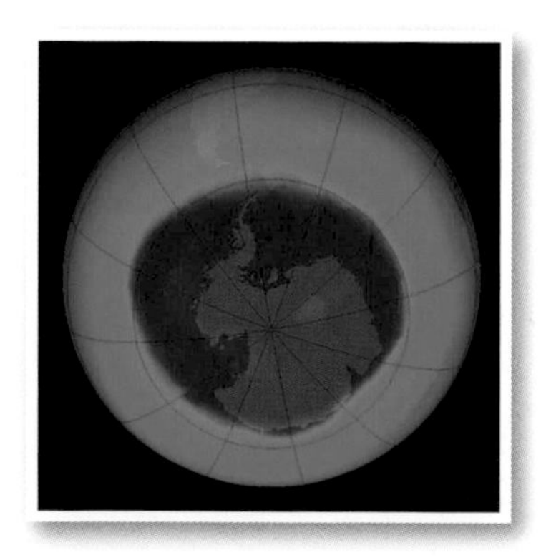

This satellite image shows the concentration of ozone over the Antarctic in 2008. The purple 'hole' in the ozone layer was one of the largest recorded, even though the concentration of CFCs has decreased steadily since 2000.

**A** Give one property of a CFC.

**B** What did the 1987 treaty signed in Montreal agree to?

## CFCs everywhere

An unusual discovery was made in 1973. A very sensitive device had just been invented to measure the concentration of compounds in the atmosphere. Wherever it was used, it detected $CCl_3F$, a commonly used CFC. Calculations showed that the total amount in the atmosphere was similar to the total amount ever made. Once they had been released into the air, CFCs were not breaking down very quickly at all. In fact, they were being removed only very slowly. The very lack of chemical reactivity that made CFCs attractive meant that they might cause problems.

At around the same time as this discovery was made, scientists realised that CFCs could break down ozone. They calculated that a single chlorine radical could cause the breakdown of 100 000 ozone molecules. With around a million tonnes of CFCs being made every year, there was huge potential to deplete the ozone layer within decades. Things were about to get even worse.

The unusual conditions over the poles mean that ozone levels there vary naturally with the seasons. Measurements in 1985 revealed a startling discovery. The concentration of ozone in the ozone layer had decreased far more than anyone expected. A link had been shown between the use and release of CFCs, and the depletion of the ozone layer. It was clear to scientists and the wider public that something had to be done urgently.

## Replacements for CFCs

Society at large has now agreed with the scientists' views about the ozone layer and the role of CFCs in depleting it. Chemists have developed safer alternatives to CFCs. They include alkanes, and compounds called hydrofluorocarbons or **HFCs**. These do not contain chlorine atoms and they do not damage the ozone layer.

```
      F
      |
  F — C — H
      |
      F
```

▲ The displayed formula of trifluoromethane, an HFC used to replace CFCs in refrigerators and air conditioning units

### Did you know...?

The concentration of ozone in the stratosphere is measured in Dobson units, named after Gordon Dobson. He was the scientist at the University of Oxford who built the first instrument to measure total ozone from the ground in the 1920s.

**C** How many ozone molecules can one chlorine radical destroy?

### Key words

**refrigerant, propellant, HFC**

### Questions

1. Why are CFCs usually gases?
2. What replacements have been developed for CFCs?

E

3. Explain why the use of CFCs in the UK and elsewhere has been banned.
4. Explain why CFCs disappear from the atmosphere very slowly.

C

5. Describe how the attitude of scientists to CFCs has changed over time.
6. Explain why CFCs will carry on depleting ozone for long time to come.

A*

# 11: Water hardness

## Learning objectives

After studying this topic, you should be able to:

- describe the properties of hard and soft water
- explain how hardness in water is caused
- describe the differences between temporary hardness and permanent hardness

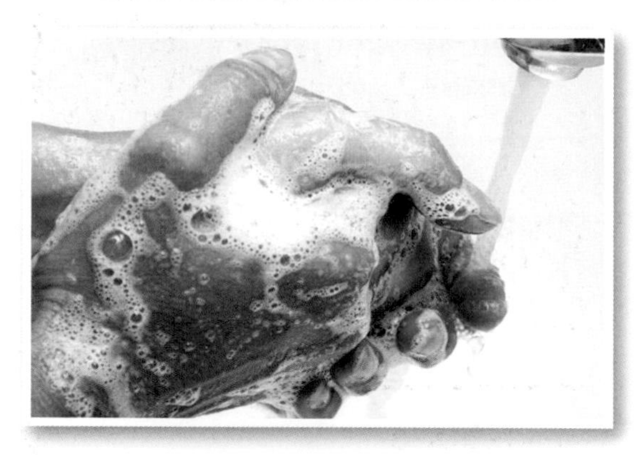

▲ Soapless detergents lather just as well in hard water as they do in soft water, but ordinary soap does not lather well in hard water

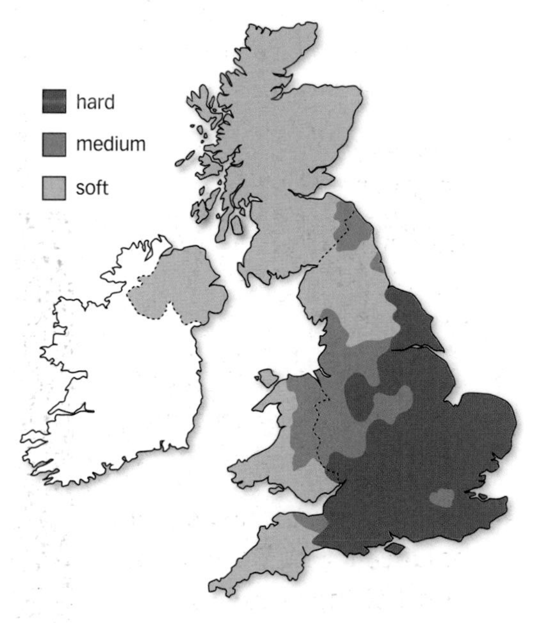


▲ How hard or soft the water from your taps is depends upon where you live

## Hard and soft water

The water that flows from your taps is not pure. It contains chlorine to kill harmful bacteria, and various dissolved mineral ions. These ions may be good for your health. For example, calcium ions help to maintain healthy bones and teeth. However, they can make it difficult to get a lather with soap. **Soft water** contains low concentrations of calcium ions and magnesium ions, and easily forms a lather with soap. **Hard water** contains higher concentrations of these ions, which stops the soap forming a lather easily.

## Acid attack

Rainwater is naturally slightly acidic. Carbon dioxide in the air dissolves in the rain, producing a weak acid. Chalk and limestone consist mainly of calcium carbonate. Rain falling on these rocks reacts with the calcium carbonate to form calcium hydrogencarbonate:

calcium carbonate + water + carbon dioxide → calcium hydrogencarbonate

Calcium carbonate is insoluble in water, but calcium hydrogencarbonate is not. The rock gradually erodes away as water from rain, rivers, lakes, and reservoirs stays in contact with it.

**A** Why is rainwater naturally slightly acidic?

The rocks in some parts of country are mostly igneous rocks like granite. These are not so susceptible to acid attack and the water there is usually soft. Chalk and limestone in other parts of the country are susceptible to acid attack, and the water there is hard. The degree of hardness depends upon the rocks the water has flowed over and through.

## Types of hardness

There are two types of hardness:

- **permanent hardness** – caused by dissolved calcium sulfate
- **temporary hardness** – caused by dissolved calcium hydrogencarbonate.

When water with permanent hardness is boiled, the hardness is not removed. However, when water with temporary hardness is boiled, the hardness is removed. This happens because calcium hydrogencarbonate decomposes when heated:

calcium hydrogencarbonate → calcium carbonate + water + carbon dioxide

This is essentially the reverse of the process that produced the hard water in the first place. It can lead to deposits of calcium carbonate as limescale in kettles, irons, and heating systems.

### Symbol equations

These are the symbol equations for the formation of hard water from calcium carbonate in rocks, and the decomposition of calcium hydrogencarbonate when water with temporary hardness is boiled:

$$CaCO_3 + H_2O + CO_2 \rightarrow Ca(HCO_3)_2$$

$$Ca(HCO_3)_2 \rightarrow CaCO_3 + H_2O + CO_2$$

## Limescale

Limescale is calcium carbonate. It damages equipment that uses hot water and increases the amount of energy used in them. Limescale removers are weak acids, like citric acid. Weak acids will not react with the metal of the equipment, but they will react with the limescale to form a calcium salt, water, and carbon dioxide. For example:

calcium carbonate + citric acid → calcium citrate + water + carbon dioxide

**C** Why are weak acids used as limescale removers?

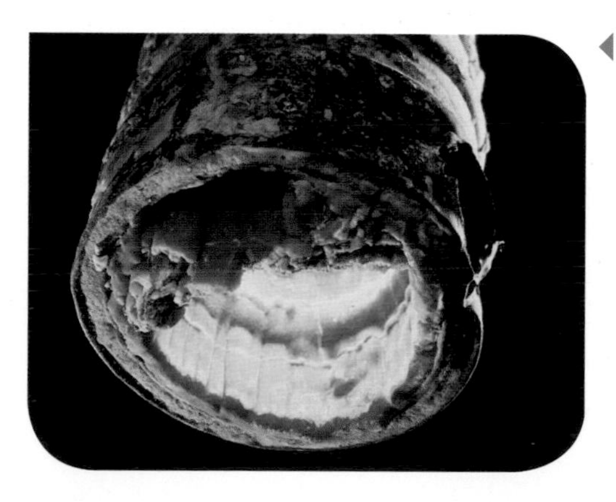

This copper hot water pipe is almost completely blocked by a thick layer of limescale, calcium carbonate

### Key words

**soft water, hard water, permanent hardness, temporary hardness**

**B** Which type of hardness can be removed by boiling?

### Did you know...?

It has been estimated that hard water costs British industry around £1 billion per year. Limescale damages pipes, boilers, and other equipment that involves hot water. It also increases energy costs. A layer just 1 mm thick can increase energy costs by around 8%.

### Questions

1. Why can you tell from the lather formed by soap that water is hard or soft? Why would this not work with a soapless detergent?
2. Why does water hardness vary across the UK?

E

3. Describe, with the help of an equation, how hard water forms.
4. Suggest why some people use boiled water in their steam irons.

C

5. Describe, with the help of a balanced symbol equation, how boiling removes temporary hardness.

A*

# 12: Removing water hardness

## Learning objectives

After studying this topic, you should be able to:

- ✔ describe how hardness in water can be removed
- ✔ explain how an ion-exchange resin can soften hard water
- ✔ interpret data from water hardness experiments
- ✔ explain how sodium carbonate can soften hard water

## Key words

**washing soda**, **ion-exchange resin**

**A** What is washing soda?

▲ Washing soda can soften hard water

## Washing soda

Temporary hardness in water can be removed by boiling, but this is not the only way to soften hard water. **Washing soda** is sodium carbonate, white crystals that dissolve in water to make an alkaline solution. It has several uses in the home, including removing grease from cookers and pans, and unblocking sinks and drains. Washing soda is added to water to soften it when washing clothes. Both temporary and permanent hardness are removed by washing soda. Without it, extra detergent would be needed in hard water areas.

### More about washing soda

Most carbonates are insoluble, including calcium carbonate and magnesium carbonate, but sodium carbonate is soluble in water. When washing soda is added to hard water, a precipitation reaction happens. For example:

calcium hydrogencarbonate + sodium carbonate → calcium carbonate + sodium hydrogencarbonate

$$Ca(HCO_3)_2(aq) + Na_2CO_3(aq) \rightarrow CaCO_3(s) + 2NaHCO_3(aq)$$

The hard water is softened because the calcium ions and magnesium ions are removed as a precipitate.

## Ion-exchange

There is another way to remove both temporary and permanent hardness from water. Ion-exchange columns contain **ion-exchange resins**, tiny beads packed into a plastic or metal tube. Hard water is softened as it passes through an ion-exchange column. This is ideal as a permanent solution to hard water in the home. The ion-exchange column is usually plumbed into the home's water supply so that all the water used in the home is softened.

Ion-exchange resins work by swapping ions in the water for ions from the resin. As the hard water passes through, calcium ions and magnesium ions from the water attach to the resin and sodium ions leave it. The water is softened because the calcium ions and magnesium ions are removed.

Eventually, all the sodium ions in the resin are replaced by calcium ions and magnesium ions. When this happens, the resin is regenerated by flushing sodium chloride solution through the column. Sodium ions are swapped back into the resin. Calcium ions and magnesium ions are pushed out and washed away in waste water. Domestic water softeners may carry out this process automatically.

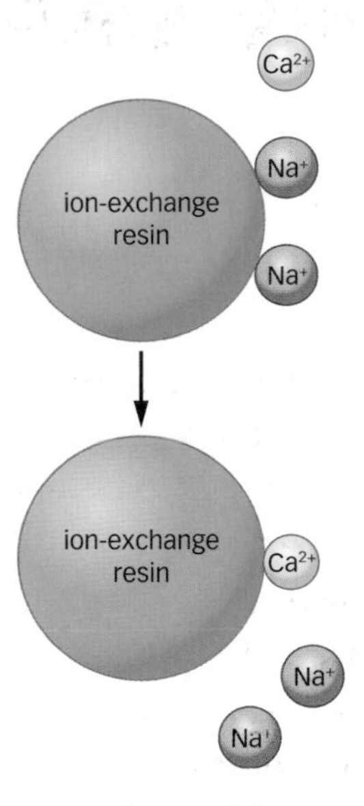


Ion-exchange resins work by swapping calcium ions and magnesium ions in water for sodium ions in the resin

## Water experiments

Hardness in water can be investigated using soap solution. This is added drop by drop from a teat pipette to a fixed volume of water. The more drops needed to get a permanent lather, the harder the water. The table shows the results of an experiment involving three different samples of water.

| Sample | Drops of soap solution needed |
|---|---|
| A | 4 |
| B | 12 |
| C | 8 |

Sample A was the softest because it needed the least amount of soap solution to get a permanent lather.

**B** Which sample of water was the hardest?
**C** Why is dishwasher salt needed?

### Did you know...?

Dishwashing machines contain an ion-exchange column to soften the water. This stops white streaks of calcium carbonate and other precipitates being left on the clean plates as they dry. Dishwasher salt has to be added regularly to regenerate the ion-exchange resin.

### Questions

1. Give two ways in which temporary and permanent hardness can be removed.
2. The three water samples in the table were evaporated to dryness. Explain why water B left a lot of white solid behind.

E

3. Explain how an ion-exchange resin works.
4. In the table, water sample C was produced by boiling sample B. What do the results show about water sample B?

C

5. Explain how washing soda can soften hard water.

A*

Salt granules being added to the ion-exchange resin in a dishwasher

# 13: Natural fats and oils

### Learning objectives

After studying this topic, you should be able to:

- ✔ describe some of the properties and uses of oils and fats
- ✔ describe the differences between saturated and unsaturated fats
- ✔ explain why bromine can be used to test for unsaturation
- ✔ explain why unsaturated fats are healthier options in the diet

▲ Edible fats and oils

glycerol

fatty acid

▲ Glycerol and a fatty acid. Most natural fatty acids contain 4–28 carbon atoms.

### Exam tip OCR

- ✔ Unsaturated fats and oils have at least one C=C bond in their displayed formula.

## Fats and oils

Animals and plants contain fats and oils. For example milk, cheese, and butter contain animal fats, while margarine and cooking oils contain vegetable oils. All these substances have a very similar chemical structure. Whether the substance is called a fat or an oil depends upon its state at room temperature:

- fats are solid at room temperature
- oils are liquid at room temperature.

Vegetable oils are not only an important part of a healthy diet, they are also important raw materials for the chemical industry. Biodiesel is an alternative to diesel made from crude oil. It is made from vegetable oils such as rapeseed oil. Unlike ordinary diesel, biodiesel is a renewable resource. Soap is also made from vegetable oils.

**A** What is the difference between a fat and an oil?

## Esters

Fats and oils are **esters**. These compounds form when a carboxylic acid reacts with an alcohol. Some esters are used in perfumes and as solvents. These are simple esters like ethyl ethanoate, made from ethanoic acid and ethanol. Fats and oils are more complex. They consist of **fatty acids** chemically joined to **glycerol**, which is an alcohol. Fatty acids have long chains of carbon atoms. Glycerol has three hydroxyl groups, –OH, unlike many other alcohols that only have one.

**B** What are the two components of a fat or oil molecule?

## Saturated or unsaturated?

Fats and oils can be saturated or unsaturated:

- in a **saturated** fat or oil, all the carbon-carbon bonds in the fatty acid parts are single covalent bonds
- in an **unsaturated** fat or oil, one or more of the carbon-carbon bonds in the fatty acid parts is a double covalent bond.

This is similar to alkanes and alkenes. Both these types of compound are hydrocarbons, but alkanes have all single bonds and are saturated, while alkenes have one or more double bonds and are unsaturated. Unsaturation in alkenes, fats, and oils can be shown using bromine water:

- bromine water stays orange when mixed with a sample of a saturated alkene, fat, or oil
- bromine water goes colourless when mixed with enough unsaturated alkene, fat, or oil.

**C** What colour is bromine water?

## More about unsaturation

Bromine reacts with the carbon-carbon double bonds in unsaturated fats and oils. An **addition reaction** happens there, producing a dibromo compound that is colourless. Saturated fats and oils do not contain carbon-carbon double bonds, so they cannot react with bromine to produce a colourless compound.

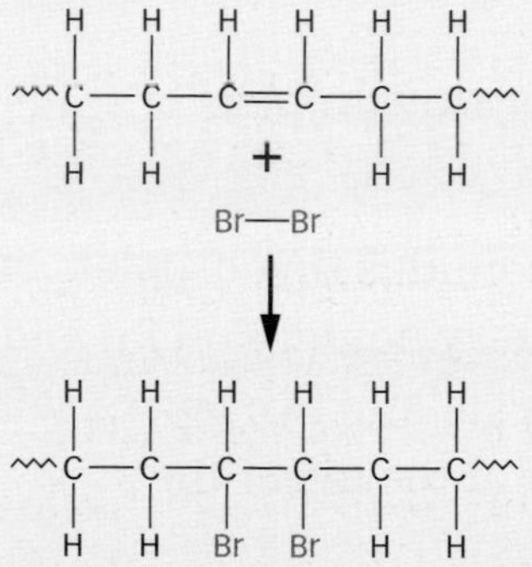

Part of an unsaturated fatty acid, showing how bromine reacts with the carbon-carbon double bond. This addition reaction produces a colourless dibromo compound.

Fats and oils are an important part of a healthy diet, but too much of them can cause health problems and make us overweight. Saturated fats, found mainly in meat and dairy products, can raise the level of cholesterol in the blood. High levels of cholesterol increase the risk of blocked arteries and heart disease.

Unsaturated fats and oils, found mainly in vegetables, fruits and nuts, tend to be more healthy choices. For example, omega-3 oils from oily fish such as mackerel are thought to help reduce the risk of heart disease.

### Key words

**ester, fatty acid, glycerol, saturated, unsaturated, addition reaction**

### Exam tip OCR

- ✔ You do not need to know any detail about the structure of glycerol and fatty acids, except that fats and oils are esters, and have chains of carbon atoms that can contain single bonds and double bonds.
- ✔ For a given mass of fat or oil, the more unsaturated it is, the greater volume of bromine water it can decolourise.

### Questions

1. Give two uses of natural fats or oils in the chemical industry.
2. What type of compound are fats and oils?

E

3. In terms of carbon-carbon bonds, what is the difference between a saturated fat and an unsaturated fat?
4. Describe how bromine water can be used to distinguish between a saturated fat and an unsaturated fat.

C

5. Explain how the bromine test for unsaturation works.
6. Explain why unsaturated fats are regarded as healthier for us than saturated fats.

A*

### Learning objectives

After studying this topic, you should be able to:

- describe oil and water emulsions
- describe how margarine is made from vegetable oils
- describe how soap is made
- explain how soap is made

### Key words

**immiscible, emulsion, oil-in-water emulsion, water-in-oil emulsion, saponification, hydrolysis reaction**

Vegetable oil and water are immiscible liquids

### Exam tip OCR

- Emulsifiers like lecithin are used as food additives, stabilising mayonnaise and other edible emulsions.

## Emulsions

If vegetable oil and water are poured into a beaker, the vegetable oil forms a layer on top of the water. This is because vegetable oil is less dense than water, and the two liquids are **immiscible** – they do not dissolve into one another. However, if they are vigorously shaken together, they form a mixture called an **emulsion**.

In an emulsion, tiny droplets of one of the liquids are dispersed throughout the other liquid. The two liquids in an emulsion will eventually settle out into separate layers again, unless the emulsion is stabilised by an emulsifier.

**A** What does immiscible mean?

## Different emulsions

There are two types of emulsion, depending on which liquid forms droplets and which liquid surrounds the droplets:

- **oil-in-water emulsions** consist of tiny droplets of oil dispersed in water
- **water-in-oil emulsions** consist of tiny droplets of water dispersed in oil.

Milk is an oil-in-water emulsion. It contains tiny droplets of butterfat dispersed throughout a watery liquid. Full fat and semi-skimmed milk eventually separate into two layers, with the butterfat forming a creamy layer at the top of the container.

Butter is a water-in-oil emulsion. It contains tiny droplets of watery liquid dispersed throughout the butterfat.

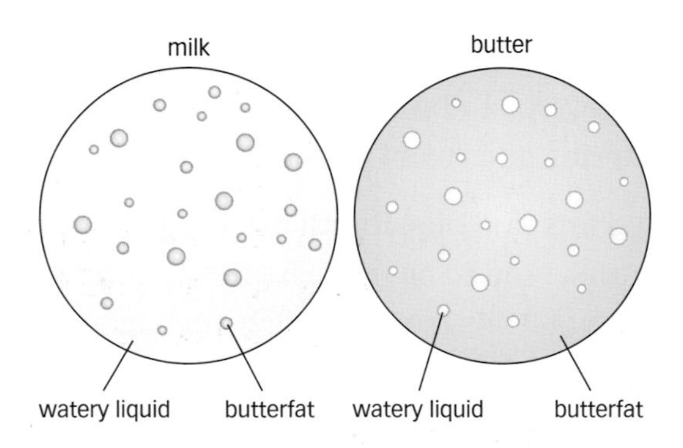


Milk is an oil-in-water emulsion and butter is a water-in-oil emulsion

## Margarine

Margarine is a water-in-oil emulsion. It consists of a blend of different vegetable oils, mixed with water. It is important that margarine from the refrigerator is soft enough to spread easily on bread and toast, but not be so soft that it runs everywhere. Unsaturated vegetable oils tend to have relatively low melting points, so they must be blended with saturated vegetable oils to achieve the desired consistency. Unsaturated vegetable oils are turned into saturated vegetable oils by reacting them with hydrogen. The hydrogen reacts at the carbon-carbon double bonds, turning them into single bonds.

Margarine is a water-in-oil emulsion containing a mixture of vegetable oils and fats

Sodium hydroxide solution reacts with oils and fats to make soap

## Making soap

Soap is made when oils or fats react with hot sodium hydroxide solution. Vegetable oils such as olive oil, coconut oil, and palm oil are often used to make soap. The reaction splits the oil into glycerol, and sodium salts of the fatty acids. These sodium salts are the soap. This process of splitting up natural fats and oils using sodium hydroxide solution is called **saponification**.

### More about saponification

Saponification is an example of a **hydrolysis reaction**. This is a type of reaction in which a compound is broken down by its reaction with water. During saponification, the hydroxide ions from the sodium hydroxide solution break down the oil or fat molecule in a similar way to a water molecule. They cause the bonds between the fatty acids and glycerol to break. Here is the word equation for the overall process:

fat + sodium hydroxide → soap + glycerol

**B** What is saponification?

### Questions

1. Describe how an emulsion is formed.
2. Give an example of an oil-in-water emulsion and a water-in-oil emulsion.

E

3. Describe how margarine is manufactured.
4. Describe how soap is made from natural fats and oils.

C

5. Explain, with the help of an equation, what happens during saponification.

A*

# C6 15: Detergents

## Learning objectives

After studying this topic, you should be able to:

- describe the function of each ingredient in a washing powder and a washing-up liquid
- explain the advantages of washing clothes at low temperatures
- describe the chemical nature of a detergent
- explain how detergents can remove fat or oil stains

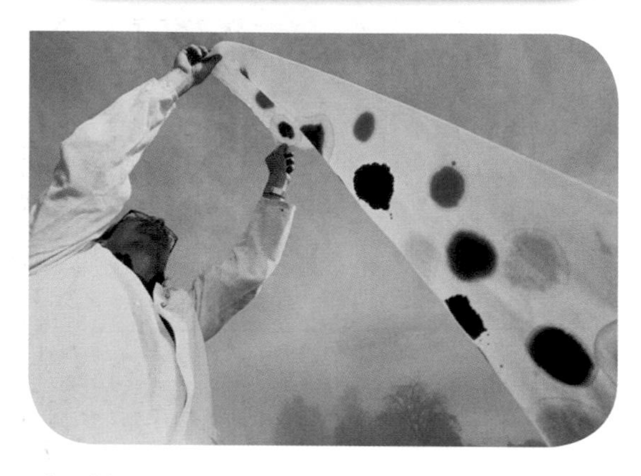

This scientist is carrying out research into the effectiveness of washing powder. The sheet has been deliberately stained with different substances to see how well the powder removes them without damaging the fabric.

| Ingredient | Function |
|---|---|
| active detergent | does the cleaning |
| bleaches | remove coloured stains by destroying the dye |
| optical brighteners | give white fabrics a 'whiter than white' appearance |
| water softener | softens hard water to avoid 'scum' on the clothes |
| enzymes | remove food stains in low-temperature washes |

## Detergents

**Detergents** are ingredients of washing powders and washing-up liquids. They are substances that can surround fat or oil molecules in stains and remove them from the clothes or plates. Some washing powders contain soap, but most do not. They are soapless and contain synthetic detergents. They have the advantage that they form a lather in hard water, as well as in soft water. However, soaps are more suited to handwashing and cause less damage to delicate fabrics.

Detergent molecules are similar in structure to emulsifier molecules. They have a **hydrophilic** or water-loving head, and a **hydrophobic** or water-hating tail.

hydrophilic head

hydrophobic tail

The displayed formula of a typical synthetic detergent molecule. The charged hydrophilic head forms bonds with water molecules, and the hydrophobic tail forms bonds with oil and fat molecules.

### More about detergents

The hydrophilic end of the detergent molecule forms strong intermolecular forces with water molecules in the wash, but not with the molecules in greasy stains. Meanwhile, the hydrophobic end of the detergent molecule forms strong intermolecular forces with the fat or oil molecules in greasy stains, but not with the water molecules in the wash. The detergent molecules can surround the fat or oil molecules in the stain, lifting them from the fabric and into the washing water.

## Washing powders

Washing powders are used to clean fabrics including clothes, sheets, and towels. They have several ingredients to improve their cleaning performance. The table on the left shows the main ones and what they do.

Washing powders that contain enzymes are called biological powders. Those that do not are called non-biological powders or just non-bio powders. Biological powders must be used in low-temperature washes in which the water is at 40 °C or less. This is because the enzymes will be denatured at higher temperatures and so will stop working. However, the ability to clean clothes in cooler water means that less energy is used to heat water for the washing machine. Some fabrics are damaged by high temperatures, so a greater range of clothing can be cleaned using biological powders.

**A** What does a biological powder have that a non-bio powder does not have?

### Key words

**detergent, hydrophilic, hydrophobic**

### Did you know...?

The enzymes in biological powders include lipases to break down fats and oils, and proteases to break down proteins. It is important that washing machines rinse the clothes thoroughly as part of the washing cycle, otherwise the enzymes may irritate your skin.

## Washing-up liquid

Washing-up liquids are used to clean plates, cutlery, and pans without using a dishwashing machine. They also have several ingredients to improve their cleaning performance. The table below shows the main ones and what they do.

| Ingredient | Function |
|---|---|
| active detergent | does the cleaning |
| water | makes the liquid thinner and less viscous, so it pours easily |
| colouring agent and fragrance | make the washing-up liquid more attractive to use |
| rinse agent | helps the water drain from the crockery |

Washing-up liquids contain several ingredients to make washing up more efficient

**B** What does the rinse agent do?

### Questions

1. Describe the function of each of the five main ingredients in washing powders.
2. Describe the function of each of the four main ingredients in washing-up liquids.

E

3. Explain the advantages of using low temperature washes for clothes.
4. Describe the structure of detergent molecules.

C

5. Explain how detergents can remove fat or oil stains.

A*

### Exam tip OCR

✔ You do not need to know the detailed structure of a detergent molecule, but you should recognise that it has a hydrophilic head and a hydrophobic tail.

# 16: Cleaning and washing

## Learning objectives

After studying this topic, you should be able to:

- identify the correct solvent to remove a particular stain
- describe dry cleaning
- interpret data from experiments on the effectiveness of washing powders and washing-up liquids
- explain how dry cleaning works
- deduce from experimental data which detergent contains an enzyme

Silk will be damaged in a washing machine so it is best to hand wash it or have it dry cleaned

These clothes have been dry cleaned

## The S words

If a substance is **insoluble** in a particular liquid, the substance will not dissolve in it. However, if a substance is **soluble** in a particular liquid, the substance will dissolve in it to form a **solution**. In a solution, the dissolved substance is called the **solute** and the liquid that does the dissolving is called the **solvent**. For example, sugar is soluble in water. Sugar dissolves in water to form sugar solution, in which sugar is the solute and water is the solvent.

Water is a very good solvent for many different substances, but it is not the only solvent. For example, nail polish is insoluble in water but it is soluble in ethyl ethanoate and propanone. Different solvents will dissolve different substances. This is very useful for cleaning clothes made with fabrics that would be damaged by being washed in water. The label on these clothes may show 'dry clean only'.

## Dry cleaning

**Dry cleaning** is a way to clean clothes without using water. A different solvent such as tetrachloroethene is used instead. Solvents like this are good at dissolving greasy stains, and other stains that do not dissolve in water.

$$Cl_2C{=}CCl_2$$

Tetrachloroethene is used as a dry cleaning solvent

At the dry cleaners, the clothing is carefully checked for anything that might dissolve in the solvent or be damaged by it. This can include plastic pens left in pockets, which may dissolve and release their ink all over the clothing. The clothes are cleaned in a machine rather like a large washing machine, but with equipment to stop solvent fumes escaping. The clothes are washed in the dry-cleaning solvent, rinsed with fresh solvent, and dried in warm air. Used solvent is distilled so that it can be reused.

**A** What are the solute and solvent in salty water?
**B** Name a solvent for nail polish.
**C** Name a solvent used in dry cleaning.

## More about dry cleaning

Removing stains using dry cleaning solvent involves a balance between intermolecular forces. There are weak intermolecular forces between molecules of grease in a stain, and also weak intermolecular forces between solvent molecules. However, the solvent molecules can also form intermolecular forces with grease molecules.

The solvent molecules surround the grease molecules, lifting them off the fabric and into the bulk of the solvent. The solvent itself becomes dirty during the cleaning cycle as the stains dissolve in it, which is why the clothes must be rinsed with fresh solvent.

**Key words**

**insoluble, soluble, solution, solute, solvent, dry cleaning**

## Using clothes labels

Clothes have wash labels sewn in them. These have various symbols to explain how the clothes should be washed. It is wise to follow the instructions if you do not want to damage your clothes. The labels include information such as:

- the water temperature to use in the washing machine
- whether the clothes should be hand washed
- the correct drying conditions, including whether to tumble dry or not
- the correct ironing temperature, including whether the item can be ironed
- the correct dry-cleaning conditions to use.

▲ These are some symbols found on clothes labels

### Questions

1 What do the words soluble and insoluble mean?

2 What information does a wash label on clothing give you?

↓ E

3 Describe what dry cleaning involves, including an example of the solvent used and the type of stain removed.

4 Suggest why used dry-cleaning solvent is distilled rather than disposed of.

↓ C

5 Explain, in terms of intermolecular forces, how dry cleaning works.

↓ A*

### Exam tip OCR

✔ You do not need to be able to recall or interpret washing symbols from clothes labels.

✔ You should be able to interpret data from experiments involving washing powders and washing-up liquids. This could be to see whether a washing powder contains an enzyme, or to see which one washed the most plates or was most effective.

# Module summary

## Revision checklist

- In electrolysis, positive ions in a liquid move to the cathode and gain electrons. Negative ions move to the anode and lose electrons.
- Molten substances decompose into elements during electrolysis.
- Hydrogen and oxygen can be formed from solutions because of the H+ and OH– ions in water.
- Fuel cells use the reaction between hydrogen and oxygen to release electrical energy.
- Fuel cells produce clean energy efficiently, but need expensive catalysts and a hydrogen supply.
- In redox reactions, reduction (gain of electrons) and oxidation (loss of electrons) happen in the same reaction.
- More reactive metals will displace less reactive metals from solutions.
- Rusting is prevented by covering iron with a protective layer or by sacrificial protection using more reactive metals.
- Ethanol is an alcohol made by fermenting sugar or by reacting ethene with steam.
- Alcohols contain an OH group.
- CFC (chlorofluorocarbon) molecules deplete the ozone layer by means of reactions involving chlorine radicals.
- Ozone absorbs damaging ultraviolet light, preventing it reaching the Earth's surface.
- Most countries have agreed to ban CFCs and replace them with less damaging alternatives.
- Hard water is caused by calcium and magnesium ions that dissolve when rainwater containing carbon dioxide flows through rocks.
- Temporary hardness is removed by boiling and is caused by dissolved calcium hydrogencarbonate. Permanent hardness is caused by calcium sulfate and isn't removed by boiling.
- Washing soda or ion-exchange resins also soften water.
- Fats and oils are naturally occurring esters.
- Fats and oils are split up by alkalis to form soap and glycerol. They can also be made into biodiesel.
- Unsaturated fats and oils contain C=C double bonds.
- Emulsifiers cause oils and water to mix as an emulsion.
- Washing powders use synthetic detergents to remove fats and oils. Enzymes and bleaches remove other stains.
- Dry cleaning uses solvents other than water to remove grease stains.

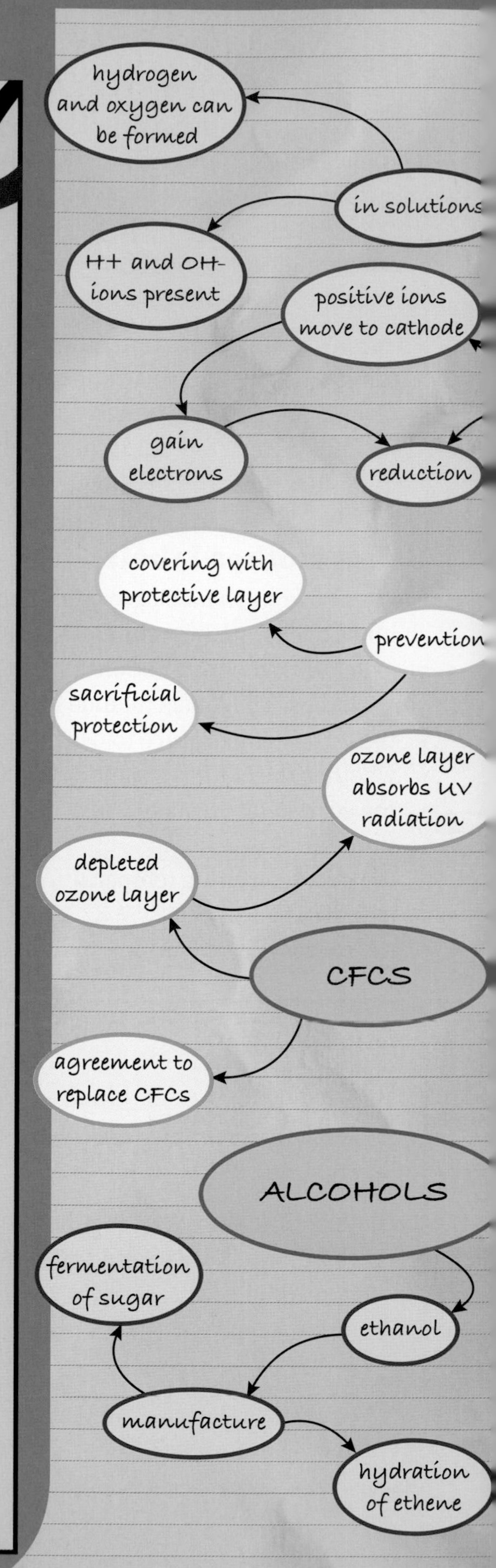




NOW USE THE C6 GRADE CHECKER ON PAGE 250

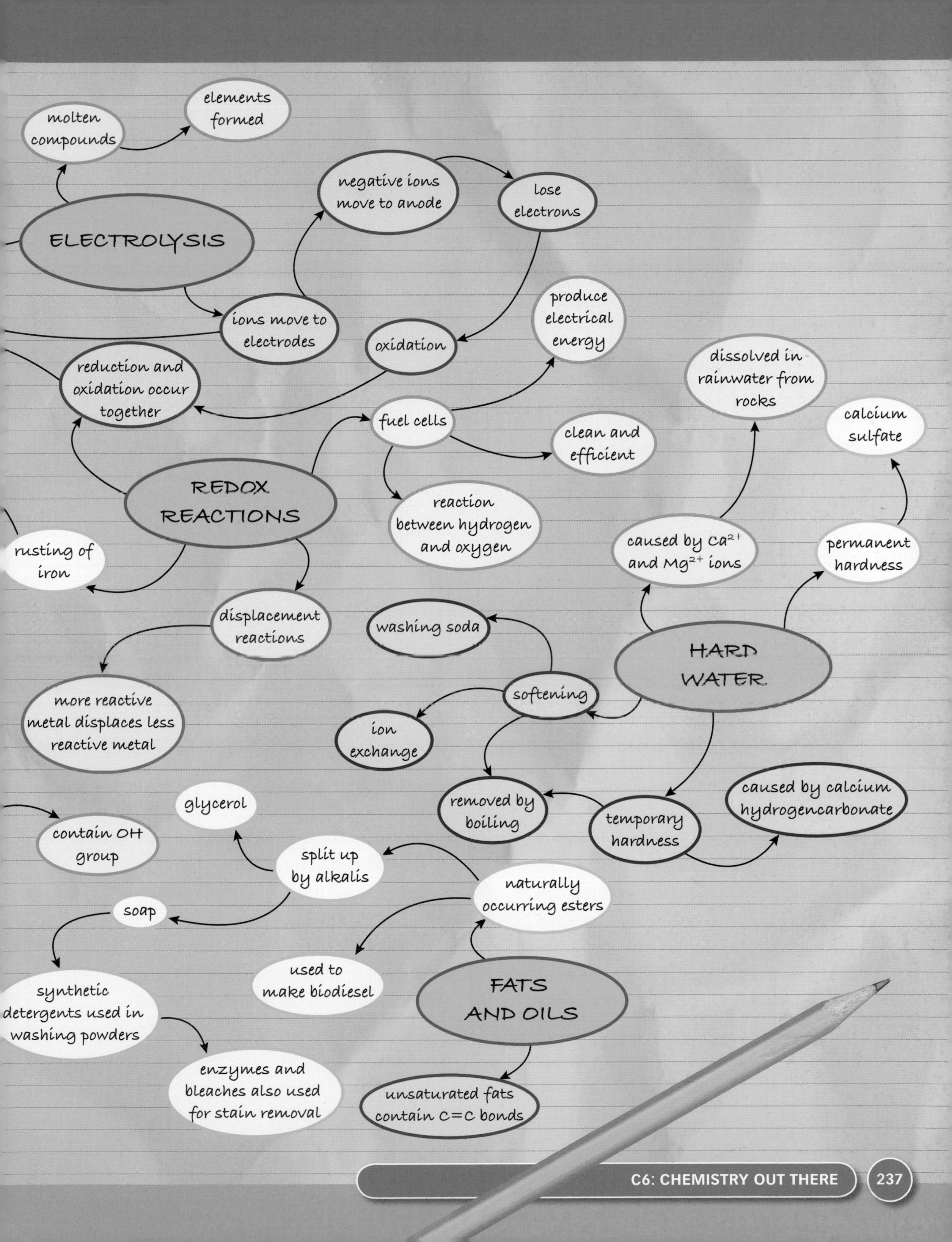

molten compounds
elements formed
ELECTROLYSIS
negative ions move to anode
lose electrons
ions move to electrodes
oxidation
produce electrical energy
reduction and oxidation occur together
fuel cells
clean and efficient
dissolved in rainwater from rocks
calcium sulfate
REDOX REACTIONS
reaction between hydrogen and oxygen
caused by $Ca^{2+}$ and $Mg^{2+}$ ions
permanent hardness
rusting of iron
displacement reactions
washing soda
HARD WATER
softening
more reactive metal displaces less reactive metal
ion exchange
removed by boiling
temporary hardness
caused by calcium hydrogencarbonate
glycerol
contain OH group
split up by alkalis
naturally occurring esters
soap
used to make biodiesel
FATS AND OILS
synthetic detergents used in washing powders
enzymes and bleaches also used for stain removal
unsaturated fats contain C=C bonds

OCR gateway Upgrade

## Answering Extended Writing questions

**QUESTION**

Explain what happens when iron rusts. Describe two ways of preventing rusting and explain how they work

**The quality of written communication will be assessed in your answer to this question.**

Water and air make iron rust. If you stop the air and water getting to the iron you can stop iron rusting greasing and painting are used to do this.

*Examiner:* This answer contains several correct points and the candidate has answered all of the requirements of the question at a basic level. The spelling is good, although the last sentence should be divided into two with a full stop after 'rusting'.

Rust is iron oxide so it happens in a redox reaction with oxygen. Galvanising is a good way of protecting iron. The zinc covers the iron and prevents air reaching it.

*Examiner:* The candidate uses some good vocabulary here, such as 'galvanising' and 'redox'. Water is also involved in rusting, but this isn't mentioned. Only one method of protecting iron is included, and although the description is partly correct, the candidate should also mention that zinc is a sacrificial metal. Spelling, punctuation, and grammar are good.

Rusting happens when iron reacts with water and oxygen to make hydrated iron oxide. This is an oxidation reaction. It can be prevented by painting the iron to stop the oxygen getting to the iron or by sacrificial protection. Iron is covered in zinc and the zinc losses electrons because it is reactive.

A*

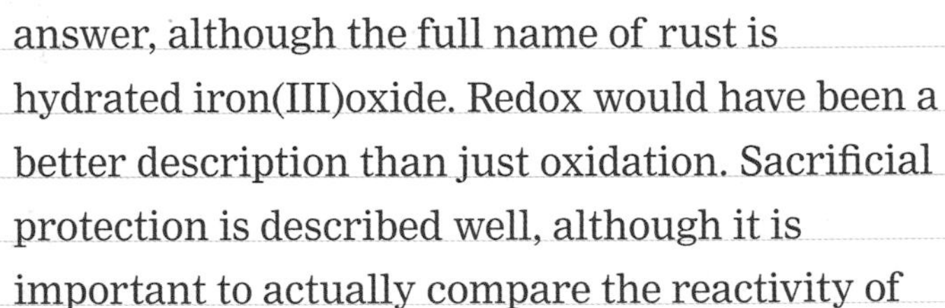

*Examiner:* This a reasonably full and correct answer, although the full name of rust is hydrated iron(III)oxide. Redox would have been a better description than just oxidation. Sacrificial protection is described well, although it is important to actually compare the reactivity of iron and zinc. 'Loses' is spelt incorrectly.

# Exam-style questions

**1** Here is the apparatus needed to carry out electrolysis on a solution of copper(II) sulfate.

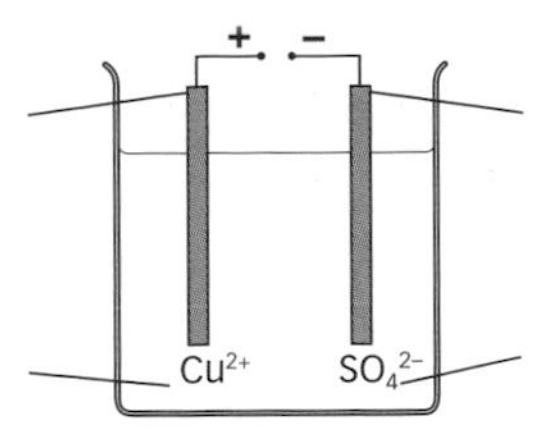


A02 **a** Complete the diagram by adding labels. Use the following words:

anode anion cathode cation

A02 **b** Give two observations that you would expect to see when the power supply is turned on.

**2** Many parts of Britain have hard water due to the presence of calcium and magnesium ions.

Calcium ions get into water because of the action of rainwater on limestone rocks.

A02 **a** Complete this equation for the reaction that dissolves the calcium ions:

$$CaCO_3 + H_2O + \_\_\_\_\_ \leftrightharpoons Ca(HCO_3)$$

A02 **b** Name the product of the reaction.

A01 **c** This process produces temporary hardness in water. How is this different from permanent hardness?

**3** Iron rusts when it is exposed to air and water.

A01 **a** **i** Give the chemical name for rust.

**ii** Rusting is a redox reaction. Explain the meaning of this term.

A01 **b** Sometimes iron is coated with a layer of zinc. This is called sacrificial protection. Explain how sacrificial protection works.

A02 **c** When zinc is added to a solution of copper(II) sulfate, a brown colour develops on the surface of the zinc.

**i** Name the brown colouring.

**ii** Complete the equation for the reaction that occurs:

$$Zn\ (s) + Cu^{2+} \leftrightharpoons \_\_\_\_\_\_\_\_$$

**iii** Which substance is oxidised? Explain your answer.

E ↓ C ↓ A*

## Extended Writing

**4** A01 Ethanol is a chemical substance with several important uses. Describe some of these uses and explain two ways in which ethanol is manufactured.

**5** A01 CFCs have been banned in most countries because scientific research has shown that they deplete the ozone layer. Describe how this depletion is thought to occur and why it has consequences for human health at the Earth's surface.

**6** A02 A03 Car manufacturers are developing cars powered by fuel cells in which hydrogen and oxygen react and release electrical energy. Some people suggest that we should use fuel cells rather than the combustion of fossil fuels because it will be more environmentally friendly. Discuss the arguments for and against this.

A01 Recall the science

A02 Apply your knowledge

A03 Evaluate and analyse the evidence

# Grade checker

**Revising module C1**

To help you start your revision, the specification for module C1 has been summarised in the checklist below. Work your way along each row and make sure that you are happy with all the statements for your target grade.

If you are not sure of any of the statements for your target grade, make a note of them as part of your revision plan. You can then work back through the relevant parts of pages 14–45 to fill gaps in your knowledge as a start to your revision.

| | To aim for a grade G–E | To aim for a grade D–C | To aim for a grade B–A* |
|---|---|---|---|
| C1a | **Recall** that crude oil, coal, and gas are fossil fuels.<br>**Describe** what non-renewable fuels are.<br>**Recognise** what fractional distillation does.<br>**Understand** the principle of fractional distillation.<br>**Recognise** the fractions obtained from crude oil.<br>**Recall** that LPG contains propane and butane gases.<br>**Describe** environmental problems associated with crude oil.<br>**Label** laboratory apparatus for cracking liquid paraffin.<br>**Describe** cracking as a process that converts large hydrocarbon molecules into smaller, more useful ones. | **Explain** why fossil fuels are non-renewable, finite resources.<br>**Describe** oil as a mixture of many hydrocarbons.<br>**Label** a diagram of a fractional distillation column.<br>**Describe** how fractional distillation works.<br>**Explain** potential environmental problems associated with transportation of crude oil.<br>**Describe** the process of cracking in terms of alkane and alkene molecules. | **Discuss** problems associated with the finite nature of crude oil.<br>**Explain** fractional distillation in terms of molecular size, intermolecular forces, and boiling point.<br>**Understand** that intermolecular bonds break during boiling, but covalent bonds do not.<br>**Explain** political problems associated with crude oil.<br>**Explain** how cracking helps refineries match supply with demand. |
| C1b | **List** the factors involved in choosing the best fuel for a particular purpose.<br>**Recall** that fuel combustion releases useful heat energy.<br>**Understand** that complete combustion requires plentiful oxygen, while incomplete combustion occurs when there is a shortage of oxygen.<br>**Understand** that complete combustion of a hydrocarbon fuel makes carbon dioxide and water, while incomplete combustion makes carbon monoxide, carbon, and water.<br>**Construct** word equations for complete and incomplete combustion of hydrocarbon fuels, given the reactants and products.<br>**Explain** why a blue Bunsen flame releases more energy than a yellow one.<br>**Understand** that a yellow flame produces soot.<br>**Recall** that carbon monoxide is a poisonous gas. | **Suggest** key factors to consider when choosing the best fuel for a particular purpose.<br>**Describe** an experiment to show that complete combustion produces carbon dioxide and water.<br>**Construct** word equations for complete and incomplete combustion of a hydrocarbon (not all reactants and products given).<br>**Explain** the advantages of complete combustion of hydrocarbon fuels over incomplete combustion. | **Evaluate** the use of different fuels.<br>**Explain** the reasons for the increase in use of fossil fuels.<br>**Construct** a balanced symbol equation for the complete combustion of a hydrocarbon, given its molecular formula.<br>**Construct** a balanced symbol equation for the incomplete combustion of a hydrocarbon, given its molecular formula and the product. |
| C1c | **Recall** that air contains oxygen, nitrogen, water vapour and carbon dioxide.<br>**Understand** that present day atmospheric oxygen, nitrogen, and carbon dioxide levels are constant. | **Recall** the percentage composition by volume of clean air.<br>**Describe** a simple carbon cycle involving photosynthesis, respiration, and combustion. | **Evaluate** the effects of human influences on the composition of air.<br>**Describe** a possible theory for how the present day atmosphere evolved over time, based on composition of present day volcanic gases. |

| | To aim for a grade G–E | To aim for a grade D–C | To aim for a grade B–A* |
|---|---|---|---|
| C1c | **Understand** that photosynthesis decreases carbon dioxide levels and increases oxygen levels in the air.<br>**Understand** that respiration and combustion increase carbon dioxide levels and decrease oxygen levels in the air.<br>**Relate** common air pollutants to their source.<br>**Understand** the function of a catalytic converter. | **Describe** how the present day atmosphere evolved.<br>**Explain** why atmospheric pollution control is important.<br>**Understand** that a catalytic converter changes carbon monoxide into carbon dioxide. | **Explain** why high temperatures in internal combustion engines allow nitrogen and oxygen to react to form oxides of nitrogen.<br>**Describe** how a catalytic converter converts carbon monoxide to carbon dioxide. |
| C1d | **Recall** the two elements chemically combined in a hydrocarbon.<br>**Recognise** a hydrocarbon from its formula.<br>**Recognise** that alkanes and alkenes are hydrocarbons.<br>**Deduce** the name of an addition polymer given the name of the monomer.<br>**Recall** that polymers are made when monomers join together in a polymerisation reaction. | **Recall** that a hydrocarbon is a compound formed between carbon and hydrogen atoms.<br>**Explain** why a compound is a hydrocarbon.<br>**Understand** that alkanes are hydrocarbons with single covalent bonds only, and that alkenes are hydrocarbons with double covalent bonds between carbon atoms.<br>**Describe** how to test for an alkene.<br>**Recognise** the displayed formula for a polymer.<br>**Describe** addition polymerisation. | **Describe** a saturated compound.<br>**Describe** an unsaturated compound.<br>**Explain** the reaction between bromine and alkenes.<br>**Draw** the displayed formula of an addition polymer from its monomer and vice versa.<br>**Explain** that addition polymerisation involves the reaction of unsaturated monomer molecules to form a saturated polymer. |
| C1e | **Recall** that nylon is used in clothing.<br>**Understand** that many polymers are non-biodegradable.<br>**Explain** the problems with disposing of non-biodegradable polymers.<br>**Recall** the ways in which waste polymers can be disposed of. | **Suggest** desirable properties for a polymer intended for a particular purpose.<br>**Explain** the suitability of a polymer for a particular purpose, based on its properties.<br>**Compare** the properties of nylon and Gore-Tex®.<br>**Explain** why chemists are developing new types of polymers.<br>**Explain** the environmental and economic issues associated with the use of polymers. | **Understand** that atoms in plastics are joined by strong covalent bonds.<br>**Relate** the properties of plastics to simple models of their structure.<br>**Explain** why Gore-Tex® is waterproof yet breathable. |
| C1f | **Recognise** the factors that mean a chemical change is taking place.<br>**Explain** why cooking is a chemical change.<br>**Relate** types of food additive to their function.<br>**Explain** how baking powder makes cakes rise.<br>**Recall** the chemical test for carbon dioxide. | **Recall** that protein molecules in eggs and meat change shape (are denatured) in cooking.<br>**Describe** emulsifiers.<br>**Recall** the word equation for the decomposition of sodium hydrogencarbonate.<br>**Construct** a balanced symbol equation for the decomposition of sodium hydrogencarbonate. | **Explain** why the texture of eggs and meat changes when cooked.<br>**Explain** why cooked potato is easier to digest.<br>**Explain** why an emulsifier helps to prevent oil and water separating.<br>**Construct** a balanced symbol equation for the decomposition of sodium hydrogencarbonate (formulae not given). |
| C1g | **Understand** that cosmetics may be either synthetic or natural.<br>**Recall** that esters are perfumes that can be made synthetically.<br>**Recall** the physical properties of perfumes.<br>**Understand** the terms solvent, solute, solution, soluble, and insoluble.<br>**Recall** that testing cosmetics on animals is banned in the EU.<br>**Explain** the need to thoroughly test cosmetics. | **Recall** that alcohols react with acids to form an ester and water.<br>**Describe** a simple experiment to make an ester.<br>**Explain** why a perfume needs certain properties.<br>**Recall** that esters can be used as solvents.<br>**Describe** a solution.<br>**Explain** why testing cosmetics on animals is banned in the EU. | **Explain** the volatility of perfumes in terms of kinetic theory.<br>**Explain** why water will not dissolve nail varnish.<br>**Explain** why people have different opinions on testing cosmetics on animals. |
| C1h | **Relate** paint ingredients to their function.<br>**Recall** the formulation of oil paints.<br>**Recall** the properties and uses of thermochromic pigments.<br>**Recall** that phosphorescent pigments can glow. | **Describe** paint as a colloid.<br>**Describe** how many paints dry.<br>**Describe** emulsion paints as water based paints that dry when the solvent evaporates.<br>**Explain** why thermochromic pigments are suited to a particular use.<br>**Explain** why phosphorescent pigments glow. | **Explain** why a colloid will not separate.<br>**Explain** how oil paints dry.<br>**Explain** how acrylic paints can be added to thermochromic pigments.<br>**Recall** that phosphorescent pigments are safer than alternative radioactive substances. |

# Grade checker

**Revising module C2**

To help you start your revision, the specification for module C2 has been summarised in the checklist below. Work your way along each row and make sure that you are happy with all the statements for your target grade.

If you are not sure of any of the statements for your target grade, make a note of them as part of your revision plan. You can then work back through the relevant parts of pages 52–83 to fill gaps in your knowledge as a start to your revision.

| | To aim for a grade G–E | To aim for a grade D–C | To aim for a grade B–A* |
|---|---|---|---|
| C2a | **Describe** the structure of the Earth, to include the crust, mantle, and core.<br>**Understand** the movement of tectonic plates.<br>**Recognise** theories on the nature of the Earth's surface.<br>**Recognise** that Earth scientists accept the theory of plate tectonics.<br>**Explain** how the size of crystals in an igneous rock is related to rate of cooling.<br>**Describe** magma and lava.<br>**Understand** that different volcanoes erupt different kinds of lava.<br>**Explain** why people live near volcanoes. | **Describe** the lithosphere.<br>**Explain** the problems associated with studying the structure of the Earth.<br>**Explain** why the theory of plate tectonics is now widely accepted.<br>**Understand** that the type of volcanic eruption depends on the composition of the magma.<br>**Explain** why geologists study volcanoes. | **Describe** the properties of the mantle.<br>**Describe** the theory of plate tectonics.<br>**Describe** the development of the theory of plate tectonics.<br>**Describe** types of igneous rocks formed from lava.<br>**Explain** why geologists are now able to better predict volcanic eruptions. |
| C2b | **Recall** that rock is used for buildings and roads.<br>**Explain** why there are environmental problems associated with quarrying or mining.<br>**Recall** that limestone and marble are both forms of calcium carbonate.<br>**Recall** that limestone thermally decomposes to make calcium oxide and carbon dioxide.<br>**Describe** how concrete is made.<br>**Describe** how concrete can be reinforced. | **Relate** some construction materials to the substances from which they are manufactured.<br>**Compare** the hardness of limestone, marble, and granite.<br>**Construct** the word equation for the decomposition of limestone.<br>**Construct** the balanced symbol equation for the decomposition of limestone.<br>**Describe** thermal decomposition.<br>**Recall** how cement is made.<br>**Recall** that reinforced concrete is a composite material. | **Explain** why marble, granite, and limestone have different hardness.<br>**Construct** the balanced symbol equation for the decomposition of limestone (formulae not given).<br>**Explain** why reinforced concrete is a better construction material than non-reinforced concrete. |
| C2c | **Understand** how copper can be extracted from its ore.<br>**Recall** that copper can be purified by electrolysis.<br>**Explain** why recycling copper is cheaper than extracting it from its ore.<br>**Recall** that alloys are mixtures of one or more metal elements.<br>**Recognise** that brass, bronze, solder, steel, and amalgam are alloys.<br>**Recall** large scale uses for alloys.<br>**Recognise** that the properties of an alloy differ from those of the metals from which it is made. | **Recall** that reduction is the removal of oxygen.<br>**Label** the apparatus needed to purify copper by electrolysis.<br>**Explain** the advantages and disadvantages of recycling copper.<br>**Recall** the main metals in amalgam, brass, and solder.<br>**Explain** why metals (including alloys) are suited to a particular use. | **Describe** the use of electrolysis in the purification of copper.<br>**Explain** why the electrolytic purification of copper involves both oxidation and reduction.<br>**Evaluate** the suitability of metals for a particular use.<br>**Explain** how the use of smart alloys has increased the number of applications of alloys. |
| C2d | **Understand** that rusting needs iron, water, and oxygen.<br>**Recall** that aluminium does not corrode in moist conditions. | **Understand** that salt water and acid rain accelerate rusting.<br>**Explain** why rusting involves oxidation.<br>**Construct** the word equation for rusting. | **Explain** advantages and disadvantages of building car bodies from aluminium or steel.<br>**Evaluate** information on materials used to manufacture cars. |

| To aim for a grade G–E | To aim for a grade D–C | To aim for a grade B–A* | |
|---|---|---|---|
| **Compare** the properties of iron and aluminium.<br>**Recall** the materials needed to build a car.<br>**Describe** the advantages of recycling materials. | **Explain** why aluminium does not corrode.<br>**Understand** that alloys can be more useful than the pure metals from which they are made.<br>**Describe** advantages and disadvantages of building car bodies from aluminium or steel.<br>**Explain** the advantages and disadvantages of recycling materials used to make cars. | | C2d |
| **Recall** that in the Haber process ammonia is made from nitrogen from the air and hydrogen from cracking oil fractions or from natural gas.<br>**Describe** the factors that determine the cost of making a new substance.<br>**Recognise** the symbol for a reversible reaction.<br>**Understand** that a reversible reaction proceeds in both directions.<br>**Recall** uses of ammonia. | **Describe** how ammonia is made in the Haber process.<br>**Construct** the balanced symbol equation for the manufacture of ammonia in the Haber process.<br>**Describe** how different factors affect the cost of making a new substance.<br>**Recognise** the importance of ammonia in relation to world food production. | **Explain** the conditions used in the Haber process.<br>**Construct** the balanced symbol equation for the manufacture of ammonia in the Haber process (formulae not given).<br>**Explain** that economic considerations determine the conditions used in the manufacture of chemicals. | C2e |
| **Describe** how universal indicator can be used to estimate the pH of a solution.<br>**Recall** that an alkali is a soluble base.<br>**Understand** that an acid can be neutralised by a base or alkali, and vice versa. | **Understand** that some indicators show sudden changes in colour.<br>**Recall** the process of neutralisation.<br>**Understand** that the concentration of an acid is determined by the concentration of $H^+$ ions.<br>**Explain** why metal oxides and metal hydroxides neutralise acids.<br>**Recall** that carbonates neutralise acids to give water, a salt, and carbon dioxide.<br>**Construct** word equations to show the neutralisation of acids.<br>**Predict** the salt produced when a named base or carbonate is neutralised. | **Explain** why an acid is neutralised by an alkali in terms of the ions present.<br>**Construct** balanced symbol equations for the neutralisation of acids by bases and carbonates. | C2f |
| **Recall** that fertilisers increase crop yield.<br>**Recall** that plants absorb minerals through roots.<br>**Recall** that nitrogen, phosphorus, and potassium are essential for plant growth.<br>**Recognise** the essential elements.<br>**Understand** that fertiliser use can be beneficial or problematic.<br>**Identify** the apparatus needed to make a fertiliser.<br>**Recall** two nitrogenous fertilisers manufactured from ammonia. | **Explain** why fertilisers must dissolve in water before they can be absorbed by plants.<br>**Identify** arguments for and against fertiliser use.<br>**Predict** the name of the acid and alkali needed to make a named fertiliser. | **Explain** how fertiliser use increases crop yield.<br>**Explain** the process of eutrophication.<br>**Describe** the preparation of a named synthetic fertiliser by the reaction of an acid and an alkali. | C2g |
| **Recall** that sodium chloride can be obtained from the sea or from salt deposits.<br>**Recall** that electrolysis of concentrated sodium chloride gives chlorine and hydrogen.<br>**Recall** the chemical test for chlorine.<br>**Understand** that sodium chloride is an important raw material in the chemical industry.<br>**Recall** that household bleach, PVC, and solvents are made from salt.<br>**Recall** that chlorine is used to sterilise water.<br>**Recall** that hydrogen is used in the manufacture of margarine.<br>**Recall** that sodium hydroxide is used in soap. | **Describe** how salt can be mined as rock salt or by solution mining.<br>**Explain** how mining for salt can cause subsidence.<br>**Recall** the products of the electrolysis of brine.<br>**Explain** the importance of using inert electrodes in the electrolysis of sodium chloride solution.<br>**Describe** how sodium hydroxide and chlorine are used to make bleach. | **Explain** how electrolysis of brine produces sodium hydroxide, hydrogen, and chlorine.<br>**Explain** why electrolysis of brine involves both reduction and oxidation.<br>**Explain** the economic importance of the chlor-alkali industry. | C2h |

# Grade checker

**Revising module C3**

To help you start your revision, the specification for module C3 has been summarised in the checklist below. Work your way along each row and make sure that you are happy with all the statements for your target grade.

If you are not sure of any of the statements for your target grade, make a note of them as part of your revision plan. You can then work back through the relevant parts of pages 90–121 to fill gaps in your knowledge as a start to your revision.

| | To aim for a grade G–E | To aim for a grade D–C | To aim for a grade B–A* |
|---|---|---|---|
| C3a | **Recognise** that some reactions can be fast and others very slow.<br>**Label** the laboratory apparatus needed to measure the rate of a reaction producing a gas.<br>**Plot** experimental results involving gas volumes or mass loss on a graph.<br>**Plot** experimental results involving reaction times on a graph.<br>**Explain** why a reaction stops. | **Understand** that the rate of a reaction measures how much product is formed in a fixed time period.<br>**Recognise** and use the idea that the amount of product formed is directly proportional to the amount of limiting reactant used.<br>**Recall** that the limiting reactant is the reactant not in excess that is all used up at the end of the reaction. | **Use** the following units for the rate of reaction: g/s or g/min, $cm^3$/s or $cm^3$/min.<br>**Explain**, in terms of reacting particles, why the amount of product formed is directly proportional to the amount of limiting reactant used. |
| C3b | **Recognise** that a chemical reaction takes place when particles collide.<br>**Describe** the effect of changing temperature on the rate of a chemical reaction.<br>**Describe** the effect of changing concentration on the rate of a chemical reaction.<br>**Describe** the effect of changing pressure on the rate of a chemical reaction of gases. | **Understand** that rate of reaction depends on the number of collisions between reacting particles.<br>**Explain**, in terms of reacting particles why changes in temperature, concentration, and pressure change the rate of reaction.<br>**Draw** sketch graphs to show the effect of changing temperature, concentration, or pressure on the rate of reaction and the amount of product formed in a reaction. | **Understand** that the rate of reaction depends on collision frequency and the energy transferred in the collision.<br>**Explain**, in terms of collisions between reacting particles, why changes in temperature, concentration, and pressure change the rate of reaction. |
| C3c | **Recall** that the rate of a reaction can be increased by the addition of a catalyst.<br>**Recall** that the rate of a reaction can be increased by using powdered reactant.<br>**Describe** an explosion as a very fast reaction that releases a large volume of gaseous products. | **Describe** a catalyst as a substance that changes the rate of reaction and is unchanged at the end of the reaction.<br>**Understand** that only a small amount of a catalyst is needed to catalyse large amounts of reactants and that a catalyst is specific to a particular reaction.<br>**Explain** the difference in rate of reaction between a lump of reactant and powdered reactant.<br>**Explain** the dangers of fine combustible powders in factories. | **Recognise** that a catalyst is specific to a particular reaction.<br>**Explain**, in terms of collisions between reacting particles, the difference in rate of reaction between a lump of reactant and powdered reactant. |
| C3d | **Calculate** the relative formula mass of a substance from its formula, given the appropriate relative atomic masses.<br>**Understand** that the total mass of reactants at the start of a reaction is equal to the total mass of products made.<br>**Use** the principle of conservation of mass to calculate mass of reactant or product.<br>**Use** simple ratios to calculate reacting masses and product masses. | **Use** relative formula masses and a symbol equation (both provided) to show that mass is conserved during a reaction.<br>**Explain** why mass is conserved in chemical reactions.<br>**Recognise** and use the idea that the mass of product formed is directly proportional to the mass of limiting reactant used. | **Use** relative formula masses and a symbol equation (provided) to show that mass is conserved during a reaction.<br>**Calculate** masses of products or reactants from balanced symbol equations using relative formula masses. |

| | To aim for a grade G–E | To aim for a grade D–C | To aim for a grade B–A* |
|---|---|---|---|
| C3e | **Understand** percentage yield as a way of comparing amount of product made to the amount expected.<br>**Recognise** possible reasons why the percentage yield of a product is less than 100%.<br>**Understand** atom economy as a way of measuring the amount of atoms that are wasted when manufacturing a chemical. | **Recall** and **use** the formula for percentage yield.<br>**Recall** and **use** the formula for atom economy.<br>**Calculate** atom economy, given a balanced symbol equation and appropriate relative formula masses. | **Explain** why an industrial process aims for the highest possible percentage yield.<br>**Explain** why an industrial process aims for the highest possible atom economy. |
| C3f | **Recall** that an exothermic reaction is one in which energy is transferred into the surroundings.<br>**Recall** that an endothermic reaction is one in which energy is taken from the surroundings.<br>**Recognise** exothermic and endothermic reactions using temperature changes.<br>**Describe**, using a diagram, a simple calorimetric method for comparing the energy transferred in combustion reactions. | **Recall** bond making as an exothermic process and bond breaking as an endothermic process.<br>**Describe** a simple calorimetric method for comparing the energy transferred per gram of fuel combusted.<br>**Calculate** energy transferred. | **Explain** why a reaction is exothermic or endothermic based on the energy changes that occur during bond breaking and making.<br>**Use** the energy transfer to calculate the mass of water heated and temperature change.<br>**Calculate** the energy output of a fuel in J/g. |
| C3g | **Describe** the differences between a batch and a continuous process.<br>**List** the factors that affect the cost of making and developing a pharmaceutical drug.<br>**Explain** why pharmaceutical drugs need to be thoroughly tested before they can be licensed for use.<br>**Understand** that the raw materials for chemicals such as pharmaceuticals can be made synthetically or extracted from plants.<br>**Explain** why it is important to manufacture pharmaceutical drugs to be as pure as possible.<br>**Describe** how melting point, boiling point, and thin layer chromatography can be used to establish the purity of a compound. | **Explain** why batch processes are often used for the production of pharmaceutical drugs, but continuous processes are used to produce other chemicals.<br>**Explain** why it is often expensive to make and develop new pharmaceutical drugs.<br>**Describe** how chemicals are extracted from plant sources. | **Evaluate** the advantages and disadvantages of batch and continuous manufacturing processes.<br>**Explain** why it is difficult to test and develop new pharmaceutical drugs that are safe to use. |
| C3h | **Explain** why diamond, graphite, and Buckminster fullerene are all forms of carbon.<br>**Recognise** the structures of diamond, graphite, and Buckminster fullerene.<br>**List** the physical properties of diamond.<br>**List** the physical properties of graphite.<br>**Recall** that nanotubes are used to reinforce graphite in tennis rackets because nanotubes are very strong.<br>**Recall** that nanotubes are used as semiconductors in electrical circuits. | **Explain** why diamond, graphite, and fullerenes are allotropes of carbon.<br>**Explain**, in terms of properties, why diamond is used in cutting tools and jewellery.<br>**Explain**, in terms of properties, why graphite is used in pencil leads and in lubricants.<br>**Explain** why diamond and graphite have a giant molecular structure.<br>**Explain** why fullerenes can be used in new drug delivery systems. | **Explain**, in terms of structure and bonding, the properties of diamond.<br>**Explain**, in terms of structure and bonding, the properties of graphite.<br>**Predict** and explain the properties of substances that have a giant molecular structure.<br>**Explain** how the structure of nanotubes enables them to be used as catalysts. |

# Grade checker

**Revising module C4**

To help you start your revision, the specification for module C4 has been summarised in the checklist below. Work your way along each row and make sure that you are happy with all the statements for your target grade.

If you are not sure of any of the statements for your target grade, make a note of them as part of your revision plan. You can then work back through the relevant parts of pages 128–159 to fill gaps in your knowledge as a start to your revision.

| | To aim for a grade G–E | To aim for a grade D–C | To aim for a grade B–A* |
|---|---|---|---|
| C4a | **Understand** that an atom has a nucleus surrounded by electrons.<br>**Recall** that a nucleus is positively charged, an electron is negatively charged, and an atom is neutral.<br>**Understand** that atoms have a very small mass and a very small size.<br>**Identify** the atomic number of an element.<br>**Recall** atomic number.<br>**Recall** mass number.<br>**Explain** why a substance is an element or a compound given its formula.<br>**Deduce** the number of occupied shells or the number of electrons of an element.<br>**Describe** the main stages in the development of atomic structure. | **Recall** that the nucleus is made up of protons and neutrons.<br>**Recall** the relative charge and relative mass of an electron, a proton, and a neutron.<br>**Describe** isotopes as varieties of an element that have the same atomic number but different mass numbers.<br>**Describe** the arrangement of elements in the periodic table.<br>**Explain** how the identity of an element can be deduced from its electronic structure.<br>**Describe** Dalton's atomic theory and how the work of certain scientists contributed to the development of the theory of atomic structure. | **Explain** why an atom is neutral in terms of its subatomic particles.<br>**Understand** the radius and mass of atoms.<br>**Deduce** the number of protons, electrons, and neutrons in a particle.<br>**Identify** isotopes from data about the number of electrons, protons, and neutrons in particles.<br>**Deduce** the electronic structure of the first 20 elements in the periodic table.<br>**Explain** the significance of the work of Dalton, J.J. Thomson, Rutherford, and Bohr in the development of the theory of atomic structure. |
| C4b | **Recall** that an ion is a charged atom or group of atoms.<br>**Recognise** an ion, an atom, and a molecule from given formulae.<br>**Compare** the electrical conductivity of sodium chloride as a solid, molten liquid, and solution.<br>**Compare** the melting points of sodium chloride and magnesium oxide. | **Understand** that atoms with an outer shell of 8 electrons have a stable electronic structure.<br>**Explain** how and why atoms form ions.<br>**Understand** ionic bonding.<br>**Deduce** the formula of an ionic compound.<br>**Recall** that sodium chloride solution conducts electricity.<br>**Recall** that magnesium oxide and sodium chloride conduct electricity when molten.<br>**Describe** the structure of sodium chloride or magnesium oxide, and their properties. | **Explain**, using the 'dot and cross' model, the ionic bonding in simple binary compounds.<br>**Explain**, in terms of structure and bonding, some of the physical properties of sodium.<br>**Explain**, in terms of structure and bonding, why the melting point of sodium chloride is lower than that of magnesium oxide.<br>**Predict** and **explain** the properties of substances that have a giant ionic structure. |
| C4c | **Recall** ionic and covalent bonds.<br>**Understand** that carbon dioxide and water do not conduct electricity.<br>**Deduce**, using a periodic table, elements that are in the same group.<br>**Describe** a group of elements.<br>**Deduce**, using a periodic table, elements that are in the same period.<br>**Describe** a period of elements.<br>**Describe** the main stages in the development of the classification of elements.<br>**Understand** that classification of elements was provisional. | **Recall** that non-metals combine together by sharing electron pairs in covalent bonding.<br>**Describe** carbon dioxide and water as simple molecules.<br>**Recognise** that the group number is the same as the number of electrons in the outer shell.<br>**Deduce** the group to which an element belongs.<br>**Recognise** that the period of an element corresponds to the shells in its electronic structure.<br>**Deduce** the period to which an element belongs from its electronic structure.<br>**Describe** the evidence or observations that contributed to the development of new models of periodic classification of elements. | **Explain**, using the 'dot and cross' model, the covalent bonding in simple binary compounds.<br>**Explain** some of the physical properties of carbon dioxide and water.<br>**Predict** and **explain** the properties of substances with a simple molecular structure.<br>**Explain** how further evidence confirmed Mendeleev's ideas about the periodic table. |

| To aim for a grade G–E | To aim for a grade D–C | To aim for a grade B–A* | |
|---|---|---|---|
| **Explain** why the Group 1 elements are known as the alkali metals.<br>**Explain** why Group 1 elements are stored under oil.<br>**Describe** the reaction of lithium, sodium and potassium with water.<br>**Construct** the word equation for the reaction of a Group 1 element with water.<br>**Recognise** sodium, lithium, and potassium as Group 1 elements.<br>**Recall** the flame test colours for lithium, sodium, and potassium compounds. | **Predict** the reactions of Group 1 elements with water.<br>**Construct** the balanced symbol equation for the reaction of a Group 1 element with water.<br>**Explain** why Group 1 elements have similar properties.<br>**Describe** how to use a flame test to identify the presence of certain compounds. | **Construct** the balanced symbol equation for the reaction of a Group 1 element with water (formulae not given).<br>**Predict** the physical properties of one Group 1 element given information about the others.<br>**Construct** a balanced symbol equation to show the formation of an ion of a Group 1 element from its atom.<br>**Explain** the trend in reactivity of the Group 1 elements with water.<br>**Recall** the loss of electrons as oxidation.<br>**Explain** why a process is oxidation. | C4d |
| **Recall** that the Group 7 elements are known as the halogens.<br>**Recognise** Group 7 elements.<br>**Describe** the uses of some Group 7 elements.<br>**Recognise** that Group 7 elements react vigorously with Group 1 elements.<br>**Construct** the word equation for the reaction between a Group 1 element and a Group 7 element.<br>**Recall** that the reactivity of the Group 7 elements decreases down the group.<br>**Construct** the word equation for the reaction between a Group 7 element and a metal halide. | **Describe** the physical appearance of the Group 7 elements at room temperature.<br>**Identify** the metal halide formed when a Group 1 element reacts with a Group 7 element.<br>**Construct** the word equation for the reaction between a Group 1 and a Group 7 element.<br>**Construct** the balanced symbol equation for the reaction of a Group 1 with a Group 7 element.<br>**Describe** the displacement reactions of Group 7 elements with solutions of metal halides.<br>**Construct** the word equation for the reaction between a Group 7 element and a metal halide (not all reactants and products given).<br>**Construct** balanced symbol equations for the reactions between Group 7 elements and metal halides (some or all formulae given).<br>**Explain** why Group 7 elements have similar properties. | **Predict** the properties of fluorine or astatine, given the properties of the other Group 7 elements.<br>**Construct** the balanced symbol equation for the reaction of a Group 1 element with a Group 7 element (formulae not given).<br>**Construct** balanced symbol equations for the reactions between Group 7 elements and metal halides (formulae not given).<br>**Predict** the feasibility of displacement reactions.<br>**Understand** that Group 7 elements have similar properties.<br>**Construct** an equation to show the formation of a halide ion from a halogen molecule.<br>**Explain** the trend in reactivity of the Group 7 elements.<br>**Recall** the gain of electrons as reduction.<br>**Explain** why a process is reduction. | C4e |
| **Identify** whether an element is a transition element from its position in the periodic table.<br>**Recognise** that all transition elements are metals and have typical metallic properties.<br>**Deduce** the name or symbol of a transition element using the periodic table.<br>**Recall** copper and iron as transition elements.<br>**Describe** a thermal decomposition.<br>**Construct** word equations for thermal decomposition reactions.<br>**Recall** the test for carbon dioxide.<br>**Describe** precipitation. | **Recall** that compounds of transition elements are often coloured.<br>**Recall** that transition elements and their compounds are often used as catalysts.<br>**Describe** the thermal decomposition of carbonates of transition elements.<br>**Describe** the use of sodium hydroxide solution to identify the presence of transition metal ions in a solution. | **Construct** the balanced symbol equations for the thermal decomposition of transition metals.<br>**Construct** balanced symbol equations for the reactions between $Cu^{2+}$, $Fe^{2+}$ and $Fe^{3+}$ with $OH^-$ (without state symbols), given the formulae of the ions. | C4f |
| **Explain** why iron is used in cars and bridges.<br>**Explain** why copper is used in electrical wiring.<br>**List** the physical properties of metals.<br>**Suggest** properties needed by a metal for a particular given use.<br>**Recognise** that the particles in a metal are held together by metallic bonds.<br>**Recall** that metals can be superconductors. | **Explain** why metals are suited to a given use.<br>**Understand** why metals have high melting and boiling points.<br>**Describe** how metals conduct electricity.<br>**Describe** what is meant by the term superconductor.<br>**Describe** the potential benefits of superconductors. | **Describe** metallic bonding.<br>**Explain**, in terms of structure, why metals have high melting and boiling points, and conduct electricity.<br>**Explain** some of the drawbacks of superconductors. | C4g |
| **Recall** different types of UK water resources.<br>**Explain** why water is an important resource for many important industrial chemical processes.<br>**List** some water pollutants.<br>**Recall** that chlorination kills microbes in water.<br>**Recall** a use of barium chloride solution. | **Explain** why it is important to conserve water.<br>**Explain** why drinking water may contain some pollutants.<br>**Describe** the water purification process.<br>**Construct** word equations for reactions.<br>**Understand** precipitation reactions. | **Explain** why some soluble substances are not removed from water during purification.<br>**Explain** the disadvantages of using distillation of to make.<br>**Construct** balanced symbol equations for some reactions. | C4h |

# Grade checker

**Revising module C5**

To help you start your revision, the specification for module C5 has been summarised in the checklist below. Work your way along each row and make sure that you are happy with all the statements for your target grade.

If you are not sure of any of the statements for your target grade, make a note of them as part of your revision plan. You can then work back through the relevant parts of pages 166–197 to fill gaps in your knowledge as a start to your revision.

| | To aim for a grade G–E | To aim for a grade D–C | To aim for a grade B–A* |
|---|---|---|---|
| C5a | **Recall** that the unit for the amount of a substance is the mole.<br>**Recall** that the unit for molar mass is g/mol.<br>**Understand** that the molar mass of a substance is its relative formula mass in grams.<br>**Calculate** the molar mass of a substance from its formula.<br>**Understand** that mass is conserved during a chemical reaction.<br>**Use** understanding of conservation of mass to carry out very simple calculations. | **Calculate** the molar mass of a substance from its formula (with brackets) using the appropriate relative atomic masses.<br>**Given** a set of reacting masses, calculate further reacting amounts by simple ratio. | **Recall** and use the relationship between molar mass, number of moles, and mass.<br>**Recall** the relative atomic mass of an element.<br>**Calculate** mass of products and/or reactants using the mole concept. |
| C5b | **Determine** the mass of an element in a known mass of compound, given the masses of the other elements present.<br>**Calculate** the molar mass of a substance from its formula. | **Understand** that an empirical formula gives the simplest whole number ratio of each type of atom in a compound.<br>**Deduce** the empirical formula of a compound, given its chemical formula.<br>**Calculate** the percentage by mass of an element in a compound. | **Determine** the number of moles of an element from the mass of that element.<br>**Calculate** the empirical formula of a compound by looking at the mass. |
| C5c | **Recall** that concentration of solutions may be measured in $g/dm^3$ or $mol/dm^3$.<br>**Recall** that volume is measured in $dm^3$ or $cm^3$.<br>**Recall** that 1000 $cm^3$ equals 1 $dm^3$.<br>**Describe** how to dilute a concentrated solution.<br>**Explain** the need for dilution. | **Understand** that the more concentrated a solution, the more solute particles there are in a given volume.<br>**Convert** volume in $cm^3$ into $dm^3$ or vice versa.<br>**Perform** calculations involving concentration for simple dilutions of solutions. | **Recall** and use the relationship between the amount in moles, concentration in $mol/dm^3$ and volume in $dm^3$.<br>**Understand** that sodium ions may come from several sources. |
| C5d | **Explain** how universal indicator is used.<br>**Identify** the apparatus used in an acid-base titration.<br>**Describe** the procedure for carrying out a simple acid-base titration.<br>**Explain** why it is important to use a pipette filler when using a pipette in an acid-base titration.<br>**Calculate** the titre given appropriate information from tables or diagrams.<br>**Understand** that the titre depends on the concentration of the acid or alkali.<br>**Describe** the colours of some indicators. | **Explain** the need for several consistent titre readings in titrations.<br>**Describe** the difference in colour change during a titration using a single indicator. | **Sketch** a pH titration curve for the titration of an acid or an alkali.<br>**Calculate** the concentration of an acid or alkali from titration results.<br>**Explain** why an acid-base titration should use a single indicator rather than a mixed indicator. |
| C5e | **Identify** apparatus used to collect the volume of a gas produced in a reaction. | **Describe** experimental methods to measure the volume and mass of gas produced in a reaction. | **Explain** why the amount of product formed is directly proportional to the amount of limiting reactant used. |

| | To aim for a grade G–E | To aim for a grade D–C | To aim for a grade B–A* |
|---|---|---|---|
| C5e | **Understand** that measurement of change of mass may be used to monitor the amount of gas made in a reaction.<br>**Explain** why a reaction stops. | **Understand** that the amount of product formed is directly proportional to the amount of limiting reactant used.<br>**Recall** that the limiting reactant is the reactant that is all used up at the end of the reaction.<br>**Explain** why a reaction stops, in terms of the limiting reactant present. | **Calculate** the volume of a known number of moles of gas.<br>**Calculate** the amount in moles of a volume of gas.<br>**Sketch** a graph to show how the volume of gas produced during the course of a reaction changes. |
| C5f | **Understand** that a reversible reaction can proceed both forwards and backwards.<br>**Recall** that the $\rightleftharpoons$ symbol is used to show that a reaction is reversible.<br>**Recognise** reactions that are reversible, given the word or balanced symbol equations.<br>**Recall** the raw materials used to make sulfuric acid by the Contact Process.<br>**Describe** the manufacture of sulfuric acid. | **Recall** that in a reversible reaction at equilibrium the rate of the forward and backward reactions are the same and concentrations do not change.<br>**Understand** that when equilibrium is on the right, the concentration of product is greater than the concentration of reactant and vice versa.<br>**Recall** that a change in temperature, pressure or concentration of reactant or product may change the position of equilibrium.<br>**Understand** that the reaction between sulfur dioxide and oxygen is reversible.<br>**Describe** the conditions used in the Contact Process. | **Explain** why a reversible reaction may reach equilibrium.<br>**Understand** in simple qualitative terms factors that affect the position of equilibrium.<br>**Explain** the effect of changing certain factors on the position of equilibrium.<br>**Explain** the conditions used in the Contact Process. |
| C5g | **Recall** that ethanoic acid is a weak acid.<br>**Recall** that hydrochloric, nitric, and sulfuric acids are strong acids.<br>**Understand** that strong acids have a lower pH than weak acids of the same concentration.<br>**Recall** that both ethanoic acid and hydrochloric acid react with magnesium to give hydrogen and with calcium carbonate to give carbon dioxide.<br>**Recall** that magnesium and calcium carbonate react slower with ethanoic acid than with hydrochloric acid.<br>**Understand** that the same amount of hydrochloric and of ethanoic acid produce the same volume of gaseous products in some reactions.<br>**Understand** that ethanoic acid has a lower electrical conductivity than hydrochloric acid.<br>**Recall** that electrolysis of both ethanoic acid and hydrochloric acid makes hydrogen at the negative electrode. | **Understand** that an acid ionises in water to produce $H^+$ ions.<br>**Understand** that a strong acid completely ionises in water and a weak acid does not fully ionise and forms an equilibrium mixture.<br>**Explain** why ethanoic acid reacts slower than hydrochloric acid of the same concentration.<br>**Explain** why the volume of gaseous products of the reactions of acids is determined by the amounts of reactants present, not the acid strength.<br>**Explain**, in terms of movement of ions, why ethanoic acid is less conductive than hydrochloric acid.<br>**Explain** why hydrogen is produced during the electrolysis of ethanoic acid and of hydrochloric acid. | **Explain** why the pH of a weak acid is much higher than the pH of a strong acid.<br>**Explain** the difference between acid strength and acid concentration.<br>**Construct** equations for the ionisation of weak and strong acids, given the formula of the mono-basic acid.<br>**Explain** why ethanoic acid reacts slower than hydrochloric acid.<br>**Explain**, in terms of transfer of charge, why ethanoic acid is less conductive than hydrochloric acid. |
| C5h | **Describe** a precipitation reaction.<br>**Understand** that most precipitation reactions involve ions from one solution reacting with ions from another solution.<br>**Describe** how lead nitrate solution can be used to test for halide ions.<br>**Describe** how barium chloride solution can be used to test for sulfate ions.<br>**Identify** the reactants and the products from an ionic equation.<br>**Recognise** and **use** state symbols.<br>**Label** the apparatus used during the preparation of an insoluble compound by precipitation. | **Understand** that ionic substances contain ions that are in fixed positions but can move.<br>**Understand** that in a precipitation reaction ions must collide with other ions to react to form a precipitate.<br>**Construct** word equations for simple precipitation reactions.<br>**Describe** the stages involved in the preparation of a dry sample of an insoluble compound by precipitation. | **Explain**, in terms of collisions between ions, why most precipitation reactions are extremely fast.<br>**Construct** ionic equations, with state symbols, for simple precipitation reactions.<br>**Explain** the concept of spectator ions. |

**Revising module C6**

To help you start your revision, the specification for module C6 has been summarised in the checklist below. Work your way along each row and make sure that you are happy with all the statements for your target grade.

If you are not sure of any of the statements for your target grade, make a note of them as part of your revision plan. You can then work back through the relevant parts of pages 204–235 to fill gaps in your knowledge as a start to your revision.

| | To aim for a grade G–E | To aim for a grade D–C | To aim for a grade B–A* |
|---|---|---|---|
| C6a | **Describe** electrolysis as the decomposition of a liquid by passing an electric current through it.<br>**Recall** that the anode is positive and the cathode negative.<br>**Recall** that cations are positively charged and anions are negatively charged.<br>**Describe** the electrolyte.<br>**Recognise** anions and cations.<br>**Identify** the apparatus needed to electrolyse aqueous solutions.<br>**Recognise** that positive ions discharge at the negative electrode and negative ions at the positive electrode.<br>**Describe** the chemical tests for hydrogen and oxygen.<br>**Describe** the observations of the electrolysis of copper(II) sulfate solution using carbon electrodes.<br>**Predict** the products of electrolytic decomposition of the molten electrolytes. | **Describe** electrolysis in terms of flow of charge by moving ions and the discharge of ions at the electrodes.<br>**Recall** the products of the electrolysis of NaOH(aq), and $H_2SO_4$(aq).<br>**Recall** the products of the electrolysis of $CuSO_4$(aq) with carbon electrodes.<br>**Understand** that the amount of substance produced during electrolysis is directly proportional to the time and to the current.<br>**Explain** why an ionic solid cannot be electrolysed, but the molten liquid can. | **Construct** half equations for the electrode processes that happen during the electrolysis of NaOH(aq), or $H_2SO_4$(aq).<br>**Explain** why the electrolysis of NaOH(aq) makes $H_2$ rather than Na at the cathode.<br>**Construct** half equations for electrode processes that happen during the electrolysis of $CuSO_4$(aq) using carbon electrodes.<br>**Perform** calculations based on current, time, and the amount of substance produced in electrolysis.<br>**Construct** half equations for the electrode processes that happen during the electrolysis of molten binary ionic compounds. |
| C6b | **Understand** that the reaction between hydrogen and oxygen is exothermic.<br>**Construct** the word equation for the reaction between hydrogen and oxygen.<br>**Describe** a fuel cell as a cell supplied with fuel and oxygen, and how it uses the energy to produce electrical energy efficiently.<br>**Recall** that hydrogen is the fuel in a hydrogen-oxygen fuel cell.<br>**Recall** an important use of fuel cells.<br>**Explain** why a hydrogen-oxygen fuel cell does not form a polluting waste product.<br>**Recall** that the combustion of fossil fuels has been linked with climate change. | **Construct** the balanced symbol equation for the reaction between hydrogen and oxygen.<br>**Construct** the balanced symbol equation for the overall reaction in a hydrogen-oxygen fuel cell.<br>**List** some advantages of using a hydrogen-oxygen fuel cell to provide electrical power in a spacecraft.<br>**Explain** why the car industry is developing fuel cells. | **Draw** and **interpret** an energy level diagram for the reaction between hydrogen and oxygen.<br>**Draw** and **interpret** energy level diagrams for other reactions given appropriate information.<br>**Explain** the changes that take place at each electrode in a hydrogen-oxygen fuel cell.<br>**Explain** the advantages of a hydrogen-oxygen fuel cell.<br>**Explain** why use of hydrogen-oxygen fuel cells will still produce pollution. |
| C6c | **Understand** that redox reactions involve oxidation and reduction.<br>**Recall** that rusting of iron and steel requires both oxygen and water.<br>**List** methods of preventing rust.<br>**Understand** how oil, grease, and paint prevent iron from rusting.<br>**Recall** the following order of reactivity (most to least): magnesium, zinc, iron, and tin.<br>**Predict**, with a reason, whether a displacement reaction will take place. | **Describe** oxidation as the addition of oxygen or the reaction of a substance with oxygen.<br>**Describe** reduction as the removal of oxygen from a substance.<br>**Understand** that rusting is a redox reaction.<br>**Construct** the word equation for rusting.<br>**Explain** how galvanising protects iron from rusting.<br>**Construct** word equations for displacement reactions between metals and metal salt solutions. | **Understand** that oxidation involves loss of electrons and reduction the gain of electrons.<br>**Explain**, in terms of oxidation and reduction, the interconversion of certain types of systems.<br>**Explain** why rusting is a redox reaction.<br>**Explain** how sacrificial protection protects iron from rusting.<br>**Explain** the disadvantage of using tin plate as a means of protecting iron from rusting.<br>**Evaluate** different means of rust prevention.<br>**Construct** symbol equations for displacement reactions between metals and metal salt solutions.<br>**Explain** displacement reactions in terms of oxidation and reduction. |

| | To aim for a grade G–E | To aim for a grade D–C | To aim for a grade B–A* |
|---|---|---|---|
| C6d | **Explain** why alcohols are not hydrocarbons.<br>**Recall** the conditions needed for fermentation.<br>**Recall** the main uses of ethanol.<br>**Recall** that hydration of ethene produces ethanol. | **Recall** the molecular formula and displayed formula of ethanol.<br>**Recall** the word equation for fermentation.<br>**Construct** the balanced symbol equation for fermentation, given all the formulae.<br>**Describe** how ethanol can be made by fermentation.<br>**Explain** why ethanol made by fermentation is a renewable fuel, and why ethanol made by hydration of ethene is not.<br>**Describe** how ethanol is produced for industry.<br>**Construct** the word and balanced symbol equations for the hydration of ethane. | **Recall** the general formula of an alcohol.<br>**Use** the general formula of alcohols to write the molecular formula of an alcohol.<br>**Draw** the displayed formulae of alcohols containing up to five carbon atoms.<br>**Construct** the balanced symbol equation for fermentation (some or no formulae given).<br>**Explain** the conditions used in fermentation.<br>**Evaluate** the merits of the two methods of making ethanol. |
| C6e | **Recall** what atoms a chlorofluorocarbon contains.<br>**Recall** the uses of CFCs.<br>**Recall** that ozone is a form of oxygen.<br>**Describe** some properties of CFCs.<br>**Describe** that increased levels of ultraviolet light can lead to medical problems.<br>**Recall** that hydrocarbons can provide safer alternatives to CFCs. | **Explain** why the use of CFCs has been banned in the UK.<br>**Describe** how CFCs deplete the ozone layer.<br>**Construct** an equation to show the formation of chlorine atoms from CFCs.<br>**Recall** that a chlorine radical is a chlorine atom.<br>**Explain** why CFCs are only removed slowly from the stratosphere.<br>**Describe** how depletion of the ozone layer allows more ultraviolet light to reach the Earth.<br>**Recall** that CFCs can be replaced with alkanes or HFCs that will not cause damage. | **Describe** and explain how scientists' attitude to CFCs has changed.<br>**Explain** how a carbon-chlorine bond can break and what it forms.<br>**Explain** why only a small number of chlorine atoms will destroy a large number of ozone molecules.<br>**Explain** why CFCs will continue to deplete ozone a long time after their use has been banned.<br>**Explain** how ozone absorbs ultraviolet light in the stratosphere. |
| C6f | **Recall** that hard water does not lather well with soap.<br>**Recall** that both hard and soft water lather well with soapless detergents.<br>**Recall** what causes hard water.<br>**Recall** that boiling destroys temporary hardness.<br>**Describe** how hardness in water can be removed. | **Describe** the origin of temporary hardness.<br>**Construct** the word equation for the reaction between calcium carbonate, water, and carbon dioxide.<br>**Recall** that temporary hardness is caused by dissolved calcium hydrogencarbonate.<br>**Recall** that permanent hardness is caused by dissolved calcium sulphate.<br>**Describe** how boiling removes temporary hardness.<br>**Explain** how an ion-exchange resin works.<br>**Plan** experiments on hardness. | **Construct** the symbol equation for the decomposition of calcium hydrogencarbonate.<br>**Explain** how washing soda can soften hard water. |
| C6g | **Understand** that natural fats and oils are important raw materials.<br>**Recall** that vegetable oils can be used to make biodiesel.<br>**Recall** that, at room temperature, oils are liquids and fats are solids.<br>**Describe** an emulsion.<br>**Recall** that milk is an oil-in-water emulsion and butter is a water-in-oil emulsion.<br>**Recall** that a vegetable oil reacts with sodium hydroxide to produce a soap. | **Recall** that animal and vegetable fats and oils are esters.<br>**Explain** whether a fat or oil is saturated or unsaturated, given its displayed formula.<br>**Describe** how unsaturation in fats and oils can be shown using bromine water.<br>**Describe** how margarine is manufactured.<br>**Describe** emulsions and how immiscible liquid can form an emulsion.<br>**Describe** how natural fats and oils can be split up by hot sodium hydroxide solution. | **Explain** why unsaturated fats are healthier as part of a balanced diet.<br>**Explain** why bromine can be used to test for unsaturated fats and oils.<br>**Explain** the saponification of fats and oils. |
| C6h | **Relate** ingredients in washing powder and washing-up liquid to their function.<br>**Understand** the terms solvent, solute, solution, soluble, and insoluble.<br>**Recognise** that different solvents will dissolve different substances.<br>**Identify** the correct solvent to remove a stain. | **Explain** the advantages of using low temperature washes.<br>**Describe** detergents as molecules that have a hydrophilic head and a hydrophobic tail.<br>**Describe** dry cleaning as a process used to clean clothes that does not involve water. | **Explain** how detergents can remove fat or oil stains.<br>**Explain**, in terms of intermolecular forces, how a dry cleaning solvent removes stains. |

# Glossary

**acid** Substance that ionises when dissolved in water to produce hydrogen ions, $H^+$, in solution.
**acid rain** Rain containing some gases from the air that make an acid solution. Acid rain dissolves some building materials such as limestone.
**acidic** Having a pH value less than 7.
**actual yield** Mass of product obtained from a reaction found by experiment.
**addition polymerisation** Polymerisation reaction that only involves addition reactions.
**addition reaction** Reaction in which two reactants combine to make only one product.
**aerobic bacteria** Bacteria that need oxygen to live.
**aggregate** Mixture of small stones, which is mixed with cement and sand to make concrete.
**alcohol** Organic compound that contains the –OH functional group.
**algal bloom** Thick layer of algae on a water surface, caused by the presence of too much fertiliser.
**alkali** Soluble base.
**alkali metal** Element in Group 1 of the periodic table (lithium, sodium, potassium, rubidium, caesium, francium).
**alkaline** Having a pH greater than 7.
**alkane** Hydrocarbon that contains no rings or double bonds in its molecule.
**alkene** Hydrocarbon that contains a double bond (two shared pairs of electrons) between two carbon atoms.
**allotrope** Form of an element that can exist in more than one form. Each allotrope has different physical properties.
**alloy** Mixture of metals that has more useful properties than the metals on their own.
**anion** Negatively charged ion.
**anode** Positively charged electrode.
**anomalous** Measurement that is significantly different from others obtained in a series of runs of the same experiment.
**antioxidant** Food additive that slows down the reaction of a food with oxygen from the air.
**aquifer** Layer of rock that stores a large quantity of water.
**atmosphere** Mixture of gases surrounding the Earth.
**atom** Smallest part of an element that can take part in chemical reactions.
**atom economy** Describes the proportion of the reactant atoms that end up in the desired product. The atom economy is given by ($M_r$ of desired product ÷ total $M_r$ of all products) × 100.
**atomic number** Number of protons in the nucleus of an atom.
**base** Substance that will neutralise an acid to form a salt and water only.
**batch process** Chemical reaction in which one batch of reactants at a time is converted to products.
**bleach** Substance that removes the colours from dyes by oxidising them.
**breathable clothing** Clothing that allows water vapour to pass through but does not allow liquid water (rain) to pass through.
**brine** Concentrated solution of sodium chloride in water.
**Buckminster fullerene** Form of carbon consisting of molecules containing 60 carbon atoms joined together to form a hollow sphere.
**bulk chemical** Chemical that is manufactured or used in large quantities, usually in a continuous process.
**burette** Glass tube with a tap at the bottom that is used to measure how much liquid has passed through the tap.
**calorimeter** Container used to measure temperature changes that occur during a chemical reaction.
**calorimetry** Method of investigating the energy changes that take place during a chemical reaction.
**carbohydrate** Compound such as sugar or starch that contains carbon, hydrogen, and oxygen with the general formula $CH_2O$.
**carbon monoxide** Poisonous gas that contains one atom of carbon and one atom of oxygen in each molecule, CO.
**catalyse** Speeding up a reaction using a catalyst.
**catalyst** Substance that speeds up a reaction without being used up in the reaction.
**catalytic converter** System that removes pollutants from car exhaust gases.
**cathode** Negatively charged electrode.
**cation** Positively charged ion.
**cement** Building material made by heating a mixture of powdered limestone and clay.
**chain reaction** Reaction in which radicals take part. At each stage one of the products is a new radical.
**chemical change** Change in which new substances are formed that have different chemical properties from the original substances.
**chemical symbol** Letter or pair of letters used to represent one atom of an element, eg H represents one atom of hydrogen; Fe represents one atom of iron.
**chlorination** Addition of chlorine to water to kill microbes.
**chlorine radical** Chlorine atom that is not combined in a molecule.
**chlorofluorocarbon (CFC)** Compound with molecules that contain the elements carbon, chlorine, and fluorine only.
**closed system** System in which no material can enter or leave.
**colloid** Mixture of substances dispersed throughout a liquid.
**combustible** Describes a reactant that can take part in a combustion (burning) reaction.
**combustion** Process of burning fossil fuels, which releases carbon dioxide into the atmosphere and provides energy.
**complete combustion** The burning of a hydrocarbon to produce carbon dioxide and water only.
**composite material** Material made up of a number of different materials mixed together.
**compound** Substance made up of two or more different elements, chemically combined.
**concentration** Quantity of solute contained in a given amount of solution, usually measured in $g/dm^3$ or $mol/dm^3$.
**concrete** Building material made by mixing cement, sand, and aggregate with water.
**conservation of mass** The principle that the total mass of the products of a reaction is the same as the total mass of the reactants.
**construction material** Material used for building.
**Contact Process** Industrial method for

making sulfuric acid by reacting sulfur dioxide with oxygen.

**continuous process** Chemical reaction that takes place by continuously supplying the reactants and removing the products as they form.

**continuous variable** Variable that can have any value.

**core** Innermost layer of the Earth, which is rich in nickel and iron.

**corrosion** Reaction of a metal with chemicals, especially water and oxygen, in the air. Corrosion weakens structures made from the metal.

**cosmetics** Substances used to change a person's appearance.

**covalent bond** Shared pair of electrons between atoms in a molecule.

**covalent compound** Compound in which the atoms are joined together in molecules by covalent bonds.

**cracking** Process of breaking large molecules into smaller ones by heating or by using a catalyst.

**cross-link** Bond that links two polymer chains together.

**crude oil** Liquid mixture of hydrocarbons, formed by the decay of the remains of creatures that lived millions of years ago.

**crust** Outer, solid layer of the Earth.

**crystal** Piece of material in which the atoms, molecules, or ions are arranged in a regular three-dimensional pattern.

**decomposition** Process of breaking up molecules into smaller ones.

**deforestation** Process of cutting down large numbers of trees to clear the ground for building or for growing crops.

**delocalised electron** Electron that is free to move throughout a structure.

**denatured** State of proteins that have had their structure altered by heating or by chemical treatment.

**denaturing** Process of changing the structure of a protein molecule by heating (cooking) or chemical treatment.

**dense** Describes something that has a large mass in proportion to its volume.

**density** Relationship between mass and volume.

**detergent** Compound with molecules that have one end soluble in oil and one end soluble in water. This means that it will lift grease from clothes.

**directly proportional** When one variable changes in the same ratio as another, they are described as being directly proportional.

**discharged** Ion that has had its charge removed by gain or loss of electrons.

**dispersed** Spread throughout.

**displace** To replace (push out) an element in a compound.

**displacement reaction** Reaction in which one element replaces another in a compound.

**displayed formula** Description of a covalently bonded compound or element that uses symbols for atoms and that also shows the covalent bonds between the atoms.

**distillation** Process by which a liquid is purified by heating it to form a gas and then condensing the gas back to a liquid in a clean container.

**dot and cross diagram** Diagram that shows how the electrons are arranged in a molecule or in ions.

**dry cleaning** Process of removing dirt from clothes using solvents other than water.

**dye** Coloured substance that dissolves in a liquid and can be used to colour cloth.

**electrical conductivity** Measure of the ability of a substance to allow an electric current to pass through it.

**electrode** Conductor through which an electric current enters or leaves a melted or dissolved ionic compound in electrolysis.

**electrolysis** Process by which melted or dissolved ionic compounds are broken down by passing an electric current through them.

**electrolyte** Liquid that will conduct an electric current and decompose.

**electron** Subatomic particle found outside the nucleus of an atom. It has a charge of –1 and a very small mass compared to protons and neutrons.

**electronic structure** Arrangement of the electrons in an atom in shells around the nucleus.

**element** Substance that consists only of atoms with the same atomic number.

**empirical formula** Simplest formula that shows the relative number of each kind of atom present in a compound.

**emulsifier** Material that enables a stable mixture to be formed between two liquids that would normally separate, such as oil and water.

**emulsion** System that contains small droplets of one liquid dispersed through a second liquid.

**emulsion paint** Paint made up of substances forming an emulsion in water.

**end point** Point in a titration at which the chemical being added has exactly reacted with the chemical in the flask or beaker.

**endothermic reaction** Reaction in which energy is transferred from the surroundings to a reacting mixture.

**energy level diagram** Diagram that shows the relative energy contents of the reactants and products in a chemical reaction.

**equilibrium** Point in a reversible reaction when the forward and backward reactions are occurring at the same rate.

**eruption** Process in which gases and molten rock come to the Earth's surface through a volcano.

**essential element** One of the elements nitrogen, phosphorus, and potassium that are required for plant growth.

**ester** Compound such as ethyl ethanoate made by reacting an organic acid with an alcohol.

**ethanol** Alcohol with the molecular formula $CH_3CH_2OH$.

**eutrophication** The effect of having too much fertiliser in a lake, which causes so much algal growth that fish and other life forms die.

**excess** A reactant that is not all used up in a reaction that uses all of another reactant is said to be in excess.

**exothermic reaction** Reaction in which energy is transferred to the surroundings, in the form of heat, light, and sound, for example.

**explosion** Very rapid reaction that releases a large amount of energy or a large volume of gas in a very short time.

**explosive** Compound capable of taking part in an explosion.

**fatty acid** An organic acid that has a –COOH functional group attached to a long hydrocarbon chain. The chain may contain both single and double carbon–carbon bonds.

**fermentation** Process in which sugars are converted to alcohols and carbon dioxide by the action of enzymes in yeast.

**fertiliser** Material added to soil to provide nutrients for plant growth.

**filter funnel** Piece of equipment that contains the filter paper in separation of a solid from a liquid.

**filtration** Process by which smaller particles of solid are removed from a liquid by passing the liquid through a filter.

**finite resource** Resource, such as a fuel, that exists in a certain amount and that is not being replaced as it is used so will eventually run out.

**flame test** Means of identifying metals by observing the colour they produce in the flame of a Bunsen burner.

**flavour enhancer** Additive that improves the taste of a food.

**food additive** Substance added to a food to improve its appearance, taste, or keeping quality.

**food colour** Food additive that improves the colour of a food, making it look more attractive.

**formula** Description of a compound or an element that uses symbols for atoms. It shows how many atoms of each type are in the substance.

**fossil fuel** Fuel that was formed by the decay of the remains of creatures that lived millions of years ago. Coal, oil, and natural gas are all fossil fuels.

**fraction** Part of a mixture that can be separated from the rest by a physical process such as boiling.

**fractional distillation** Separation of a mixture into fractions that boil at different temperatures.

**fuel** Substance that releases energy when it is burnt.

**fuel cell** Device that transforms chemical energy directly into electrical energy but does not involve combustion.

**galvanising** Coating iron with a layer of zinc to protect the iron from corrosion.

**gas syringe** Piece of equipment used to collect and measure the volume of a gas.

**geologists** Scientists who study the Earth.

**giant ionic lattice** Regular, three-dimensional pattern of ions held together by strong electrostatic forces of attraction.

**global warming** Effect caused by greenhouse gases building up in the atmosphere and causing a rise in the global temperature.

**glycerol** Organic compound with molecular formula $CH_2OHCHOHCH_2OH$.

**greenhouse gas** Gas in the atmosphere that absorbs infrared radiation from the Earth's surface and so reduces the loss of energy from the Earth into space. This has the effect of warming the Earth.

**group** Vertical column in the periodic table. Elements in a group have similar chemical properties.

**Group 1** Group (vertical column) of elements on the left hand side of the periodic table. An atom of each element has one electron in its outer shell.

**Group 7** Group (vertical column) of elements on the right hand side of the periodic table. An atom of each element has seven electrons in its outer shell.

**guideline daily amount (GDA)** Amount of energy or foodstuff to be consumed each day that should be sufficient to maintain health.

**Haber process** Industrial reaction in which nitrogen and hydrogen react to form ammonia.

**half equation** Equation that shows the loss or gain of electrons by an atom or ion in a redox reaction.

**halogen** Element in Group 7 of the periodic table (fluorine, chlorine, bromine, iodine, astatine).

**hard water** Water that does not readily form a lather with soap.

**HFC** Abbreviation for hydrofluorocarbon, a compound containing carbon, hydrogen, and fluorine only.

**hydration reaction** Reaction in which water is chemically added to a compound to form new molecules.

**hydrocarbon** Compound whose molecules contain atoms of hydrogen and carbon only.

**hydrogen ion** Hydrogen atom that has lost one electron, $H^+$.

**hydrolysis reaction** Reaction of a compound with water so that it splits into two products.

**hydrophilic** The part of a molecule that interacts well with water.

**hydrophobic** The part of a molecule that does not interact well with water but does interact well with oil.

**igneous rock** Solid rock formed when molten rock cools.

**immiscible** Description of liquids that form separate layers when they are in the same container.

**impermeable** Layer of material that does does not allow a liquid to pass through it.

**incomplete combustion** Burning of a hydrocarbon to produce substances that could be oxidised further, such as carbon (soot) and carbon monoxide.

**indicator** Substance that changes colour in solutions with different pH values.

**insoluble** Describes a substance that will not dissolve in a liquid.

**intermolecular force** Relatively weak force between a molecule and its neighbours.

**ion** Atom, or group of atoms, that has gained or lost one or more electrons and so is electrically charged.

**ion-exchange resin** Substance with molecules that contain ions that can be exchanged with others in a solution that is passed through the resin.

**ionic bond** Strong electrostatic force of attraction between oppositely charged ions. Ionic bonds act in all directions.

**ionic compound** Compound made up of ions.

**ionic equation** Equation that summarises a reaction between ions by showing only the ions that take part in the reaction.

**ionise** Breaking a molecule into two charged particles (ions), one with a positive charge and one with a negative charge.

**isotope** Atoms of the same element that have different numbers of neutrons in their nuclei are called isotopes.

**laminated** Made up of layers joined together.

**landfill site** Large hole in the ground in which rubbish is dumped.

**lava** Molten rock that has been released through a volcano.

**limescale** Deposit of calcium carbonate that builds up in pipes carrying hard water or in kettles and other containers in which hard water is heated.

**limestone** Rock that consists mainly of calcium carbonate. Much limestone contains the shells of marine animals that lived in prehistoric times.

**limewater** Solution of calcium hydroxide in water.

**limiting reactant** Reactant that is completely used up in a reaction and so determines how much product can be made.

**lithosphere** Part of the Earth composed of the crust and the outer part of the mantle.

**lubricant** Substance that helps moving parts slide over each other easily.

**lustrous** Having a shiny surface.

**magma** Molten rock that comes to the surface of the Earth through weak spots in the crust.

**malleable** Describes something that can be beaten into sheets.

**mantle** Thick solid layer of the Earth between the core and the crust. Although it is solid it can flow very slowly.

**marketing** Process of selling a particular chemical or process to customers.

**mass number** Total number of protons and neutrons in the nucleus of an atom.

**measuring cylinder** Piece of equipment that can be used to measure the volume of a liquid.

**metal halide** Compound made up of a metal and a halogen.

**metallic bond** Force of attraction between positively charged metal ions and delocalised electrons in a metal.

**metamorphic rock** Rock that was formed as a sedimentary rock but has been changed by the effects of high temperature and pressure.

**mixed indicator** Indicator that contains more than one substance and so shows more than one colour change during a titration.

**molar mass** Mass in grams of one mole of substance.

**molar volume** Volume occupied by one mole of a gas.

**mole** Amount of substance that contains the same number of particles (about $6 \times 10^{23}$) as exactly 12 grams of the isotope carbon-12.

**molecular formula** Description of a compound or an element that uses symbols for atoms. It shows the relative number of atoms of each type in the substance.

**molecule** Particle made up of two or more atoms chemically bonded together.

**monomer** Material whose molecules are the individual structural units of a polymer.

**mortar** Building material made by mixing cement and sand with water.

**nanotube** Sheets of carbon atoms arranged in hexagons that are wrapped around each other to form a cylinder with a hollow core.

**neutral** Having a pH value equal to 7.

**neutralisation** Reaction of an acid with a base to form a neutral solution.

**neutralisation reaction** Reaction in which an acid reacts with a base to leave no excess of acid or base.

**neutron** Subatomic particle found in the nucleus of an atom. It has no charge and a relative mass of 1.

**nitrogenous fertiliser** Fertiliser that supplies nitrogen to the soil.

**non-biodegradable** Not able to be broken down by the action of bacteria.

**non-renewable resource** Resource, such as a fuel, that was made a long time ago and is being used up faster than it is being made now.

**nucleus** Relatively heavy central part of an atom, made up of protons and neutrons.

**oil paint** Paint in which materials are dispersed in oil.

**oil slick** Layer of oil floating on water in the sea or in a lake.

**oil well** Shaft drilled into the Earth to allow crude oil to be brought to the surface.

**oil-in-water emulsion** Emulsion in which droplets of oil are dispersed through water.

**ore** Mineral that contains so much of a metal that the metal can be extracted from it at a profit.

**organic acid** Acid that has molecules containing a carbon chain.

**oxidation** Chemical reaction in which an element or compound gains oxygen, or an atom or an ion loses electrons.

**oxidising agent** Chemical that causes oxidation in a redox reaction and is itself reduced.

**ozone** Form of oxygen in which each molecule contains three atoms, $O_3$.

**ozone layer** Layer in the atmosphere that is rich in ozone. It shields the Earth's surface from ultraviolet radiation from the Sun.

**paint** Mixture of substances used to decorate or protect surfaces.

**payback time** Time taken by a company to earn enough profit from making and selling a new product to pay for its development.

**percentage yield** Actual yield × 100 ÷ maximum theoretical yield.

**perfume** Volatile substance that has a pleasant smell.

**period** Horizontal row in the periodic table. The atomic number of each element increases by one, reading from left to right.

**periodic table** Table in which the elements are arranged in rows (periods) and columns (groups) in order of their atomic number.

**permanent hardness** Water hardness that cannot be removed by boiling the water.

**petrochemical** Chemical obtained from crude oil.

**pH** Measure of the concentration of hydrogen ions in a solution.

**pH curve** Graph that shows the change in the pH of a solution as acid is added to alkali or vice versa.

**pH scale** Scale that describes the concentration of hydrogen ions in a solution.

**pharmaceutical** Chemical that is used in medicine and so needs to be obtained as a very pure product, usually by a batch process.

**phosphorescent** Describes something that gives out light that has been absorbed earlier.

**photochemical smog** Form of atmospheric pollution caused by the action of sunlight on pollutants from exhaust gases from cars.

**pigment** Coloured material used in paints.

**pipette** Piece of apparatus that delivers accurately a fixed volume of liquid.

**pipette filler** Safety device for filling a pipette.

**pollutant** Substance released into the environment that can cause harm to living things, including humans.

**polymer** Material whose molecules are made up of many repeated units.

**polymerisation** Reaction in which monomer molecules join together to make a polymer.

**position of equilibrium** Description of the relative amounts of reactants and products in a reaction at equilibrium.

**precipitate** Suspension of small, solid particles, spread throughout a liquid or solution.

**precipitation reaction** Reaction in which a precipitate forms.

**predicted yield** Mass of product that would be obtained from a given mass of reactant if the reaction exactly followed the equation with no losses.

**product** Substance produced in a chemical reaction. Products are written on the right hand side of a chemical equation.

**propellant** Chemical used in an aerosol can to eject the useful contents.

**protein** Very large molecule containing carbon, hydrogen, oxygen, and nitrogen that is found in living things.

**proton** Subatomic particle found in the nucleus of an atom. It has a charge of +1 and a relative mass of 1.

**quarry** Area in the Earth's surface from which large quantities of rock are dug.

**radiation** Transfer of energy across a distance, sometimes in the form of particles given off by the nuclei of radioactive atoms.

**radioactive** Describes something that emits radiation from the nucleus of an atom.

**rate of reaction** The amount of product made ÷ time, or the amount of reactant used ÷ time.

**reactant** Substance that takes part in a chemical reaction. Reactants are written on the left hand side of a chemical equation.

**reaction time** The time between the start of a reaction and its completion.

**reactivity series** List of metals in decreasing order of reactivity.

**redox reaction** Chemical reaction that involves oxidation of one substance and reduction of another.

**reducing agent** Chemical that causes reduction in a redox reaction and is itself oxidised.

**reduction** Chemical reaction in which oxygen is removed from a compound, or an atom or ion gains electrons.

**refrigerant** Chemical used to transport heat energy in a refrigerator.

**reinforced concrete** Cement that is made stronger by allowing it to set round steel bars or mesh.

**relative atomic mass** The relative atomic mass of an element compares the mass of atoms of the element with the mass of atoms of the $^{12}C$ isotope. It is an average value of the isotopes of the element. Its symbol is $A_r$.

**relative formula mass** The relative formula mass of a substance is the mass of a unit of that substance compared to the mass of a $^{12}C$ carbon atom. It is worked out by adding together all the $A_r$ values for the atoms in the formula. Its symbol is $M_r$.

**resistance** Measure of how difficult it is for an electric current to flow through a material.

**reversible reaction** Reaction that can go either way so that the products of one reaction act as the reactants of the opposite reaction.

**rtp** Abbreviation for 'room temperature and pressure'. At rtp one mole of any gas has a volume of 24 $dm^3$.

**rusting** Corrosion of iron when it is exposed to air and water.

**saponification** Process of treating an oil or fat with an alkali to produce the salt of a fatty acid and glycerol.

**saturated** Compound that contains no double bonds in its molecules.

**sedimentary rock** Rock such as limestone that was formed when the remains of sea creatures fell to the floor of the oceans and built up into a thick layer.

**sedimentation** Process by which larger particles of solid in a liquid sink to the bottom over time.

**shape-memory alloy** Alloy that can return to its original shape when it has been bent or twisted out of shape.

**shell** Region of space in an atom that can hold a number of electrons.

**simple molecule** Molecule containing only a few atoms joined by covalent bonds.

**single indicator** Indicator that shows only one colour change at the end point of a titration.

**soft water** Water that readily forms a lather with soap.

**soluble** Describes a substance that will dissolve in a liquid.

**solute** Substance dissolved in a liquid to form a solution.

**solution** Mixture of a substance (solute) dissolved in a liquid (solvent).

**solution mining** Process of obtaining a chemical such as salt from underground by dissolving it in water so that it can be pumped to the surface.

**solvent** Liquid that dissolves a substance to form a solution.

**soot** Deposit of carbon formed by the incomplete combustion of compounds containing carbon.

**speciality chemical** High value chemical that is required for a particular reaction or customer.

**spectator ions** Ions that are present in a reaction mixture but not involved in the reaction.

**starch** Large carbohydrate with molecules that are polymers of sugars.

**state symbol** Symbol used to represent the physical state of a substance taking part in a chemical reaction (aq, dissolved in water; g, gas; l, liquid; s, solid).

**strong acid** Acid that has only some of its molecules ionised when in solution.

**subatomic particle** Particles from which atoms are made, including protons, neutrons, and electrons.

**subduction** Process of one tectonic plate being pushed under another as they move.

**subsidence** Lowering of the ground level as a result of underground mining.

**successful collision** Collision between reactant particles that results in a reaction.

**superconductor** Metal that has a very low (almost zero) electrical resistance.

**surface area** The area of a solid that is in contact with the other reactants, in a reaction involving a solid.

**synthetic** Describes a substance that does not occur in nature but has to be made in a laboratory.

**tectonic plate** One of the large pieces of the Earth's crust that gradually moves across the surface of the mantle.

**temporary hardness** Water hardness that can be removed by boiling the water.

**tensile strength** Resistance of a material to breaking when stretched.

**texture** The feel of a material.

**thermal decomposition** Process of breaking a compound down into simpler substances by heating it.

**thermal decomposition reaction** Reaction in which a reactant is broken up into simpler products by the action of heat.

**thermochromic** Describes a substance that changes colour according to the temperature.

**titration** Procedure in which one solution is added to a known volume of another to measure the volume of the first solution required to react with the second.

**titre** The measured volume of the solution added during a titration to reach the end point.

**toxicity** Extent to which a substance can damage a living creature.

**transition element** Element that occurs between Groups 2 and 3 in the periodic table.

**universal indicator** Mixture of indicators that shows a range of colours in solutions with different pH values.

**unsaturated** Compound that contains at least one double bond in its molecules.

**upward displacement** Method of collecting a gas by displacing a liquid in a closed inverted tube.

**volatility** Ability of a substance to evaporate.

**volcanic activity** Eruption of volcanoes, releasing large amounts of gases into the atmosphere.

**washing soda** The compound sodium carbonate, $Na_2CO_3$.

**water-in-oil emulsion** Emulsion in which droplets of water are dispersed through oil.

**waterproof clothing** Clothing that does not allow rain to pass through to the inside.

**weak acid** Acid that has all its molecules ionised when in solution.

**yeast** Name given to single-celled fungi that cause fermentation.

**yield** Amount of product made in a reaction.

# Index

## Periodic table

| 1 | 2 | | | | | | | | | | | 3 | 4 | 5 | 6 | 7 | 0 |
|---|---|---|---|---|---|---|---|---|---|---|---|---|---|---|---|---|---|
| | | | | | | | 1<br>**H**<br>hydrogen<br>1 | | | | | | | | | | 4<br>**He**<br>helium<br>2 |
| 7<br>**Li**<br>lithium<br>3 | 9<br>**Be**<br>beryllium<br>4 | | | | | | | | | | | 11<br>**B**<br>boron<br>5 | 12<br>**C**<br>carbon<br>6 | 14<br>**N**<br>nitrogen<br>7 | 16<br>**O**<br>oxygen<br>8 | 19<br>**F**<br>fluorine<br>9 | 20<br>**Ne**<br>neon<br>10 |
| 23<br>**Na**<br>sodium<br>11 | 24<br>**Mg**<br>magnesium<br>12 | | | | | | | | | | | 27<br>**Al**<br>aluminium<br>13 | 28<br>**Si**<br>silicon<br>14 | 31<br>**P**<br>phosphorus<br>15 | 32<br>**S**<br>sulfur<br>16 | 35.5<br>**Cl**<br>chlorine<br>17 | 40<br>**Ar**<br>argon<br>18 |
| 39<br>**K**<br>potassium<br>19 | 40<br>**Ca**<br>calcium<br>20 | 45<br>**Sc**<br>scandium<br>21 | 48<br>**Ti**<br>titanium<br>22 | 51<br>**V**<br>vanadium<br>23 | 52<br>**Cr**<br>chromium<br>24 | 55<br>**Mn**<br>manganese<br>25 | 56<br>**Fe**<br>iron<br>26 | 59<br>**Co**<br>cobalt<br>27 | 59<br>**Ni**<br>nickel<br>28 | 63.5<br>**Cu**<br>copper<br>29 | 65<br>**Zn**<br>zinc<br>30 | 70<br>**Ga**<br>gallium<br>31 | 73<br>**Ge**<br>germanium<br>32 | 75<br>**As**<br>arsenic<br>33 | 79<br>**Se**<br>selenium<br>34 | 80<br>**Br**<br>bromine<br>35 | 84<br>**Kr**<br>krypton<br>36 |
| 85<br>**Rb**<br>rubidium<br>37 | 88<br>**Sr**<br>strontium<br>38 | 89<br>**Y**<br>yttrium<br>39 | 91<br>**Zr**<br>zirconium<br>40 | 93<br>**Nb**<br>niobium<br>41 | 96<br>**Mo**<br>molybdenum<br>42 | [98]<br>**Tc**<br>technetium<br>43 | 101<br>**Ru**<br>ruthenium<br>44 | 103<br>**Rh**<br>rhodium<br>45 | 106<br>**Pd**<br>palladium<br>46 | 108<br>**Ag**<br>silver<br>47 | 112<br>**Cd**<br>cadmium<br>48 | 115<br>**In**<br>indium<br>49 | 119<br>**Sn**<br>tin<br>50 | 122<br>**Sb**<br>antimony<br>51 | 128<br>**Te**<br>tellurium<br>52 | 127<br>**I**<br>iodine<br>53 | 131<br>**Xe**<br>xenon<br>54 |
| 133<br>**Cs**<br>caesium<br>55 | 137<br>**Ba**<br>barium<br>56 | 139<br>**La***<br>lanthanum<br>57 | 178<br>**Hf**<br>hafnium<br>72 | 181<br>**Ta**<br>tantalum<br>73 | 184<br>**W**<br>tungsten<br>74 | 186<br>**Re**<br>rhenium<br>75 | 190<br>**Os**<br>osmium<br>76 | 192<br>**Ir**<br>iridium<br>77 | 195<br>**Pt**<br>platinum<br>78 | 197<br>**Au**<br>gold<br>79 | 201<br>**Hg**<br>mercury<br>80 | 204<br>**Tl**<br>thallium<br>81 | 207<br>**Pb**<br>lead<br>82 | 209<br>**Bi**<br>bismuth<br>83 | [209]<br>**Po**<br>polonium<br>84 | [210]<br>**At**<br>astatine<br>85 | [222]<br>**Rn**<br>radon<br>86 |
| [223]<br>**Fr**<br>francium<br>87 | [226]<br>**Ra**<br>radium<br>88 | [227]<br>**Ac***<br>actinium<br>89 | [261]<br>**Rf**<br>rutherfordium<br>104 | [262]<br>**Db**<br>dubnium<br>105 | [266]<br>**Sg**<br>seaborgium<br>106 | [264]<br>**Bh**<br>bohrium<br>107 | [277]<br>**Hs**<br>hassium<br>108 | [268]<br>**Mt**<br>meitnerium<br>109 | [271]<br>**Ds**<br>darmstadtium<br>110 | [272]<br>**Rg**<br>roentgenium<br>111 | Elements with atomic numbers 112–116 have been reported but not fully authenticated | | | | | | |

**Key**

| relative atomic mass<br>**atomic symbol**<br>name<br>atomic (proton) number |
|---|

* The lanthanoids (atomic numbers 58–71) and the actinoids (atomic numbers 90–103) have been omitted.

## Acknowledgements

The publisher and authors would like to thank the following for their permission to reproduce photographs and other copyright material:

**p8** Laurence Gough/Istockphoto; **p9** Laurence Gough/Istockphoto; **p10T** Laurence Gough/Istockphoto; **p10B** 81a/Alamy; **p11** Chris Pearsall/Alamy; **p13** Roger Asbury/Shutterstock; **p14L** Paul Rapson/SPL; **p14R** George Clerk/Istockphoto; **p16R** Richard Folwell/SPL; **p16L** Randy Brandon/Photolibrary; **p18** MaxFX/Shutterstock; **p19L** Andrew Lambert Photography/SPL; **p19R** Andrew Lambert Photography/SPL; **p20M** Martyn F. Chillmaid/SPL; **p20R** Tommounsey/Istockphoto; **p20L** Andsem/Istockphoto; **p21** Martyn F. Chillmaid/SPL; **p22** Andrey Prokhorov/Istockphoto; **p23** Theodore Scott/Istockphoto; **p24** Bronskov/Shutterstock; **p25L** David R. Frazier/SPL; **p25R** Simon Fraser/SPL; **p27** Andrew Lambert Photography/SPL; **p28BL** Jim DeLillo/Istockphoto; **p28BR** Ionescu Bogdan Cristian/Istockphoto; **p28T** Leonid Nyshko/Istockphoto; **p30** Ljupco/Istockphoto; **p31** Thomas Trötscher/Istockphoto ; **p32** Rob Hill/Istockphoto; **p33** Robert Brook/SPL; **p34TL** JackJelly/Istockphoto; **p34BL** JackJelly/Istockphoto; **p34R** Dr Keith Wheeler/SPL; **p35** Bluestocking/Istockphoto; **p36** Fertnig/Istockphoto ; **p37T** Varela/Istockphoto; **p37B** ViktorFischer/Istockphoto; **p38T** Wojciech Gajda/Istockphoto; **p38B** Aluxum/Istockphoto; **p40TL** Martyn F. Chillmaid/SPL; **p40R** Ruzanna/Istockphoto; **p40BL** ScantyNebula/Istockphoto; **p41** J.C. Revy, ISM/SPL; **p42L** Jonya/Istockphoto; **p42R** Gannet77/Istockphoto; **p43** Charles D. Winters/SPL; **p44T** Andrew Lambert Photography/SPL; **p44B** Cordelia Molloy/SPL; **p45R** Graffizone/Istockphoto; **p45L** John Solie/Istockphoto; **p51** PHB.cz (Richard Semik)/Shutterstock; **p53** JulienGrondin/Istockphoto; **p54L** Dirk Wiersma/SPL; **p54R** Dirk Wiersma/SPL; **p55L** Tyson Paul/Istockphoto; **p55R** Simon Fraser/SPL; **p56L** Mark Thomas/SPL; **p56R** Martin Bond/SPL; **p57** Martin Bond/SPL; **p58** Charles D. Winters/SPL; **p59** Maximilian Stock Ltd/SPL; **p60** Karol Kozlowski/Istockphoto; **p61** Matej Krajcovic/Istockphoto; **p62R** Andreas Reh/Istockphoto; **p62L** Markus Gann/Shutterstock; **p63** Garysludden/Istockphoto; **p64** John Guard/Istockphoto; **p65R** Lesley Jacques/Istockphoto; **p65L** Paul Rapson/SPL; **p66L** Ricardo Azoury/Istockphoto; **p66R** Car Culture/Corbis; **p67T** Ian Hamilton/Istockphoto; **p67B** Roger Job/Reporters/SPL; **p68** Jonathan Maddock/Istockphoto; **p69** SPL; **p71** Charles D. Winters/SPL; **p72R** Andrew Lambert Photography/SPL; **p72TL** Martyn F. Chillmaid/SPL; **p72BL** Gustoimages/SPL; **p73** Charles D. Winters/SPL; **p74T** Andrew Lambert Photography/SPL; **p74B** Charles D. Winters/SPL; **p76L** Redmal/Istockphoto; **p76R** Federico Rostagno/Istockphoto; **p77** Heike Kampe/Istockphoto; **p78R** Maximilian Stock Ltd/SPL; **p78L** Robert Brook/SPL; **p79** Andrew Lambert Photography/SPL; **p80** OlesiaRu&IvanRu/Shutterstock; **p81** Carlos Dominguez/SPL; **p82** Bob Edwards/SPL; **p83** Andrew Lambert Photography/SPL; **p89** Maximillian Stock LTD/SPL; **p90R** Photos.com; **p90L** Andrew Lambert Photography/SPL; **p92** Photos.com; **p94L** Andrew Lambert Photography/SPL; **p94R** Martyn F. Chillmaid/SPL; **p96** MARKA/Alamy; **p98** Photos.com; **p99** Martyn F. Chillmaid/SPL; **p100R** Malcolm Fielding, Johnson Matthey Plc/SPL; **p100L** Charles D. Winters/SPL; **p101** Charles D. Winters/SPL; **p102** Philippe Plailly/SPL; **p104R** Martyn F. Chillmaid/SPL; **p104L** Andrew Lambert Photography/SPL; **p105** Andrew Lambert Photography/SPL; **p106** Andrew Lambert Photography/SPL; **p107** Health Protection Agency/SPL; **p108** Ria Novosti/SPL; **p109** Photos.com; **p110RT** Ruslan Gilmanshin/Istockphoto; **p110RB** Michael Durham/Getty; **p110L** pablo del rio sotelo/Istockphoto; **p111L** Cordelia Molloy/SPL; **p111R** Charles D. Winters/SPL; **p112** Andrew Lambert Photography/SPL; **p114R** Martin Bond/SPL; **p114L** Peter Menzel/SPL; **p116** Maximilian Stock Ltd/SPL; **p118L** Ingvar Tjostheim/Shutterstock; **p118R** Ryan Balderas/Istockphoto; **p120L** Geoff Tompkinson/SPL; **p120R** Kevin Britland/Shutterstock; **p121** Eye of Science/SPL; **p127** Adam Hart-Davis/SPL; **p128** Martin Bond/SPL; **p130** Mark Dunn/Alamy; **p132** John Chumack/SPL; **p134** Oleg Mitiukhin/Istockphoto; **p137** Tek Image/SPL; **p138R** Photos.com; **p138L** Gustoimages/SPL; **p140L** Martyn F. Chillmaid/SPL; **p140M** Andrew Lambert Photography/SPL; **p140R** Andrew Lambert Photography/SPL; **p142** Martina Barbist/Istockphoto; **p143L** David Taylor/SPL; **p143M** David Taylor/SPL; **p143R** David Taylor/SPL; **p144TR** Hakan Caglav/Istockphoto; **p144BR** A. McClenaghan/SPL; **p144L** Andrew Lambert Photography/SPL; **p145** Martyn F. Chillmaid/SPL; **p146** Andrew Lambert Photography/SPL; **p148R** Martyn F. Chillmaid/SPL; **p148L** CC Studio/SPL; **p149** Andrew Lambert Photography/SPL; **p150** Paul Rapson/SPL; **p151L** Andrew Lambert Photography/SPL; **p151R** Andrew Lambert Photography/SPL; **p152R** Amra Pasic/Shutterstock; **p152L** Charles D. Winters/SPL; **p154** Charles D. Winters/SPL; **p155R** Photos.com; **p155L** Andy Crump/SPL; **p156R** Photos.com; **p156L** Ranplett/Istockphoto; **p158R** Massimo Brega/Eurelios/SPL; **p158L** Charles D. Winters/SPL; **p159** Andrew Lambert Photography/SPL; **p165** Ted Clutter/SPL; **p166** Andrew Lambert Photography/SPL; **p168** Dirk Wiersma/SPL; **p170** Carlos Neto/Shutterstock; **p172** Martyn F. Chillmaid/SPL; **p173** Andrew Lambert Photography/SPL; **p174** Ian Hooton/SPL; **p176** Andrew Brookes, National Physical Laboratory/SPL; **p177** Paul Rapson/SPL; **p178** Andrew Lambert Photography/SPL; **p180TL** Jerry Mason/SPL; **p180R** Andrew Lambert Photography/SPL; **p180BL** Charles D. Winters/SPL; **p182L** American Honda Motor Co., Inc.; **p182R** Andrew Lambert Photography/SPL; **p183** Andrew Lambert Photography/SPL; **p184** Charles D. Winters/SPL; **p186** Kuttig/Alamy; **p188T** Photo Researchers/SPL; **p188B** Andrew Lambert Photography/SPL; **p190** Betty Finney/Alamy; **p191** Andrew Lambert Photography/SPL; **p192L** Christopher Nash/Alamy; **p192R** Martyn F. Chillmaid/SPL; **p194L** Andrew Lambert Photography/SPL; **p194R** SPL; **p196T** Sovereign, Ism/SPL; **p196B** Andrew Lambert Photography/SPL; **p197** Charles D. Winters/SPL; **p203** NASA/SPL; **p204** Trevor Clifford Photography/SPL; **p205** Sheila Terry/SPL; **p206T** Andrew Lambert Photography/SPL; **p206B** Trevor Clifford Photography/SPL; **p210T** NASA/SPL; **p210B** NASA; **p211** Martin Bond/SPL; **p212** Charles D. Winters/SPL; **p214** John Mole Photography/Alamy; **p215T** Photos.com; **p215B** Photos.com; **p216** Charles D. Winters/SPL; **p217** Peter Menzel/SPL; **p218** Peter Bowater/SPL; **p220** George Steinmetz/SPL; **p221** Sue Ford/SPL; **p222L** Photos.com; **p222R** NASA/SPL; **p224** Matthew Benoit/Shutterstock; **p225** Sheila Terry/SPL; **p226** Pink Sun Media/Alamy; **p227** DWD-photo/Alamy; **p228** Cordelia Molloy/SPL; **p230** Charles D. Winters/SPL; **p231** Cc Studio/SPL; **p231** BigPileStock/Alamy; **p232** Andrew Lambert Photography/SPL; **p233** Klaus Guldbrandsen/SPL; **p234T** Photos.com; **p234B** Ted Foxx/Alamy

Cover image courtesy of T-SERVICE/SCIENCE PHOTO LIBRARY.

Illustrations by Wearset Ltd and Peter Bull Studios.

Although we have made every effort to trace and contact all copyright holders before publication this has not been possible in all cases. If notified, the publisher will rectify any errors or omissions at the earliest opportunity.

Great Clarendon Street, Oxford OX2 6DP

Oxford University Press is a department of the University of Oxford. It furthers the University's objective of excellence in research, scholarship, and education by publishing worldwide in

Oxford New York

Auckland Cape Town Dar es Salaam Hong Kong Karachi Kuala Lumpur Madrid Melbourne Mexico City Nairobi New Delhi Shanghai Taipei Toronto

With offices in
Argentina Austria Brazil Chile Czech Republic France Greece Guatemala Hungary Italy Japan Poland Portugal Singapore South Korea Switzerland Thailand Turkey Ukraine Vietnam

First published 2011



British Library Cataloguing in Publication Data

Data available

ISBN 978-0-19-913573-8

10 9 8 7 6 5 4

Printed in China by Printplus

Paper used in the production of this book is a natural, recyclable product made from wood grown in sustainable forests. The manufacturing process conforms to the environmental regulations of the country of origin.